THE DICTYOSTELIDS

The Dictyostelids

By Kenneth B. Raper

With the collaboration of Ann Worley Rahn

PRINCETON UNIVERSITY PRESS

CONTENTS

This book is about a relatively small but quite remarkable group of organisms. During a 2- to 4-day cycle the dictyostelids show both animal-like (protozoan) and plant-like (fungal) characteristics. Clearly primitive, the vegetative phase consists of small, independent amoeboid cells, termed myxamoebae, that feed upon bacteria, grow, and multiply until the available food supply is exhausted. They then stream together in great numbers, forming compact, multicellular associations, or pseudoplasmodia, from which subsequently develop fructifications, or sorocarps, that consist of supportive stalks and unwalled sori containing propagative spores. The vacuolate cells that comprise the stalks, and the spores which they support, are formed from individual myxamoebae that were once similar and equipotent before differentiating along divergent paths; and the temporal separation of growth and morphogenesis, coupled with the fact that completed sorocarps contain cells of only two types, has led many biologists to choose dictyostelids as simple eukaryotic models for investigative researches of basic types and at different levels from organismal to molecular. The book tells also where to find the dictyostelids in nature, how to isolate them from natural habitats, and how to cultivate, study, and conserve them in the laboratory to achieve different goals. Fortunately, many techniques used in studying prokaryotic bacteria are equally applicable to the myxamoebae.

Dictyostelids have been known for more than a century. It was not until recently, however, that these organisms, now commonly called cellular slime molds, began to attract particular interest as objects suitable for investigating problems in cellular and developmental biology. Attention in recent years has been focused not upon *D. mucoroides*, although it is especially common in nature, but upon a species less than half its age, *D. discoideum* Raper. There are valid reasons for this choice, since the latter possesses a number of attributes, structural and behavioral, that make it particularly attractive for experimental work. Prominent among published reports, now exceeding two thousand in number, are studies on cell growth, intercellular communication and response, progressive cell integration, and, under optimal conditions, divergent cell differentiation to yield sorocarps of near constant pattern and proportions. It pleases me that hundreds of investigators in scores of laboratories throughout the world are now studying this slime mold. Nonetheless, I am aware that *D. discoideum* is but one of many described dictyostelids, and I doubt not that some of the less publicized species possess untapped potentials for equally imaginative research.

It was with the dual aim of reviewing current developments in dictyostelid biology against an historical background, and of bringing together in one place information concerning all the known dictyostelids, that preparation of this monograph was undertaken. The endeavor has not been entirely successful, for we still lack sufficiently firm criteria to construct a satisfactory natural system of classification. Two objectives have been achieved, however. Descriptions of known species together with illustrations, original and supplemental, are presented in the following pages; and a dichotomous key has been prepared, albeit somewhat artificial in construction, that should enable researchers to determine the names most appropriate for their isolates, unless perchance they are previously undescribed. A second, semi-synoptic key, less seriously regarded at first and so dubbed "Helpful Hints," may in practice prove to be as useful as the first.

The importance of the book, we believe, rests equally upon the historical and state-of-the-art accounts of Part I and the more comprehensive taxonomic coverage of Part II. If the first part appears to contain an excess of personal reference, it is because of the author's early involvement in exploring the research potential of the dictyostelids and his continuing efforts to enlarge such possibilities. In writing about *Dictyostelium discoideum* in *Harvard Magazine* (July-August 1982) under the title "Marching along with the social amoeba," Jonathan Weiner said, "Someday someone should write its biography." This the author has attempted to do.

With regard to the second part, not since the publication of E. W. Olive's *Monograph of the Acrasieae* in 1902 has there been an attempt to include descriptive information about all published dictyostelids. The closest approaches have been the chapter entitled "Dictyostelia" in L. S. Olive's more inclusive book, *The Mycetozoans* (1975); H. Krzemieniewska's paper of 1961, entitled simply "Acrasieae" and self-described as a sketchy monograph; and J. T. Bonner's two widely read editions of *The Cellular Slime Molds* (1959, 1967).

The dictyostelid story as told in these pages is incomplete, and, for reasons of space and the author's lack of technical expertise, large areas of research on these slime molds have been covered only partially or omitted. This is particularly true concerning *D. discoideum*. Fortunately, such deficiencies are mitigated in substantial part by the very recent publication of a book edited by Professor W. F. Loomis (1982) and entitled "The Development of *Dictyostelium discoideum*," which contains chapters by recognized authorities on genetics, cell motility, cell adhesion, chemotaxis, morphogenetic signaling, cell proportioning, and pattern formation, among other related topics. The Loomis book thus complements and supplements information presented here, and readers who wish to pursue researches at the genetic, enzymatic, or molecular level should consult that work.

The pronouns "we" and "our" are used throughout the text, not so much in a rhetorical sense as in partial recognition of the participation of graduate students, postdoctoral fellows, and professional associates who have worked with me in research on the dictyostelids. It is a pleasure to acknowledge specifically our indebtedness to Professor James C. Cavender of Ohio University, Athens, who first as a graduate student and subsequently as an independent investigator generously shared isolates from many parts of the world, together with illustrations from several of his published species descriptions. To Dr. Hiromitsu Hagiwara of the National Museum of Science, Tokyo, we are grateful for type cultures, original line drawings, and photomicrographs of his several new species of *Dictyostelium, Polysphondylium,* and *Acytostelium*; without these this monograph would be far less complete. To Professor Hans R. Hohl and Dr. Franz Traub of the University of Zurich, we express sincere appreciation for type cultures, copies of unpublished manuscripts, illustrations, and stimulating discussions relating to taxonomy of the Dictyosteliaceae. We wish also to thank Dr. David R. Waddell of Princeton University and Dr. Daniel Mahoney of California State University, Los Angeles, for cultures that have added new dimensions to the genus *Dictyostelium*. We are grateful to Dr. Brian Shaffer of Cambridge University and to Professor Peter Newell of Oxford University for copies of and permission to use illustrations from their publications; to Professor Michael Filosa of the University of Toronto and Dr. U.-P. Roos, University of Zurich, for copies of electron micrographs of macrocyst formation and nuclear divisions in *Dictyostelium*, respectively; to Professor James H. Gregg, University of Florida, for photomicrographs depicting stalk formation in *D. discoideum*; to Professor Ian K. Ross, University of California, Santa Barbara, for pictures of cell division in *D. discoideum*; to Professor David Francis and Dr. Robert Eisenberg, University of Delaware, for diagrams relating to pseudoplasmodial migration and the distribution of dictyostelids in soil; and to Dr. B. N. Singh of Luchnow, India, for use of his diagram showing comparative consumption of bacteria in soil by the myxamoebae of *D. mucoroides* and *D. giganteum*. We wish also to thank Dr. Claude E. Vézina, Montreal, for translating our description of the new species *D. caveatum* into Latin.

To Professor John T. Bonner of Princeton University we are especially indebted, not only for providing illustrative materials, but for his abiding faith in the merit of this endeavor, for having read the text piecemeal, and for offering valuable suggestions as preparation of the manuscript progressed. We are grateful also to Professor L. S. Olive, University of North Carolina, and to our colleague, Professor R. L. Dimond, for reading selected chapters of the completed text.

Support for our dictyostelid research has come from the U.S. Department of Agriculture, the Wisconsin Alumni Research Foundation, University of Wisconsin, the National Institutes of Health (U.S. Public Health Service), and the National Science Foundation. Without this support many of the studies reviewed in this monograph could not have been undertaken.

We wish to thank the Department of Bacteriology, The College of Agriculture and Life Sciences, and the University of Wisconsin, Madison, for laboratory facilities and services, for continued clerical assistance, and for quiet but invaluable personal support. We are especially grateful for a grant from the Brittingham Trust provided by the Chancellor of the University of Wisconsin which enabled us to complete essential research and bring preparation of *The Dictyostelids* to a successful conclusion.

Very special thanks are extended to Ann Worley Rahn for her continuing interest in the dictyostelids, for her indispensable help in preparing cultures and illustrations and in compiling the extensive bibliography, and for suggesting improvements in the text as it took form; to Nancy J. Gouker for her patience, good humor, and craftsmanship in typing the manuscript through revisions until finally stored in the word processor; and to my wife, Louise, for her encouragement, forbearance, and willingness to forgo some long anticipated travel so that the book could be written.

Kenneth B. Raper

BIOGRAPHICAL NOTE

KENNETH B. RAPER, William Trelease Professor of Bacteriology and Botany, Emeritus, University of Wisconsin, Madison, was actively engaged in teaching and research for twenty-five years prior to his official retirement in 1979. Earlier he worked for the U.S. Department of Agriculture, first in Washington, D.C., and later at the Northern Regional Research Laboratory in Peoria, Illinois. He received his A.B. degree from the University of North Carolina (1929); A.M. degrees from George Washington University (1931) and Harvard University (1935); a Ph.D. from Harvard (1936); and a D.Sc. from the University of North Carolina (1961).

Professor Raper has been a student of the dictyostelids for a full half-century. At first concerned with their presence and possible roles in soils and forest litter, he soon turned his attention to their unique developmental cycles. Such studies were immensely abetted by his discovery of *Dictyostelium discoideum* in 1933 (described in 1935), for this species possesses a number of characteristics that make it particularly suitable for experimental studies.

During the early part of World War II, Professor Raper participated in the search for and isolation of better penicillin-producing molds. While working for the Department of Agriculture, he was closely associated with the eminent mycologist Dr. Charles Thom, with whom he published two reference books on common molds, one on the genus *Aspergillus* (1945) and the other on *Penicillium* (1949). Professor Raper also established and initially directed (1940-1953) the NRRL Culture Collection at the Northern Utilization Research Center of the USDA in Peoria, Illinois. This unmatched collection of industrially and agriculturally important microorganisms has had a profound effect upon the ever increasing use of molds, yeasts, bacteria, and actinomycetes not only as agents producing substances useful to man but also as laboratory tools for investigating the basic processes of metabolism and biosynthesis.

MRS. ANN WORLEY RAHN, a native of Madison, Wisconsin, is a microbiologist in the Department of Bacteriology, University of Wisconsin, from which she received B.A. and M.S. degrees, the latter in 1974. She has conducted research on many different cellular slime molds and is codescriber of the new genus and species *Fonticula alba*; of the genus *Copromyxella* with four new species; and of three new species in the genus *Dictyostelium*.

PART I
GROWTH AND MORPHOGENESIS

CHAPTER 1

Historical Background

Slime molds of the dictyostelid type have been known since Oskar Brefeld isolated and described *Dictyostelium mucoroides* in 1869. While examining the fungus flora of horse dung he observed spores of the same form but much smaller than those of *Mucor mucedo*. Unable at first to trace their origin, he fortunately, and soon thereafter, obtained on rabbit dung in relatively clean culture abundant fruiting structures bearing similar spores. Superficially these structures resembled the sporangia and sporangiophores of a *Mucor*, hence the specific name. Several unique features, however, refuted any probable relationship: the "sporangia" had no enveloping walls and the spores were suspended freely in droplets of thin slime; no vegetative hyphae were connected with the bases of the fructifications; the spores upon germination produced not hyphae but small amoeboid cells; the myxamoebae, following a period of growth and substantial multiplication, aggregated to certain points from which fructifications arose directly; and finally the stalks that supported the pseudosporangia consisted not of hyphae, as in the true fungi, but of closely packed parenchyma-like cells. The generic name *Dictyostelium* was chosen to indicate their net-like appearance (Fig. 1-1).

Using cooked horse dung as a solid substrate, and a decoction of the same for slide cultures, Brefeld succeeded in following the entire life cycle of his slime mold; and, except for a few misinterpretations, he presented a generally accurate picture of its growth and development as this is still known. He recognized the free-living myxamoebae as constituting the vegetative phase and noted a parallel between these and an early stage in the life cycle of the Myxomycetes with which he concluded *Dictyostelium* was related. He described the presence of a thin, open-ended sheath surrounding the cellular stalk and noted the cellulosic character of this and of the walls of spores and stalk cells as well. He may have sensed but did not elaborate upon a possible role for the open-ended tube in stalk construction. His major misinterpretation, and one that troubled him later, was his assumption that the myxamoebae upon aggregation fused to form a true plasmodium, albeit a somewhat transitory one. This error was corrected in a second and more elaborate paper several years later (1884), but not before van Tieghem (1880) had demonstrated that the aggregated myxamoebae never actually fused.

It is to Ph. van Tieghem that we owe our first clear insight into the true and unique nature of the so-called cellular slime molds, including the

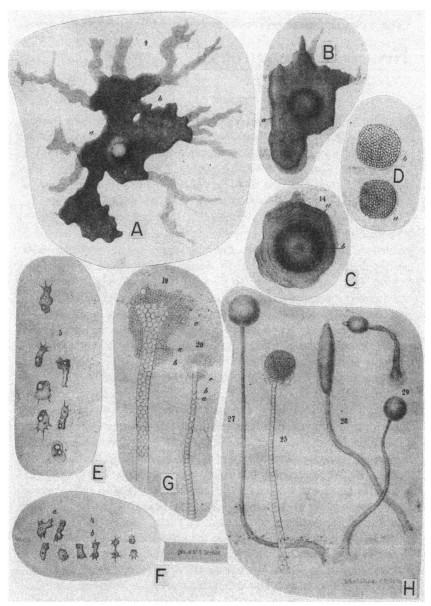

FIG. 1-1. Selected illustrations assembled from Plates I-III of Brefeld's 1869 paper describing *Dictyostelium mucoroides*. *A-C*. Progressive stages in "plasmodium" formation. *D*. "Dwarf sporangia," believed to be macrocysts. *E,F*. Myxamoebae, the latter figure depicting cell division. *G*. Terminal portions of two stalks showing cellular structure, one ending in a characteristic funnel-like expansion. *H*. Fructifications in different stages of development. Scale variable.

dictyostelids. In two brief and regrettably unillustrated papers (1880, 1884) he showed that the myxamoebae formed their characteristic fructifications through the orderly assemblage, coordination, and subsequent differentiation of large populations of separate, collaborating cells. In the first of his brief communications (1880) he described a new genus and species, *Acrasis granulata*, two new species of Cienkowski's *Guttulina* (1873), and two new species of *Dictyostelium*, *D. roseum* and *D. lacteum*, in addition to including observations on strains of Brefeld's *D. mucoroides* that he had isolated. Whereas *Acrasis granulata* was the slime mold in which the absence of a true plasmodium was first established, a fact reflected in the generic name selected for it, he clearly demonstrated that nonfusion of aggregated myxamoebae was equally characteristic of the other genera included in his "Plasmode Agrégé," or Acrasiées. In the second paper (1884) van Tieghem described a new and more complex genus and species, *Coenonia denticulata*, and in a few simple experiments demonstrated that the position of a myxamoeba within the fructifying mass determined whether it would form a stalk cell or a spore. Unfortunately, neither *Coenonia* nor *Acrasis granulata* has been rediscovered.

In the same year (1884), but without reference to van Tieghem's earlier work (1880), Brefeld published a long and comprehensive paper that included further observations on *Dictyostelium mucoroides* together with a description and superb illustrations of an additional new genus and species, *Polysphondylium violaceum* (Figs. 1-2 and 1-3). He had by this time recognized the unique character of the aggregated cell associations in these genera and introduced the term *Scheinplasmodium*, or pseudoplasmodium, which was appropriately descriptive and has since been generally accepted.

Three additional reports appeared prior to 1900: one by E. Marchal (1885) wherein a fourth species of *Dictyostelium*, *D. sphaerocephalum*, was described; one by von M. Grimm (1895) on the structure and development of *D. mucoroides* that in the main confirmed Brefeld's account of 1884 with some additional information on nuclear division; and a third by G. A. Nadson (1899) in which, for the first time, a dictyostelid, *D. mucoroides*, was reported to be growing with a known species of bacteria, *Bacillus fluorescens liquefaciens*. He concluded that the two organisms were symbionts and that the bacteria favored the slime mold by creating an alkaline reaction in the culture medium. The myxamoebae, he believed, fed solely upon nutrients in solution.

Early in the next century very important papers were published by E. W. Olive in the United States (1901, 1902) and George Potts in Germany (1902), while works of lesser scope were reported in France by P. Vuillemin (1903) and E. Pinoy (1903, 1907).

In the first of his papers, entitled "A Preliminary Enumeration of the Sorophoreae" (1901), Olive listed, with brief notations, all the known slime molds that he believed were assignable to Zopf's "Gruppe I" of the

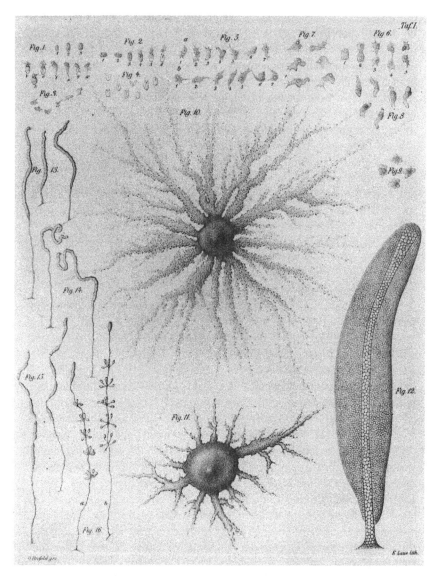

FIGS. 1-2 AND 1-3. Plates I and II of Brefeld's second paper (1884) illustrating stages in the development of his new genus and species, *Polysphondylium violaceum. 1-9.* Spore germination and vegetative myxamoebae. *10,11.* Stages in pseudoplasmodium formation. *12-19.* Different stages in sorocarp construction. *20-29.* Mostly branching patterns and details of cellular structure in sorocarps of different size. *30-33.* Stalk formation in young fructifications. *34,35.* Small fruiting bodies. Scale variable.

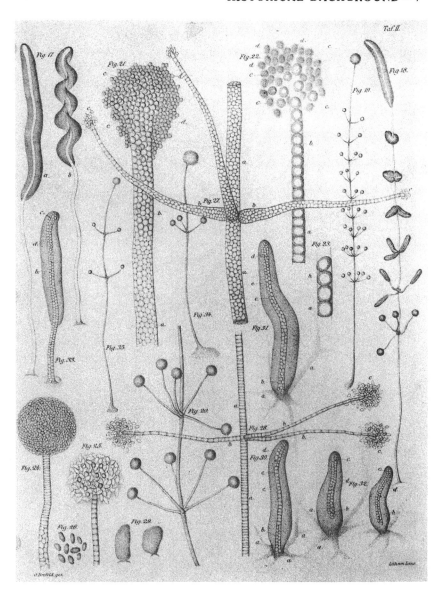

Eumycetozoa (1885), together with a new genus, *Guttulinopsis* (with three species), three new species of *Dictyostelium*, and two new species of *Polysphondylium*. This was followed a year later (1902) by his much longer "Monograph of the Acrasieae" (Fig. 4). For this he used an Anglicized version of van Tieghem's more restrictive "Acrasiées," and excluded Barker's *Diplophrys* (1868) and Cienkowski's *Labyrinthula* (1867) because

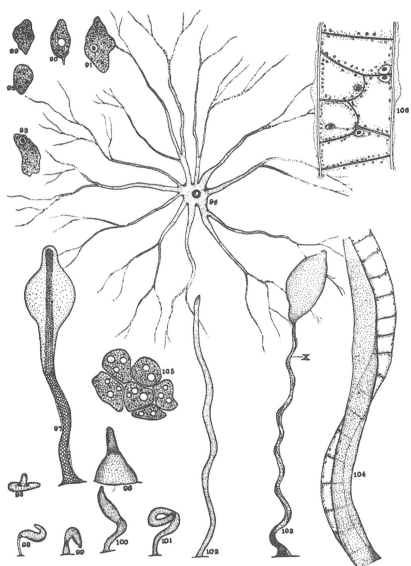

FIG. 1-4. Plate 7 of E. W. Olive's *Monograph of the Acrasieae* (1902) showing stages in the development of three dictyostelids: *Dictyostelium mucoroides* (89-93, 95-97, 106), *D. purpureum* (98-105), and *Polysphondylium violaceum* (94). The myxamoebae were removed from pseudoplasmodia, the latter somewhat compressed. Nuclei in his figures 91 & 93 are inaccurately depicted as having centrally positioned nucleoli. Scale variable.

they lacked cell associations of true aggregative origin. Based upon his studies at Harvard University under the watchful eyes of Roland Thaxter, Olive attempted to present a comprehensive picture of what he and earlier investigators had learned about the growth, development, and possible relationships of the genera and species then known. His slime molds were primarily of coprophilous origin, and, like his predecessors, he relied heavily upon horse dung and dung decoctions, either liquid or "stiffened with agar," as culture substrates. Media containing peptone and potato decoction were used also, but concentrations were not given. Unidentified bacteria were constantly present. Their possible role in the nutrition of myxamoebae was noted but not explored.

Olive (1902) did emphasize the temporal separation of the vegetative (myxamoebal growth and division) and fruiting (cell aggregation and fructification) stages, and he surmised that the myxamoebae aggregated in response to chemotropic stimuli since *Dictyostelium mucoroides* and *D. purpureum* fruited separately when cells of the two species were intermixed. He included some further information concerning stalk and spore formation and, consistent with the times, dwelt at some length upon the cytology and nuclear states of the myxamoebae and of differentiating cells. His major contribution, however, related to the systematics of the group. For over three-quarters of a century his monograph has remained the one inclusive taxonomic work on the Acrasieae despite the publication of five new genera and many new species since that time.

Although differing in subject matter, and less frequently cited, Potts's paper "Zur Physiologie des *Dictyostelium mucoroides*" (1902) was no less significant. He cultivated *Dictyostelium mucoroides* upon sterilized dung and dung infusion, upon infusion agar made from the stems of *Vicia faba*, and upon an extract of corn (maize) grains which he employed either as a liquid or solid medium. Employing corn gelatin, which he found to be especially favorable, he observed that *Dictyostelium* developed only in the presence of bacteria, and he succeeded in isolating the slime mold with a single bacterial species which he described as *Bacterium fimbriatum*.[1] He further studied the relationship between these organisms by cultivating them upon a large number of synthetic media, each of which contained some inorganic or organic source of nitrogen together with a sugar, organic acid, or other source of carbon. Although he obviously misinterpreted the manner in which the slime mold fed upon the accompanying bacteria, believing they were digested extracellularly, he made no mistake in his conclusion that the *Dictyostelium* was dependent upon bacteria for nutriment. Furthermore, he was not unmindful of the influence the bacteria might exert through altering the culture medium, and he reported that *D.*

[1] In presenting current work, the nomenclature of bacterial species conforms with the 8th Edition of Bergey's Manual (1974); whereas, in reviewing the work of earlier investigators, the nomenclature they used has been retained.

mucoroides grew well in media of slightly acid to mildly alkaline reaction. Potts succeeded in growing the myxamoebae with pure cultures of three additional bacteria, *Bacillus subtilis, B. megatherium,* and *B. fluorescens liquefaciens,* and noted that the list could probably be extended. He refuted Nadson's claim (1899) of symbiosis between the last of these bacteria and *D. mucoroides.* Exhaustion of the nutrient was believed to stimulate cell aggregation, for Potts succeeded in keeping cells in a vegetative stage for 3½ months by transferring them frequently to fresh substrata. Concerning cell aggregation, he concluded that the ray-like arrangement of the inflowing arms of myxamoebae around the midpoint appeared "almost to point to a chemotactic influence." Normal fruiting was reported to be strictly aerobic and strongly influenced by humidity, which in turn seemingly controlled transpiration of the developing fructifications, a process that Potts regarded as being of prime importance.

As the title "Une Acrasiée Bacteriophage" suggests, Vuillemin (1903) clearly perceived the relationship between *Dictyostelium mucoroides* and *Bacillus fluorescens nonliquefaciens,* the bacteria with which his slime mold was isolated and cultivated. The *Dictyostelium* grew only in the presence of bacteria, and he states unequivocally that the bacteria were ingested and digested by the myxamoebae. He appreciated also the interrelationship between the bacteria, the substrate, and the presence or absence of slime mold growth, for he concluded that the slime mold failed to grow on maltose-peptone medium with *Bacillus pyocyaneus,* a more proteolytic species, because of excess alkalinity.

Vuillemin's studies were extended in several ways by E. Pinoy (1903, 1907). He confirmed the engulfment and digestion of bacteria and succeeded in isolating from myxamoebae of *D. mucoroides* a preparation that liquefied gelatin and dissolved bacterial cells (*B. coli communis*) that had been killed by chloroform. To the active substance, or enzyme contained in this extract, he applied the name "acrasidiastase," and noted that it was similar in its action to a substance, or enzyme, earlier isolated by Mouton (1902) from a soil amoeba and termed an "amibodiastase." Dismissing symbiosis as a possibility, Pinoy (1907) considered the slime mold as living parasitically upon the colonies of associated bacteria. He cultivated the myxamoebae successfully with several Gram-negative bacteria including *Bacterium coli, Bacillus kieli, Bacillus violaceus,* and *Vibrio cholorae,* and noted that Gram-positive species such as *Bacillus subtilis, Bacillus megatherium,* and *Bacillus anthracis* were not consumed. Of the various substrates investigated, an agar medium based upon cooked flax seed proved most useful.

Pinoy made some comparisons between *Dictyostelium mucoroides* and *D. purpureum* and *Polysphondylium violaceum,* having obtained the latter cultures from Professor Thaxter. Sorus color in *D. mucoroides,* but not in *D. purpureum* or *P. violaceum,* was reported to vary with the identity of the bacterial associate and the composition of the underlying substrate. Since the fluorescent pigment of *B. fluorescens liquefaciens* could account for a yellow sorus in *D. mucoroides*, he surmised that van Tieghem's *D. roseum* (1880) might have been the same species in the presence of a red-pigmented bacterium such as *Bacillus kieli*.

A decade later F. X. Skupienski initiated studies on sexuality among the slime molds (1917, 1918). Although concerned primarily with the true Myxomycetes, such as *Didymium nigripes*, he published on *Dictyostelium mucoroides* briefly in 1918, and at greater length in his "Recherches sur le Cycle Évolutif de Certain Myxomycètes," published privately in Paris in 1920. A sexual cycle was reported where (+) and (−) myxamoeba-gametes fused to form zygotes, and these then merged to produce a plasmodium of short duration. This in turn was transformed into a single sporangium, rarely two. The spores and cells of the stalk(s) were thought to be derived by progressive division of the whole plasmodial mass as in the fruiting process of *Didymium nigripes*. This interpretation could not be substantiated by other investigators, and the discovery of a true sexual stage in *Dictyostelium* was delayed for a half-century until the origin, structure, and function of the macrocysts were revealed (Erdos, Nickerson, and Raper, 1972 and 1973; Clark, Francis, and Eisenberg, 1973; Erdos, Raper, and Vogen, 1973). Skupienski's cultural studies were informative and generally correct, except he concluded, like Nadson (1899) before him, that *D. mucoroides* and *Pseudomonas fluorescens liquefaciens* existed in a symbiotic relationship.

At about the same time (1922) R. Oehler in Würzburg reported that *Dictyostelium mucoroides* could be isolated from compost and forest soil in addition to dung, and that it represented an ideal subject for laboratory demonstrations of spore germination and the nutrition of amoeboid organisms growing in association with living bacteria. Of such bacteria Gram-negative species were most suitable and "xerose-bacteria" (*Corynebacterium xerosis*?) were less so, while Gram-positive bacilli were not recommended.

Using cultures obtained from Oehler, W. von Schukmann took up the study of *D. mucoroides* and in 1924 and 1925 published papers on its biology and morphogenesis. In the first of these he reported that substrains which had ceased to produce pseudosporangia, or even pseudoplasmodia, could be restored to normal development by implanting myxamoebae on sterilized horse dung together with an indigenous dung bacterium. Once the ability to produce pseudosporangia had been restored, the revitalized slime mold could be cultivated successfully on agar media with *B. coli*. In the second paper he, like Potts (1902), recognized nutrient exhaustion as a primary stimulus for fructification, but claimed that negative hydrotropism also played a significant role. Enhanced fruiting on roughened agar surfaces created by the incorporation of inert objects such as glass wool and glass beads was cited as evidence. In still other experiments he attempted to immunize rabbits by injecting them with myxamoebae of *D. mucoroides*. Although not dramatic by current standards, his results were nonetheless interesting. Spores of *D. mucoroides* were not influenced by the immune serum.

Except for the pioneering investigations of E. W. Olive, already cited, the *Dictyosteliaceae* attracted little or no attention in America prior to the mid-twenties, when R. A. Harper of Columbia University began publishing his analytical studies of sorocarp construction in *Dictyostelium mucoroides* and *Polysphondylium violaceum*. Having previously studied processes of morphogenesis in coenobic algae such as *Pediastrum* and *Hydrodictyon* (1918), he was drawn to the Acrasieae because of their capacity to build from previously independent myxamoebae erect, multicellular fructifications of considerable symmetry. In the first of three papers, "Morphogenesis in *Dictyostelium*" (1926), he introduced the very useful term *sorocarp* for the completed fructification, and in great detail analyzed steps in its formation. Much emphasis was placed upon photographs of living material, and although not very clear they still supported his thesis of intercellular communication and regulation leading to fairly constant levels of proportionality in completed sorocarps and in the cells that comprised them. Recognizing that many other factors influenced the construction and pattern of sorocarps, Harper believed that cellular contacts and pressures, combined with stimuli of maximal weight resistance and diminishing load, were of prime importance in effecting the readjustments in cellular form and position required during fructification. When one considers the culture techniques employed (primarily hanging drop cultures in dung extract), it is remarkable that he achieved so much.

Harper's second paper, "Morphogenesis in *Polysphondylium*" (1929), reconfirmed and extended the observations and conclusions recorded for *Dictyostelium*. For this study the cultures were grown on dung agar and the illustrations (photographs) were far better. Harper emphasized the complete separation of growth (an increase in the number and mass of myxamoebae) and morphogenesis (the aggregation and subsequent differentiation of these cells) and contrasted this with the situation for higher plants and animals, among which these processes are concurrent and inextricably mixed. The regulation of size and pattern in developing sorocarps was again examined. In *Polysphondylium* this involved the further consideration of intercellular communication and the adjustments required, first, to cause posterior masses of cells periodically to lag behind on the sorophore in their vertical ascent, and second, to cause such arrested masses to cleave and to develop from each segment a lateral branch roughly perpendicular to the central axis—each branch in its structure and anchorage representing the equivalent of a miniature sorocarp. *Polysphondylium* thus demonstrated even more clearly than did *Dictyostelium* the need for continuing sensitive adjustments between stalk and spore-forming cells. Here too Harper stressed the matter of proportionality. He cited three types of cellular phenomena that were involved, namely: coordinated creeping movements to form the pseudoplasmodial masses; specific "morphallactic" adjustments in cellular form to construct the tapering stalk and branches; and intracellular histogenic metamorphosis of cells in situ to form the cellular elements of the stalk and spores, respectively. In this paper another new and useful term, the *sorogen*, was introduced to designate the pseudoplasmodial mass after it left the substrate during sorocarp construction.

Harper's third paper, "Organization and Light Relations in *Polysphondylium*" (1932), analyzed in great detail the comparative effects of alternating daylight and darkness versus continuous relative darkness upon the form and dimensions of developing sorocarps. Cultures were incubated on a laboratory shelf near a window with equal numbers of dishes under similar bell jars. One of these jars was covered by a black cloth which "did not exclude all light," the rationale being to simulate conditions in nature. Briefly stated, Harper found that fewer and larger sorocarps were formed in darkness than in alternating light and darkness. However, the total number of whorls and of branches in the two sets were about the same—the greater number per sorocarp in darkness being offset by a greater number of sorocarps in light alternating with darkness. Dimensions and patterns of cell aggregations were not recorded.

Stimulated by Harper's papers and encouraged by him personally,[2] K. B. Raper began isolating and studying *Dictyostelium* in 1930, and with Charles Thom published a short paper on its distribution in soils soon thereafter (1932). This was followed three years later by the description of *Dictyostelium discoideum*, a new species isolated from soil of a deciduous forest. This slime mold possessed a number of singular characteristics (Raper, 1935). Foremost among these were a freely migrating pseudoplasmodial stage and the development under optimal culture conditions of sorocarps of essentially uniform construction and constant proportions irrespective of their size. The sorocarps consisted of beautifully tapered stalks (sorophores) which arose from flattened, cellular discs and bore at their apices unwalled, globose-to-slightly-citriform, white-to-yellowish spore masses (sori).

Other papers followed. In the first of these (1937) it was shown that *D. discoideum* could be grown on hay infusion agar with a great variety of bacterial associates including Gram-negative and Gram-positive forms,

[2] Raper was at that time working in the Division of Soil Microbiology, U.S. Department of Agriculture, in the laboratory of Dr. Charles Thom. Harper, Thom's former teacher at Lake Forest College and continuing good friend, would visit the laboratory each spring when he came to Washington (D.C.) for the annual meeting of the National Academy of Sciences, and while there he would inquire about our work on the Acrasieae. His salutation was always a question, and always the same: "Have you found *Acrasis*?" Regrettably, the answer too was always the same and always, "No."

rods and cocci, and spore-formers and non spore-formers. Of these the Gram-negative rods were most satisfactory. Among spore-formers the vegetative cells were ingested and consumed, whereas the spores were ejected intact. The question of symbiosis versus nonsymbiosis between the slime mold and asociated bacteria was laid to rest, and the relationship described as one of prey and predator. Finally, a method of expressing the growth of *Dictyostelium discoideum* with different bacteria and on different agar substrates in a semi-quantitative manner was introduced.

Two papers appeared in 1939. Of these one compared plant, animal, and human pathogens as bacterial associates (Raper and Smith, 1939); it was found that these served equally as well as their saprophytic cousins as nutrient for the myxamoebae of *D. discoideum*. The second paper (Raper, 1939) covered in substantial detail the influence of culture conditions upon the growth and development of *D. discoideum*, and in a very real sense laid the foundations for later laboratory investigators with this slime mold. The interrelationships between the composition of the substrate, the fermentative capabilities of the associated bacteria, and the amount and normality of slime mold growth and fructification were carefully evaluated. Working with *Escherichia coli*, and to a lesser extent with *E. coli* var. *communior* and *Pseudomonas fluorescens*, and two much-used diagnostic media, Endo Agar and Levine Eosin-Methylene Blue agar, Raper found that satisfactory substrates for two-membered cultures of *E. coli* and *D. discoideum* could be compounded by balancing an organic nitrogen source, 1.2% peptone, and a utilizable carbon source, 1.2% lactose or glucose, so that the pH leveled off and remained within a favorable range, the optimum being pH 6.0-6.5. Alternatively, KH_2PO_4 and Na_2HPO_4 could be used as buffers to establish and maintain favorable pH levels in substrates where the bacterial degradation of nutrient components did not achieve the desired balance. With limited modifications, and the introduction of *Aerobacter aerogenes* and *E. coli* B/r as alternate bacterial hosts, this basic substrate and dilutions thereof has been used quite generally for the past 40 years.

Additional papers followed, and of these "Pseudoplasmodium Formation and Organization in *Dictyostelium discoideum*" (Raper, 1940a) assumed particular importance, since it established this species as uniquely suited to laboratory cultivation and study. Shifts from one bacterial associate to another were made easy because of the bacteria-free spores borne in isolated sorocarps, and the freely migrating pseudoplasmodia enabled one to perform various types of dissections and grafts with highly significant results. It was found that the migrating structures were strongly polarized, that the extent and direction of migration was controlled by their apical tips, and that at this early stage demarcation into presumptive stalk- and spore-forming areas had already occurred. Despite this tentative commitment there still remained in all parts of the multicellular body a potential for cellular reorganization and the formation of a normal sorocarp from any isolated fragment. Such experiments significantly influenced future work in this laboratory and elsewhere.

Other papers of the period reported further investigations with *Dictyostelium discoideum*. More importantly, they demonstrated that previously described species could be studied and analyzed in the same general manner. Thus, much information of a comparative nature was obtained concerning myxamoebal growth, cell aggregation, pseudoplasmodial behavior, and morphogenesis in *D. mucoroides, D. purpureum, Polysphondylium violaceum*, and *P. pallidum* (Raper, 1940b and 1941a). Another study (Raper and Thom, 1941) involved the growth, behavior, and fructification of different species when their myxamoebae were intermixed. Two approaches were employed: (1) different species were implanted so their expanding colonies of vegetative myxamoebae became intermixed at common frontiers, and (2) myxamoebae in different stages of fructification were disassociated and subsequently intermixed. In the first case *D. discoideum* and *D. mucoroides* formed common wheel-like aggregates but subsequently pulled apart and fruited separately. In the second case, where existing pseudoplasmodia of *D. discoideum* and *D. purpureum* were disassociated at common sites, a variety of sorocarps developed, some representing each species and others intermixed in various ways. No common aggregates were formed or mixed sorocarps produced under any condition in mixtures of *D. discoideum* and *Polysphondylium violaceum* myxamoebae. Thus were suggested similarities and differences in chemotactic attractants among different species and genera that would be confirmed a quarter-century later.

Parallel with Raper's early investigations on *D. discoideum*, but unknown to him at that time, Arthur Arndt in Rostock, Germany, was making a comprehensive study of growth and development in *D. mucoroides*. His work, published posthumously in 1937 under the title "Untersuchungen über *Dictyostelium mucoroides* Brefeld," was remarkable in several ways, not only for what the paper contained but also for what might have been included had Arndt lived to publish his total researches. Special attention was given to the formation of the pseudoplasmodium and the early stages of sorocarp construction. To distinguish recognizable stages in this continuous process he introduced a series of terms, including "präplasmodium," "cumulus," "colliculus," "agmen," "conus," and "clava," that have gained limited acceptance in subsequent literature. He was the first to use time-lapse cinematography, and was able with this technique to reveal a dramatic pattern of inflowing waves in the streams of aggregating cells. He observed also that the centers of such aggregations (colliculi and coni) did not remain stationary during pseudoplasmodium formation. Upon this and other evidence he concluded that the stimulus for aggregation arose within the myxamoebae, as Olive (1902) and Potts (1902) had suggested, and not from some external source, such as negative hydrotropism, as von Schuckmann (1924) had claimed. He stopped short of suggesting what he thought that stimulus might be.

Although an historical account such as this could be extended into the present, one must, I believe, select some date as a benchmark from which to look backward and forward. If the year 1941 is chosen, I think it may be said that the rebirth of interest in the Dictyosteliaceae can be attributed to the researches of Harper, Arndt, and Raper. This work has been greatly expanded by the imaginative researches of John Bonner at Princeton University beginning in 1944, of Maurice Sussman since 1951 at Northwestern and Brandeis Universities and the University of Pittsburgh, of Brian Shaffer at Cambridge University from 1953, of Gunther Gerisch since 1959 at laboratories in Tübingen, Freiburg, Basel, and Munich, and of many more who at one time or another were associated with them, or who discovered *Dictyostelium* in the published literature and took up its study independently. It would be quite impractical to cover in this place even a fraction of the investigations reported over the years from more than a hundred laboratories throughout the world. Instead, many of these will be cited in connection with specific topics in the chapters that follow.

CHAPTER 2

Occurrence and Isolation

The dictyostelid slime molds were long considered to be primarily coprophilous, for it was from the dungs of animals, commonly herbivores, that the early investigators obtained most of their cultures (Brefeld, 1869 and 1884; Nadson, 1889; Olive, 1901 and 1902; Potts, 1902; Pinoy, 1903; and others). Even then it was recognized that other habitats could be involved as well. Van Tieghem found *Acrasis granulata* growing on a paste of beer yeast (1880), *Coenonia denticulata* on decaying beans partially covered by water (1884), and *Dictyostelium lacteum* on decomposing agarics (1880). Some decades later, Oehler (1922) recorded compost and forest soil, along with dung, as sources of *D. mucoroides*, and shortly thereafter H. Krzemieniewska and S. Krzemieniewski (1927) in Poland reported that *D. mucoroides* could be obtained regularly from cultivated soils and *Polysphondylium violaceum* somewhat less commonly from forest soils. Following their lead, Harper isolated cultures of the latter slime mold from soils collected in the parks of New York City (unpublished); and Raper and Thom (1932) confirmed the presence of these slime molds in American soils in a pattern similar to that reported from Poland. It was not, however, until the more intensive studies of Raper (1935 et seq.), Cohen (1953), and more particularly those of Cavender and Raper (1965a,b,c) and Cavender (1969 et seq.) that soils from forest areas, especially deciduous forests, were shown to represent an especially rich source of dictyostelids, both in number and diversity of species. Among these *D. mucoroides* was the most abundant.

This did not imply that dung should be ignored as a natural habitat for *Dictyostelium* and *Polysphondylium*, particularly the larger and more common species. It did indicate, however, that dung could no longer be regarded as the principal habitat, and that dictyostelids are quite commonplace in nature. In fact, one might expect to find them wherever microbial decomposition of organic matter, particularly plant residues, occurs under aerobic conditions of high moisture and moderate temperature. Furthermore, such conditions need not prevail continuously, for all the cellular slime molds have some type of resting stage, either spores or cysts, that can carry them through limited periods of adversity. If one considers the environmental requirements for growth and assumes the presence of sufficient nutrient in the form of bacterial cells, one can readily understand why soil is a singularly productive source. Whereas there may not be large masses of bacteria in soil at any particular time, they are always present and ready to proliferate under favorable conditions; and the slime molds, we should recall, need not be widely dispersed but need only be able to grow and fruit in localized niches where and when conditions are favorable. We suspect that the dictyostelids tend to be concentrated in such microenvironments.

In addition to various types of soil (forest, cultivated, prairie, desert, and marshland) and many kinds of dung, it may be of interest to record that we have isolated dictyostelids from spoiled potatoes and vegetables, decomposing grass, musty hay, rotting wood, decaying fungi of various types, and blades of grass floating in a stagnant pond. The list may be extended from the published reports of other researchers: leguminous pods, ears and tassels of corn, decaying breadfruit, dead flowers of the cannon ball tree (Olive, 1960), and the rhizospheres of many plants (Agnihothrudu, 1956).

Many different methods may be used to isolate dictyostelids in the laboratory. Where fruiting structures project above or otherwise stand apart from fungi and other microorganisms on natural habitats, it is a simple matter to touch a sorus with a fine needle and implant the adhering spores onto some agar medium, with or without added bacteria. As produced on natural substrates such as dung, or in an isolation plate with an indigenous microflora, the spores in a developing sorus are almost invariably accompanied by some type(s) of nutritive bacteria, and upon the regrowth of these the emergent myxamoebae can feed following germination. This quite obviously represents the method employed by Brefeld (1869, 1884), van Tieghem (1880, 1884), Potts (1902), Vuillemin (1903), Pinoy (1903, 1907), Skupienski (1918, 1920), and Oehler (1922); and where the accompanying bacteria were identified, in more cases than not they were reported to be *Bacillus fluorescens liquefaciens (Pseudomonas fluorescens)*, a common soil bacterium. Once they had isolated *D. mucoroides* with an indigenous bacterium, Potts, Pinoy, and Oehler succeeded in growing it in association with a few other species as well.

Such methods of isolation were at best dependent on chance encounters with dung or other objects upon which the slime molds were already present and apparent. Once it was known that dictyostelids could be isolated from soil, several methods were developed that involved an actual search for species of *Dictyostelium* and *Polysphondylium* and such other forms as might be found. Of these methods, one of the first was that of Raper and Thom (1932). Samples of soil or decaying vegetation were ground in a clean mortar and diluted with approximately ten volumes of sterile water. The resulting suspension was then streaked upon plates of mannite agar (Ashby's formula) and incubated for 2-3 weeks at 16-18°C. The nitrogen-free mannite agar, commonly used for growing *Azotobacter*, and the relatively low incubation temperature were selected to limit rampant fungal growth while supporting enough indigenous bacteria to nourish whatever dictyostelids might be present. Once sorocarps of *Dictyostelium* or *Polysphondylium* had developed, spores from isolated sori were removed and implanted on fresh substrate, either as spot or streak inoculations. By repeated transfer two-membered cultures were obtained as quickly as possible with what appeared to be the dominant bacterial associate. This varied from one slime mold to another, although species of *Pseudomonas* were the primary associates in about a fourth of the slime mold cultures (Raper, 1937). It was in this way that *Dictyostelium discoideum* (Fig. 2-1) was isolated in 1933 from decaying leaves collected in a hardwood forest of beech, birch, oak, and horse chestnuts, or buckeyes (*Aesculus Hippocastanum* L.), at Little Butts Gap (Fig. 2-2) in the mountains of western North Carolina (Raper, 1935). As can be seen from a paper published two years later (Raper, 1937), strain NC-4, the type culture of *D. discoideum*, was isolated and cultivated in association with *Vibrio alkaligenes* for many months before being shifted to *Escherichia coli*, with which it has since been maintained.

Improvements have been made with time. Mannite agar ceased to be used and instead a dilute hay infusion agar (Raper, 1937, 1951) was substituted in primary isolation plates; and samples were no longer comminuted but shaken in sterilized water during half-hour periods to dislodge and disperse slime mold spores and myxamoebae. Small aliquants of the resulting suspension were then streaked on plates of hay and dilute hay infusion agar and incubated at 20-24°C for several days, the plates being examined periodically for the presence of wheel-like pseudoplasmodia and sorocarps. As sorocarps developed (Fig. 2-3), spores were removed from isolated sori and transferred to plates of hay agar prestreaked with *E. coli* or another selected host bacterium. Inoculations were made at the intersection of crossed streaks or at one end of plate-long streaks. Each technique has its advantage: with the crossed streaks one has four possible choices from which to reisolate the slime mold after it has grown through the streaks; with the longer streak the slime mold has a greater distance over which to free itself of fungi and unwanted bacteria. If fungi, usually mucorales, are a problem, one can brush aside the aerial growth with a thin platinum needle and permit new sorocarps to develop from which fungus-free spores can be removed the following day. Ridding the slime mold of an unwanted bacterium characterized by thinly spreading growth is more difficult, but this too can be accomplished as a rule in two or three passages on a nutrient-poor substrate that is weakly acidic. If desired, a dictyostelid may be shifted from one bacterial associate to another in the same manner. A paper published some years later in the *Quarterly Review of Biology* (Raper, 1951) summarized the effectiveness of these methods for isolating cellular slime molds from soils. As many as four or five different species were oftentimes recovered from single samples consisting of a few grams of forest soil.

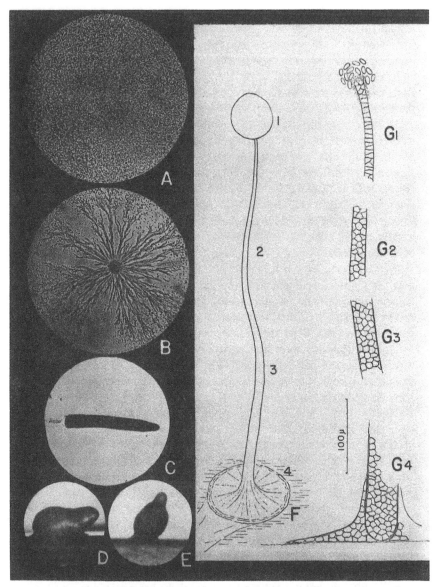

FIG. 2-1. Stages of the life cycle of *Dictyostelium discoideum*. *A*. Vegetative stage showing uniform distribution of free-living myxamoebae, × 14. *B*. Pseudoplasmodium formation by convergence of myxamoebae in streams, × 14. *C*. Migrating pseudoplasmodium, × 18. *D,E*. Transition from migration to early culmination, × 18. *F*. Diagrammatic representation of a mature sorocarp. *G1-G4*. Camera lucida drawings to show cellular structure of segments corresponding to regions indicated in *F*. (From Raper, 1951)

Fig. 2-2. Little Butts Gap, Craggy Mountains, western North Carolina, the site from which decaying leaves were collected that yielded the type of *Dictyostelium discoideum*, strain NC-4. Collection and photograph made in June 1933; species described two years later. (Raper, 1935)

Using essentially the same isolation procedures, but somewhat later, Hagiwara began an investigation of the Acrasiales in Japan (1971 et seq.) that has revealed the presence of such common species as *Dictyostelium mucoroides*, *D. minutum*, *D. sphaerocephalum*, *D. purpureum*, *Polysphondylium violaceum*, and *P. pallidum*, together with a number of newly described species that, in considerable part, seem to differ from but resemble taxa recently described in America and in Europe.

For studying the distribution of Acrasieae in the soils of Great Britain, B.N. Singh of the Rothamsted Experiment Station employed a technique he had developed earlier for isolating and studying soil protozoa (1946). Bacteria were grown for 2-7 days on nutrient agar and then one or two loopfuls were removed and spread on nonnutrient agar to form a disk, or "bacterial circle," about 1 inch in diameter. Several circles were placed in a single Petri dish. The base medium contained 1.5% washed agar and 0.5% NaCl, the latter being added to retard the growth of unwanted microorganisms. To each such circle was added small crumbs of soil or aliquots of a diluted soil suspension. Gram-negative bacteria proved es-

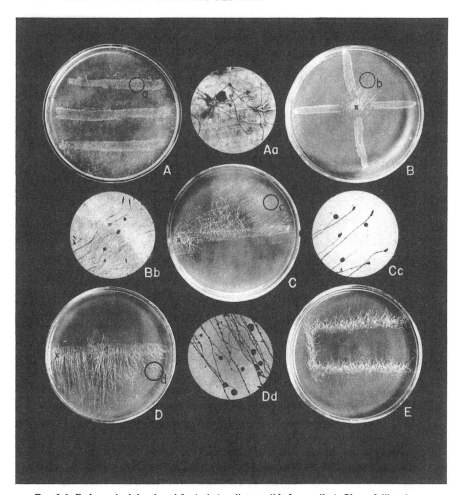

FIG. 2-3. Early method developed for isolating dictyostelids from soil. *A*. Plate of dilute hay infusion agar is streaked with soil suspension. Pseudoplasmodia consisting of inflowing myxamoebal streams develop, *Aa*, from which cells may be removed and implanted in preinoculated streaks of *Escherichia coli, B*. When pseudoplasmodia migrate or developing sorocarps extend beyond the indigenous bacterial and fungal growth, *Bb*, spores or myxamoebae are again removed and implanted in a preestablished bacterial streak *C*. Spores are removed from isolated sori near the opposite end of this streak *Cc*, and the procedure repeated a third time, *D,Dd,E*, to ensure purity of the two-membered culture. Inoculation sites are indicated by **x**. The dictyostelid illustrated is probably *Dictyostelium giganteum*. (Recast from Raper, 1951)

pecially suitable, and much of his work centered upon *Aerobacter* strain #1912, subsequently diagnosed in this laboratory as *Aerobacter aerogenes* by Ralph Wiseman.

Using the bacterial circle technique, Singh (1947a) surveyed 67 samples of arable and grassland soils from different areas of Great Britain. "*Dic-*

tyostelium spp.'' was found in 33 of 38 arable soils, whereas they were obtained from only 3 of 29 grassland soils. All apparently were *D. mucoroides*. Additionally, a new species, *Dictyostelium giganteum*, was isolated from an actively decomposing compost heap; it was described as differing from *D. mucoroides* primarily in its much longer, often creeping sorocarps. Singh expressed regret that he was unable to make quantitative estimates of the numbers of slime molds in soils, either as spores or myxamoebae. In an accompanying paper (1947b), however, he demonstrated quite convincingly that the myxamoebae of *D. mucoroides* and *D. giganteum* could proliferate, spread through, and fruit in sterilized soils to which 2-3 day cultures of nutritive bacteria were added.

At about the same time Eugene Kitzke (1950), studying soils and decomposing plant materials from Milwaukee County, Wisconsin, reported the isolation of *Dictyostelium purpureum* and *Polysphondylium violaceum*. Materials from the field were macerated, suspended in sterile water, and streaked on mannite agar. Incubation was for 2 weeks at 16-18°C. As sorocarps developed, spores from mature sori were transferred to hay infusion agar plates preinoculated with *E. coli*. Dictyostelids were isolated from only 4 of 33 samples. Three strains of *D. purpureum* and one of *P. violaceum* were represented. No mention was made of the ubiquitous *D. mucoroides*, and one wonders if this might have been overlooked or disregarded. Although *D. purpureum* is not rare, it would be most unusual to find three cultures of this species and none of the other. The study was extended the following year to include soils of several kinds collected in and near Madison, Wisconsin, and for this survey a low-nutrient soil extract agar was used as the primary isolation medium (Kitzke, 1951). Randomly selected soil particles were placed on the agar directly instead of being added in suspensions. Thirty-six of 60 soil samples yielded dictyostelids, and of the 157 cultures isolated *Dictyostelium mucoroides* represented 76, *Polysphondylium violaceum* 60, *P. pallidum* 12, and *D. minutum* 9—about the proportions one might expect. Of the different soils tested, those from deciduous forests were most productive (see also Raper, 1951). Kitzke in 1952 proposed a very different if not better procedure: soils were loosely pulverized and placed in 25 cc specimen bottles to which was added an amount of bacterial suspension (*E. coli*) sufficient to moisten but not inundate the soil, the bottles closed with loose-fitting caps, and the mixtures allowed to incubate at 20-23°C. Sorocarps appeared in 3 days, from which spores were removed and implanted on agar plates (medium not stated) in pregrown streaks of *E. coli*. Experience tells us that this might be used for isolating the larger species, such as *D. mucoroides* and *P. violaceum*, but one would hardly find an *Acytostelium* with this method.

As part of a continuing interest in the Mycetozoa, Myxomycetes, and Acrasieae, Arthur Cohen in 1953 published short reports on the isolation and cultivation of "opsimorphic" organisms, of which Part I concerned the Acrasieae specifically. (Part II, by Johanna Sobels and Cohen, related to the Myxomycetes.) Soils were collected in small plastic vials (22 mm) with screw caps, care being taken to avoid any possible cross contamination. Collections were made in Georgia, Florida, Michigan, and Massachusetts for a total of 87 samples representing forests, bottom land, savannah, pasture, and cultivated fields. Mannitol agar to which sterile rabbit dung pellets were added prior to gelation was used as a primary isolation medium; five pellets were added to each plate, equally spaced and only partially embedded. Water was added to the vial, shaken vigorously, allowed to settle, and 2-3 ml of the resulting suspension added per plate. Excess water was later drained off and the plates incubated at 20-24°C. They were observed periodically. After sorocarps appeared, spores were removed from isolated sori by needle contact and implanted at the center of PO_4 buffered nonnutrient agar plates previously spread with a washed suspension of *E. coli*. Three successive monosorial transfers generally sufficed to yield a two-membered culture. Transfers were then made to an appropriate nutrient agar for maintenance as stock cultures. Six species of *Dictyostelium* and *Polysphondylium* were isolated among a total of 52 dictyostelids. The greatest number of species (4) came from a soil collected in a dense pine forest in southern Georgia.

Working at the University of Madras, V. Agnihothrudu (1956) conducted a survey of the Dictyosteliaceae in the soils of southern India with special reference to the rhizospheres of growing plants. Isolation methods patterned after those reported by Raper (1951) were employed, supplemented by bacterial "circles" using *Aerobacter aerogenes* as recommended by Singh (1947a). A total of 83 samples were examined: 20 cultivated soils, 15 uncultivated soils, and rhizosphere soils from 48 species of crop plants and common weeds. Not surprisingly, *D. mucoroides* was present in a majority of the samples; what is more significant was its presence in 36 of the 48 rhizosphere soils. Also noteworthy was the high incidence of *Polysphondylium pallidum* (18 of 48) in these soils, as was the presence of *D. discoideum* in 8 of the total number and in 6 rhizosphere soils. Insofar as we are aware, this is one of only three reports of *D. discoideum* being isolated from sources outside the Americas, the others being that of Lee (1971) and Kawabe (1980), who obtained it from soil and decaying leaves from Japan. In addition to the above, cultures of *D. minutum, D. purpureum,* and *P. violaceum* were isolated, and for each species the highest incidence occurred in the rhizosphere soils. Four species (unnamed) were isolated from one rhizosphere sample of pigeon-pea (*Cajanus cajan*), while from a rhizosphere sample of peanut (*Rachis hypogea*) five species were obtained. It would seem that further and detailed study of such associations, now long overlooked, would be imminently worthwhile.

Also in India, but somewhat later (1961), Rai and Tewari at Lucknow University undertook the isolation of cellular slime molds from soil to fill what they believed to be a lack of record of this group from Indian soils,

". . . except that Raper in a personal communication informed Singh (1947) that he had isolated species of *Dictyostelium* from soils of Australia, India, Mexico, Brazil, Cuba and numerous stations throughout the United States." Hay infusion agar (Raper, 1937 et seq.) and nonnutrient agar (Singh, 1946, 1947a) were employed: the former by placing particles of soil on the surface of plates supporting a day's growth of suitable bacteria; the latter by adding pregrown bacteria in the form of disks or circles to the center of which soil particles or drops of a soil-water suspension were added. Soils were collected "from a depth of six inches" in randomly selected areas in the environs of Lucknow. Plates were incubated at 22°C. When sorocarps developed within a few days, spores were implanted at the intersection of crossed bacterial streaks on an appropriate substrate and subcultured as necessary to secure two-membered cultures. *Escherichia coli* and *Aerobacter aerogenes* were used as sources of bacterial nutrient. Three points deserve mention: (1) only two species were isolated, *Dictyostelium mucoroides* and *Polysphondylium violaceum*, in contrast to six in Agnihothrudu's study; (2) *D. mucoroides* was isolated from all but 2 of the soils examined (52 of 54), and *P. violaceum* from 30, a high incidence for each species; and (3) this incidence seems especially high for slime molds that are strictly aerobic and for soils said to be collected to (?) a depth of 6 inches.

In a review paper entitled "The Biology of the Cellular Slime Molds," Maurice Sussman (1956) reported that from 52 samples of soil taken from areas supporting herbaceous plants, trees, and grass in the environs of Northwestern University, the Cook County Forest Preserve, and the Winnetka Golf Course he obtained 24 isolates of the *Dictyostelium mucoroides* complex, 16 of *Polysphondylium violaceum* and *P. pallidum*, one of *D. purpureum*, and one that might be similar to E. W. Olive's *D. aureum* (1901). The method of isolation consisted, apparently, of spreading a few drops of an *E. coli* or *A. aerogenes* suspension over plates of plain agar and then grinding small amounts of soil between thumb and forefinger, permitting the finely divided sample to settle on the plates. Alternatively, suspensions of soil could be made, and a few drops of this could be combined with a drop of bacterial culture and the mixture spread over the agar. Similar methods were reported to have been used by Shaffer and Bonner. In any case, dictyostelids developed on the plates in a clone-like fashion, and Sussman (1956) surmised that "it should be possible to spread homogenized samples with perhaps a variety of bacterial associates on dilute nutrient or plain agar and obtain an accurate picture of the incidence of these forms." A modification of such an approach was introduced within a decade by Cavender and Raper (1965a) that yielded semi-quantitative results; it has been much used by Cavender (1969 et seq.), Traub (1977), Benson and Mahoney (1977), Sutherland and Raper (1978), and others since that time.

The so-called "Cavender Method" for isolating dictyostelids and other cellular slime molds from soils is presented herewith in essentially the form published by Cavender and Raper in 1965 (Fig. 2-4).

COLLECTION OF SAMPLES. Samples are collected at the soil surface or from the humus layer of forests, and in the latter case include a portion of the fermentation layer where decomposition is most active. An adequate number of soils are taken to ensure representative samples of a particular forest population. Soils may be conveniently collected in small plastic vials with wide mouths and snap-on covers, or in plastic Whirl-Pak bags. For the former method samples are taken by scraping the soil surface with the mouth of the vial, for the latter by means of a spoon or spatula. If soils cannot be plated out immediately after collection, they should be stored at a temperature slightly above freezing. Numbers of Acrasieae diminish after collection, there being a 15-25% decrease after 8 weeks storage at 4°C in a moist condition. Freezing is not recommended; it causes a sharper initial decrease in numbers than storage at 4°C. In no case should the soils be allowed to become dry.

PREPARATION OF THE MEDIUM. A dilute organic medium that stimulates growth and development of dictyostelids, but which is too dilute to allow appreciable growth of bacteria or fungi, is made from leached, dried hay (largely *Poa* spp.) weathered for several weeks and collected from fields or roadsides. An infusion is made from 10 g of hay per liter of distilled water by autoclaving at 120°C for 20 minutes, filtering the infusion through cotton on cheesecloth, and making up to volume with distilled water. The infusion is buffered with 1.5 g $KH_2 PO_4$ and 0.62 g $Na_2HPO_4 \cdot 7H_2O$ per liter to yield pH 6.0 $\pm$. Fifteen grams of agar are added per liter. The medium is sterilized at 15 lbs. for 15 minutes at 121°C.

PLATING TECHNIQUE. Petri plates are prepared containing the hay medium to a depth of 4-5 mm. To ensure a uniform distribution of the soil and bacterial suspensions when these are added, the plates are marked and retained in the same position on the table as when the agar was poured. Ten grams of soil are measured into a 500 ml flask containing 90 ml of sterile distilled water, giving an initial dilution of 1:10. Flasks are then placed on a rotary shaker at 280 rpm for 2 minutes to break up soil particles and to distribute spores and myxamoebae. A dilution of 1:25 is made by adding 5 ml of the above suspension to 7.5 ml sterile water, and 0.5 ml of this dilution is added per plate, giving a final dilution of 1/50 g soil/ plate. At the same time, about 0.4 ml of a heavy suspension of a suitable pregrown bacterium is deposited to provide nutrient for the dictyostelids. *Escherichia coli*, strain #281, has proven highly satisfactory for isolating the Dictyostelidae. Other bacteria, yeasts, or a combination of different bacteria may be substituted when isolating slime molds with different host

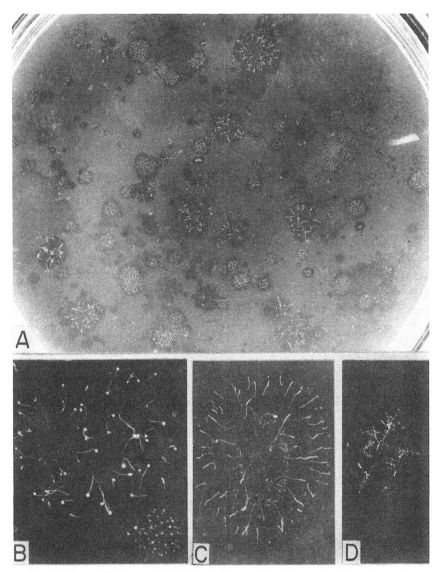

FIG. 2-4. Cavender's improved method for isolating cellular slime molds. This method yields semi-quantitative estimates of propagules (spores, myxamoebae, and microcysts), hence is useful for comparative studies. *A*. Isolation plate showing many clones of identifiable dictyostelids, ca. 5 days. *B*. A single large colony of *Dictyostelium mucoroides* and a smaller colony of *D. minutum* (lower right). *C*. A developing colony of *Polysphondylium violaceum*. *D*. A small colony of *P. pallidum*. (From Cavender and Raper, 1965a)

requirements such as *Acrasis rosea* and certain species of *Protostelium* (Olive and Stoianovich, 1960). *Escherichia coli* is grown on a tryptose 0.5%, glucose 0.1%, yeast extract 0.5%, agar medium in Roux bottles for a period of 24 hours at 30°C. The bacteria are then suspended in 12 ml of water, which provides sufficient nutriment for plating ten samples if three replicates per sample are run. The bacterial and soil suspensions are mixed together and distributed over the surface of the plates by tilting them back and forth. Porous clay covers may be used until free water has evaporated, or plates may be poured a day in advance in order for the agar to dry sufficiently to absorb the excess liquid. Plates are then incubated at 20-23°C, the latter being suitable for most dictyostelids whereas 20°C favors some alpine species.

POPULATION COUNTS. Clones appear after 3-4 days (Fig. 2-4A), and their positions should be marked early if quantitative information is desired. Enumeration and identification can be completed after sorocarps have developed. *Polysphondylium violaceum* (Fig. 2-4C), *Dictyostelium discoideum*, and certain members of the *D. mucoroides* complex (Fig. 2-4B, center) will usually appear sooner than *P. pallidum* (Fig. 2-4D), *D. minutum* (Fig. 2-4Bb, lower right), *D. lacteum,* and *D. polycephalum,* for example. This difference in rate of development necessitates rechecking the plates periodically for a week after plating.

CALCULATIONS. In recording quantitative information, clones of each species are totaled for all the samples from a given forest and the average number of clones per plate is calculated. The *density* of each species per gram of soil is determined by multiplying the average number of clones per plate for that species by 50, which represents the fractional part of a gram of soil added per plate. The total of such densities gives the *absolute density* of all dictyostelids per gram of sample.

The Cavender method represents, in some measure, a combination of the isolation methods previously described by Raper (1951) and by Sussman (1956). The basic medium is a hay infusion agar that is more dilute but otherwise comparable to that employed by Raper, whereas the addition of pregrown bacterial cells represents an adaptation of the technique proposed by Sussman. In practice, it is more rewarding than either of the earlier methods in the number of slime molds that may be isolated. Additionally, it has the distinct advantage of allowing, for the first time, some quantitative measure of (1) the dictyostelid population present in the soil of a given forest and (2) the relative abundance of different slime molds in such soils (Fig. 3-2A-E); and it provides a means of comparing this information with that obtained from other soils collected from similar or markedly different forests.

Some refinements of the Cavender method have been proposed by Kuserk, Eisenberg, and Olsen (1977), which include the collection of uniform soil cores, repeated inversion or gentle swirling of the soil-water suspension rather than vigorous shaking (to reduce injury to myxamoebae by the shearing action of soil particles), and freezing a portion of the soil suspension to kill trophic cells and thus reveal the percentage of propagules present as spores and microcysts (Cotter and Raper, 1968).

Whereas such methods afford the opportunity to isolate great numbers of strains of the more common species, their true significance lies in (1) making ecological comparisons, which will be considered in another chapter, and (2) isolating that single clone of a rare or perhaps previously unknown species.

It is especially significant that when soils are appropriately diluted and plated, the cellular slime molds develop as isolated colonies, each representing a clone, having developed, presumably, from a single myxamoeba, spore, or microcyst. It is equally important, and to us surprising, that the plates are not overrun by less desirable spore-forming bacteria, free-living soil amoebae, and filamentous fungi, which must be abundantly present in any 1:50 dilution of surface forest soil. One possible explanation is that the dense layer of pregrown bacteria upon a low-nutrient agar suitable for slime mold fructification affords a type of enrichment culture that is especially favorable for the cellular slime molds but not for most other soil organisms.

Beginning in 1965, Cavender and others have investigated the distribution and ecology of these singular organisms in various parts of the globe, including the forests of eastern North America (Cavender and Raper, 1965b and c), tropical and subtropical America (Cavender and Raper, 1968), Europe and East Africa (Cavender, 1969a, b, respectively), Southeast Asia (Cavender, 1976b), and Alaska (Cavender, 1978). With minor modifications the method has been used also by Traub (1972), Traub, Hohl, and Cavender (1981b), and Frischknect-Tobler, Traub, and Hohl (1979) for quite detailed surveys of the cellular slime molds present in the soils of Switzerland; by Benson and Mahoney (1977) for a comparable investigation of the dictyostelids in soils of southern California; by Sutherland and Raper (1978) for a study of the occurrence and distribution of cellular slime molds in the prairie soils of Wisconsin; and by Kawabe (1980) and Kanda (1981) for surveys of cellular slime molds of the Southern Alps of Japan and the Kushiro moor in Hokkaido, respectively.

Out of such studies has come a new but still incomplete perception of the potential role these amoeboid organisms may play in influencing or maintaining balanced soil microfloras; from them also has come the discovery and description of many new species and one new genus that greatly expands our concepts of the Dictyostelidae as a unique and experimentally useful group of microorganisms.

CHAPTER 3

Ecology

It is now well established that the dictyostelids represent a normal component of the microflora of most soils. It is also generally accepted that the numbers and types of dictyostelids present in any particular soil are strongly influenced, if not actually determined, by a variety of environmental and nutritional factors. Of these, ambient temperature, soil moisture, pH, plant cover and available decomposing organic matter, and the amount and character of its bacterial content are very important.

As early as 1947, B. N. Singh demonstrated in dramatic fashion the interrelationship of two of these factors upon the presence and growth of *Dictyostelium* spp. Working with *D. mucoroides* and a new species, *D. giganteum*, which he had recently isolated and described, Singh (1946, 1947b) showed that in plates of sterilized soils adjusted to different moisture contents, and to which ample pregrown bacteria had been added, sorocarps of *D. mucoroides* would develop in three days following inoculation in soils containing 15% moisture but not in soils containing lower levels. Furthermore, in plates at 15% moisture the slime molds grew and fruited but remained confined to the immediate area of inoculation, whereas at higher moisture levels they spread progressively farther, encompassing the entire plate within 10-12 days at 30% moisture. No growth occurred in plates of soil at 10% moisture, but when the water level was brought up to 40% after 17 days, large numbers of sorocarps developed within a fortnight. The bacterium (#4002) used for these soil moisture experiments was a strain known to provide excellent nutrition for *Dictyostelium* spp. when spread or streaked on nonnutrient agar plates. In further experiments, Singh dispensed equal portions of sieved barnyard soil in Petri dishes and sterilized them, following which each plate was inoculated with approximately equal numbers of one of nine different bacteria, some common and some rare but all known to attain high numbers in soil. Spores of *D. mucoroides* were introduced in central points at approximately 2×10^4/g soil. The moisture content after sterilization was 33%, and incubation was at 20-21°C. Parallel cultures with the same bacterium pregrown and added to nonnutrient agar were prepared for comparison. Results varied widely: three strains of bacteria, including #4002, were completely consumed on nonnutrient agar and supported abundant and normal fruiting in soil; one that was partially consumed on agar yielded few but normal fructifications in soil; two strains that induced abnormal fruiting on agar yielded very few sorocarps in soil; whereas three strains that induced abnormal fruiting on agar supported large numbers of normal sorocarps in soil. Thus it was shown that even at favorable levels of moisture and temperature, and in the presence of ample bacteria, the amount and pattern of fructifications of *D. mucoroides* in soil varied with the bacterial host. Such variation, however, was not consistently parallel with fruiting behavior in agar plate cultures. Two other points of interest emerged from Singh's studies: (1) the numbers of bacterium #4002 recoverable from soil in the presence of *D. mucoroides* at 3 and 10 days were far less (ca. 15%) than in similar soil plates seeded with bacteria alone, indicating the very substantial phagocytic effect of the feeding myxamoebae (Fig. 3-1); and (2) the myxamoebae of *D. mucoroides*, which had been isolated originally from soil, were far more effective in limiting or reducing the numbers of bacteria than were those of *D. giganteum*, first isolated from a decomposing compost heap.

The variation in growth and fruiting that Singh encountered among different bacteria has had its parallels in other studies as well. Whether a particular slime mold grows with a particular bacterium on a particular medium depends upon the interplay of multiple factors, nutritional and environmental. For example, whereas Raper (1937) compared the growth and fruiting of *Dictyostelium discoideum* (NC-4) associated with many species of indigenous bacteria on hay infusion agar, it must be emphasized that the semi-quantitative results he reported apply specifically to the substrate employed and the physical conditions under which the tests were conducted. The same can be said of a much more recent study by Depraitère and Darmon (1978), who investigated the growth of *D. discoideum* (strain Ax-2) upon bacteria of many species. Their results apply specifically to the substrate they chose for the tests, the much used SM medium (see Chapter 4), and cannot be taken to indicate similar behavior on substrates of different composition or lesser nutrient content.

Although less extensive than those of Singh, some early studies by Eugene Kitzke (1951) were focused not on laboratory experiments but upon the occurrence and distribution of dictyostelids in nature insofar as such information could be determined by plating and enumerating the acrasian flora of different soils. Confirming the work of Raper then in progress (1951), Kitzke clearly demonstrated the frequent occurrence of four species in forest soils, ranked in this order of occurrence: *D. mucoroides*, *Polysphondylium violaceum*, *P. pallidum*, and *D. minutum*. He observed no correlation with percentage of soil colloids but did note an increase in the number of dictyostelids with an increase in soil organic matter. The occurrence of dictyostelids did not seem to correlate with soil pH, except for lower numbers in soils near pH 5.0 on the one hand and pH 8.0 on the other. There was a positive correlation with the type of tree and shrub cover, the greatest number of slime molds being found in soils

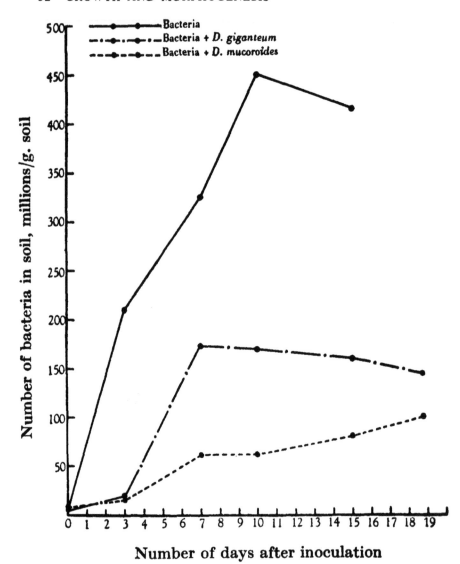

FIG. 3-1. Restriction of bacterial growth in sterilized soil by two species of *Dictyostelium*. The bacterium #4002, but otherwise unidentified, was described as "a common soil bacterium which is readily and completely eaten by the myxamoebae." Note the substantial reduction in numbers of bacteria in the presence of *D. giganteum* myxamoebae and the even greater reduction with *D. mucoroides*. (From Singh, 1947b)

from a tract known as Stewart's Woods (Verona, Wisconsin), which was characterized by large white oak, black oak, and black cherry trees and a dense understory of shrubs and herbaceous plants.

Although studies by Kitzke (1951), Raper (1951), and others had shown species of the Dictyosteliaceae to be common inhabitants of forest soils, especially surface soils of deciduous forests, the first comprehensive ecological study of these organisms was that initiated by J. C. Cavender as the basis for his dissertation research. Since he was located at the University of Wisconsin, his attention first centered upon deciduous forests in the environs of Madison. Subsequently it widened to encompass adjacent states, then eastern North America, and eventually selected areas in five other continents. The results of these studies dating from 1965 to the present have been published in a series of reports in the *American Journal of Botany* and elsewhere (q.v.). Together with somewhat parallel studies in Japan by Hagiwara (1971 et seq.), Kawabe (1980), and Kanda (1981), and recent detailed investigations in Switzerland by Frischknecht-Tobler, Traub, and Hohl (1979) and Traub, Hohl, and Cavender (1981b), these reports are believed to provide a realistic, if incomplete, picture of the global distribution of the dictyostelids.

Central to the collection and interpretation of meaningful ecological data has been (1) the formulation of procedures adequate for determining approximate numbers of cellular slime molds in soils, and (2) the refinement of methods and terminology for expressing such information in quantitative terms. The first of these objectives was accomplished through the use of the isolation procedure previously described, whereas the second was attained by determining the frequency of occurrence as well as the density of a particular slime mold within a forest area. As reported by Cavender and Raper (1965a), this was accomplished in the following way: samples were usually collected from ten different sites within a forest, and frequency, a measure of the prevalence of a species, was determined by dividing the number of occurrences of a species by 10 and multiplying by 100. Relative density, used to compare the relationship of one species to another, or to the whole population, was calculated by dividing the number of clones of each species by the total number of clones in a forest population and multiplying by 100. With such information it has been possible to compare the composition of dictyostelid populations of different forests, or of the forest at different seasons of the year, in terms of relative density (%), frequency (%), and absolute density (total clones/g dry soil), and to illustrate the first two parameters in the form of circular graphs (Fig. 3-2).

Several significant points emerged in our study of soils from the prairie-forest succession in southern Wisconsin (Cavender and Raper, 1965b). The diversity of species and the total numbers of dictyostelids increased as one progressed from (a) a tall grass prairie (two species) through (b) a

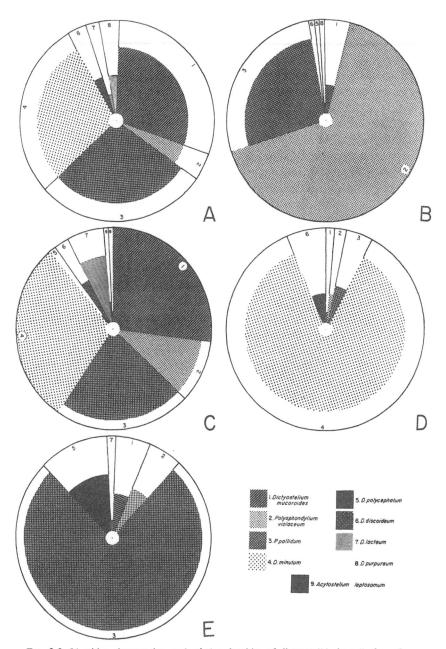

FIG. 3-2. Identities, frequencies, and relative densities of dictyostelids in soils from five forest types within the deciduous forests of temperate North America. *A*. Mixed mesophytic forest, Cumberland Falls State Park, Ky. *B*. Maple–basswood forest, Kaplan Woods State Park, Minn. *C*. Oak–hickory forest, Observatory Woods, Pine Bluff, Wis. *D*. White pine–northern hardwood forest, Cranberry Lake, N. Y. *E*. Silver maple–American elm bottomland forest, Arena, Wis. *Relative density* is shown by the portion of a graph occupied by a species; *frequency* is shown by the extent of the pattern along a radius. (From Cavender and Raper, 1965c)

bur oak opening (four species) and (c) an open oak forest (six species) to (d) a closed oak forest of large white, bur, and black oaks, with some smaller oaks and shagbark hickory (nine species). The floor of the last forest (Fig. 3-2 C) located in the Observatory Woods Scientific Area near Pine Bluff, Wisconsin, was covered by a thick blanket of slowly decomposing oak leaf litter that shielded the surface soil from rapid changes in either moisture or temperature. This covering undoubtedly contributed to both the diversity and the high incidence of dictyostelids, which in single samples sometimes ran as high as 5000/g dry soil.

In another series of tests, samples from various depths were assayed for the presence of dictyostelids, albeit their aerobic habit and the strong positive response of some species to light argued against their frequent occurrence much below the soil surface. Two forests were selected: an upland white oak and black oak woods in the Dane County School Forest, Verona, Wisconsin, and a bottomland silver maple and American elm forest on the Sugar River near Avon, Wisconsin. The soil profiles of the two stations differed markedly, as did the dictyostelid populations. Samples of the upland forest were taken from the dry leaf layer, the wet leaf layer, the fermentation layer, the surface of the humus layer, and at 1, 3, and 10 inches in depth. *Acrasis rosea* Olive and Stoianovitch (1960) was recovered from the dry leaves, but no dictyostelids were found there. Dictyostelids were found where the leaves became moist and started to decay, supporting the growth of bacteria upon which the myxamoebae fed. The largest numbers of different species, six and seven, occurred in the fermenting leaves and at the surface of the humus layer, respectively. Only two species occurred at the 3 inch level and none at 10 inches. In the bottomland soil, organic materials decay much more rapidly because of greater moisture and the chemical composition of the less acid maple and elm leaves; hence little or no litter accumulates from year to year. The largest number of dictyostelid species, four, was obtained from the surface humus, while only *Polysphondylium pallidum* was recovered from the 4 inch level, and it could have been washed downward from above. No slime mold was isolated from the 12 inch level. Significantly, *Dictyostelium polycephalum* Raper (1956) was isolated from the bottomland forest but not from the upland forest soil.

In a third series of tests, conducted to determine seasonal variations in total numbers and/or species of dictyostelids, six forests with varied soil moisture and atttendant floras were selected for careful analysis, namely (1a) dry: black oak and white oak; (1b) dry: white oak and shagbark hickory; (2) dry mesic: red oak and white oak; (3) mesic: sugar maple and slippery elm; (4) wet mesic: silver maple and American elm; and (5) wet: willow and cottonwood (Cavender and Raper, 1965b). In the dry and dry mesic forests the dictyostelid populations in spring and autumn were about double those in summer and winter, and in most cases the maximum numbers were substantially higher than those for the mesic, wet mesic, and wet forests at any season. While there was some seasonal fluctuation in the wetter soils, numbers in these tended to remain more nearly constant. Of the species that comprised these populations, *Dictyostelium minutum* was the most prevalent and occurred at high frequency in all forests; *Polysphondylium pallidum* was quite prevalent in the dry and dry mesic forests but not in the others; *D. discoideum* occurred in the 60-70% frequency range in the dry, dry mesic, and mesic forests only, whereas *D. polycephalum* was largely confined to soils of the wet mesic and wet forests.

The foregoing results tend to confirm what one familiar with the growth requirements of the Dictyostelidae might anticipate. One would expect fewer and hardier species to occur in grasslands subject to sun and drying winds, and by the same reasoning one would expect a greater and more diverse population to occupy the moist, temperate, leaf-protected surface soils and humus layers rich in bacteria. Additionally, one would expect forest soil to be less homogeneous and to contain numerous microenvironments within which one or another species could grow and fruit, even if on a very limited scale.

Ecological studies of the dictyostelids were soon extended to include all of eastern North America (Cavender and Raper, 1965c), and the results previously obtained were fully confirmed as samples were collected from Minnesota to New York and from Arkansas to Virginia. Throughout this area we discovered no genera or species that had not been found in Wisconsin (1965b) or that Raper had not previously reported (1951, 1956a, 1956b). What we found was further evidence in support of conclusions tentatively drawn from the Wisconsin studies. Such information is presented in Table 3-1, where identities and frequencies of three genera and nine species from nine contrasting types of forest are tabulated, and in Fig. 3-2, where similar information, together with species densities, for five of the collection sites is visually depicted (Cavender and Raper, 1965c).

In a subsequent study, Cavender (1972) collected extensively in eastern Canada and included nine boreal forests among the 22 sites sampled. For the most part, the dictyostelid populations consisted of the same species previously encountered in the more northern regions of the United States. The boreal forests, previously sampled only from high altitudes, were of special interest and contained in some cases very high densities, numbering 20,000/g in one soil, 12,750 in another, and 9150 in a third, perhaps reflecting less competition from other phagocytic microorganisms. In addition to *D. mucoroides*, *D. minutum*, *P. violaceum*, and *P. pallidum*, as could have been expected, *D. lacteum* and *D. discoideum* were also found in these forests.

Turning to subtropical and tropical America, Cavender collected soils from southern Texas, the Florida Everglades, several stations in Mexico, and four areas in Costa Rica. Included were forests designated as desert,

TABLE 3-1. Frequency of dictyostelids (%) in forest soils of the temperate zone of eastern North America (for explanation see text; data from Cavender & Raper, 1965c)

Type of Forest	1	2	3	4	5	6	7	8	9
Mixed Mesophytic									
Cumberland Falls (Ky.)—Fig. 3-2A	70	70	80	80	—	40	20	40	—
Buckeye Cove (Tenn.)	100	40	30	90	—	20	10	—	—
Cox Woods (Ind.)	90	30	60	100	10	—	70	—	10
Beech–Maple									
Turkey Run (Ind.)	70	50	70	90	—	—	20	30	—
Maple–Basswood									
Kaplan Woods (Minn.)—Fig. 3-2B	30	100	80	—	10	10	—	10	—
Oak–Hickory									
Observatory Woods (Wis.)—Fig. 3-2C	100	90	90	100	10	50	70	10	20
Rich Mt. Gap (Ark.)	90	80	100	50	—	—	90	80	—
Oak–Chestnut									
Wilson Creek (Va.)	20	40	80	100	—	30	40	20	—
Oak–Pine									
Huntsville (Tex.)	70	40	60	—	—	20	60	20	20
Hemlock–White Pine									
Cranberry Lake (N.Y.)—Fig. 3-2D	10	30	40	80	—	30	—	—	—
Flambeau River (Wis.)	20	30	30	100	—	40	10	—	—
Boreal Forest									
Cornell Mt. (N.Y.)	10	—	—	—	—	—	—	—	—
Mt. Collins (Wis.)	—	—	30	—	—	—	—	—	—
Bottomland Hardwoods									
Arena (Wis.)—Fig. 3-2E	40	50	90	—	60	—	10	—	—
Big Oak Tree (Mo.)	70	100	100	10	—	—	90	50	—

SPECIES: (1) *Dictyostelium mucoroides*, (2) *Polysphondylium violaceum*, (3) *P. pallidum*, (4) *D. minutum*, (5) *D. polycephalum*, (6) *D. discoideum*, (7) *D. lacteum*, (8) *D. purpureum*, (9) *Acytostelium leptosomum*

mesquite scrub, thorn, tropical deciduous, hammock, seasonal evergreen, rain, and cloud. Of these sources, the most productive soils came from seasonal evergreen forests, and one of these, from Pozo Rica, Mexico, yielded a total of 11 species, of which 4 were new: *Dictyostelium rhizopodium* Raper and Fennell (1967), *D. vinaceo-fuscum* Raper and Fennell (1967), *D. deminutivum* Anderson, Fennell, and Raper (1968), and *D. rosarium* Raper and Cavender (1968), in addition to some isolates then believed to represent E. W. Olive's *D. aureum* (1901, 1902), not previously encountered in our studies. Additional new species, *D. coeruleostipes* Raper and Fennell (1967) and *D. lavandulum* Raper and Fennell (1967), came, respectively, from seasonal evergreen forests in Mexico and rain forests in Costa Rica, while a new variety of *D. mucoroides*, *D. m.*

var. *stoloniferum* (Cavender and Raper, 1968), was isolated from the rain forests also. Of special significance was the large number of species contained in the soils from some tropical stations, but not in all. Even more significant was the number of new taxa, including four species of *Dictyostelium* with crampon-like bases (Raper and Fennell, 1967), a structure previously reported only by van Tieghem for *Coenonia denticulata* (1884). Another finding of special interest was the common occurrence of *D. purpureum* Olive (1901, 1902) in tropical and subtropical forest soils. Although present in temperate forests of North America, its incidence there is far less than the 80-100% frequency registered for many of the Mexican and Central American soils. In marked contrast, *D. minutum* Raper (1941b) was virtually absent from the tropical soils, although it is often dominant in the soils of northern deciduous forests as it may be also in coniferous forests. Of special interest was the relatively common occurrence of *D. polycephalum* Raper (1956) and *Acytostelium leptosomum* Raper and Quinlan (1958), species that tend to be limited in the northern temperate zone. Although fairly common in the deciduous forests of the United States, *D. discoideum* Raper (1935) was found only rarely in the soils of tropical America. *D. mucoroides*, *P. violaceum*, and *P. pallidum*, on the other hand, seem to be unrestricted and occur at high frequencies throughout Middle America, in the southern part of the United States, and in Mexico and Central America as well. In fact, Cavender (1973) regards these three species as cosmopolitan in distribution based upon his extensive collections and analyses.

If the yields of dictyostelids in soils from tropical America were surprisingly high, those from Europe (Cavender, 1969a) and East Africa (Cavender, 1969b) were disappointingly low. In soils collected from 17 areas in eight European countries, and from 13 stations in three countries in East Africa, only one undescribed cellular slime mold was found. In fact, the populations in Europe were much like those from temperate North America except for the absence of *D. discoideum* and *D. purpureum* (the latter has since been reported from Mallorca by Franz Traub, 1972). In East Africa the story was somewhat different: *D. purpureum* was isolated frequently, along with a form reported as "*D. mucoroides* variant." This form had been encountered fairly frequently in soils from Mexico and Costa Rica and has since been described as *Dictyostelium tenue* sp.n. by Cavender, Raper, and Norberg (1979).

In a study entitled "Geographical Distribution of Acrasieae," Cavender in 1973 summarized his studies up to that time in which he found striking correlation between latitude (and altitude) and the number of dictyostelid species recoverable from soil, the number decreasing progressively from the tropics to the arctic. This is graphically shown in Fig. 3-3. The same figure was updated in 1978 but without appreciable change. Striking also

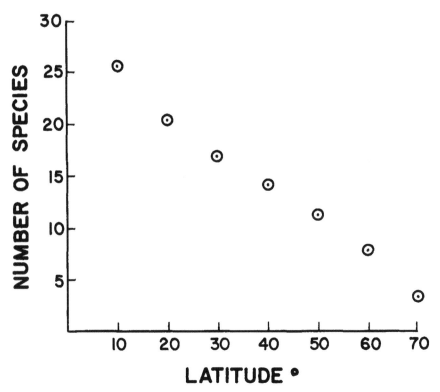

Fɪɢ. 3-3. Relationship between numbers of dictyostelid species isolated and latitudes of forests surveyed. A somewhat comparable plot could be made for altitudes of forests surveyed as well. (From Cavender, 1973)

was the similarity between the dictyostelid populations of tropical America and those of Southeast Asia and the East Indies.

Collections and analyses of forest soils from 13 stations in five countries of Southeast Asia revealed not only the common occurrence of dictyostelids in that part of the world but also the presence of some forms not previously known. Four new species and one variety were described by Cavender (1976a), namely: *Acytostelium subglobosum, D. multi-stipes, D. inter-medium, D. bifurcatum,* and *D. lacteum* var. *papilloideum.* Present among the dictyostelids also were representatives of the four crampon-based species first isolated and described from forest soils of tropical America. *D. purpureum* was generally present, while *D. discoideum* was not found in any soil. Other species commonly encountered included *D. mucoroides, P. violaceum* and *P. pallidum,* and, not infrequently, *D. polycephalum* (Cavender, 1976b).

Shifting attention from the tropics to the Arctic, Cavender recently (1978) collected and analyzed soils from the Alaskan tundra, the interior forested areas of aspen, birch, and spruce, and the rain forest of Sitka spruce and Western hemlock near Juneau. Nine species of dictyostelids were identified; *D. mucoroides* was by far the most abundant, followed by *D. sphaerocephalum* and *D. giganteum*. *Polysphondylium violaceum* was not seen, and *P. pallidum* and *D. minutum* were found only in the aspen-birch-spruce forests. Perhaps significantly, many of the *D. mucoroides* and *D. minutum* isolates produced macrocysts, the sexual stage of the Dictyosteliaceae, that may give these slime molds additional survival advantage. Of special interest in the rain forest was a singular *Dictyostelium*, erect and robust of form, unable to grow and fruit above 20°C and with an optimum temperature of 15-17°C. It was described and appropriately named *D. septentrionalis* sp.n. by Cavender in 1978.

Less extensive studies relating to regional ecology of the dictyostelids have been published by several investigators in recent years. In one of these, undertaken to complement the earlier study of Wisconsin forest soils (Cavender and Raper, 1965b), Sutherland and Raper (1978) selected for study five different types of prairie based primarily upon moisture conditions and designated as wet, wet mesic, mesic, dry mesic, and dry prairies. *Polysphondylium violaceum* was the most prevalent species, being especially abundant in wet prairies but decreasing in numbers to near extinction in the dry prairie. In contrast, *D. giganteum* showed the greatest numbers in mesic prairies and *D. sphaerocephalum* in dry mesic sites. *P. pallidum* and *D. aureo-stipes* (then reported as a yellow-stalked *Dictyostelium*) showed bimodal distributions, with the greatest numbers in the dry and wet prairies. Differing from virtually all other surveys, *D. mucoroides*, although present in all prairies, was not especially numerous in any of them—perhaps in part a matter of specific diagnoses.

In another investigation, Benson and Mahoney (1977) surveyed soil samples from 34 different vascular plant communities scattered throughout southern California. Collections were made from single sites in some communities to as many as eight or nine in others. Clones of dictyostelids varied from 0 in about 10% of the samples to 837/g of soil in one sample taken from a "moist" oak woodland. For the most part the counts were very low, averaging 94 clones/g of soil, a low density probably attributable to the usually dry hot climate and the xerophytic vegetation characteristics of many collection sites. Of very special interest and possible significance, however, was the recovery of *D. rosarium* Raper and Cavender (1968) from 22 of the 26 communities that yielded dictyostelids. Prior to their study the species was thought to be rare, having been reported only a few times; from deer dung (Washington State), horse dung (Mexico), from semi-arid mesquite forest (Texas) by Raper and Cavender (1968), and from desert soil (Washington State) by Mishou and Haskins (1971). The second most abundant species was *D. mucoroides* followed closely by *D. sphaerocephalum*. Also present in substantial numbers was a form with hyaline to lemon-yellow sorophores bearing pale to bright lemon-yellow sori that Benson and Mahoney diagnosed as *D. aureum* Olive. *Polysphondylium pallidum* and *D. minutum* were present occasionally, whereas *P. violaceum* was found beneath only 2 of the 34 plant communities.

A few years earlier, Mishou and Haskins (1971) had conducted a survey of comparable scope in the state of Washington. In most cases their soils came from quite different climatic environments and plant communities. The state was divided into ten regions based upon geography, climate, and vegetation, ranging from alpine to desert habitats. *Dictyostelium mucoroides*, the most abundant species, was found in all collection sites, followed by *D. minutum* from nine of the ten regions. A branched form (possibly *D. aureo-stipes*: KBR) reported as *D. mucoroides* variant was present in six of ten sites, while *D. lacteum, P. pallidum*, and *P. violaceum* were found in four, three, and two sites, respectively. *D. discoideum* was found only in soil from the foothills of the Cascade Mountains, whereas *D. rosarium* was seen only in soil from the arid Columbia Basin. With all values generally low, the greatest numbers of dictyostelids/g of soil were obtained from environments of moderate temperature and adequate moisture. Few dictyostelids were isolated from very wet and/or very acid soils of boreal forests, confirming an earlier report by Cavender and Raper (1965c).

Two more limited studies merit attention. In studying the distribution of the Acrasieae in Kansas grasslands, Smith and Keeling (1968) found *D. minutum, D. purpureum*, and *P. pallidum* to be present in addition to *D. mucoroides* and *P. violaceum*. Numbers were greatest in areas of ungrazed big and little blue stem, followed by grazed areas of the same grasses, with fewer still in grazed short grasses. Counts of individual species were not reported. Prevalence of dictyostelids was correlated with the amount of ground cover. With some amplification to include moisture and a favorable pH, the same factor had been reported by Jergensen and Long (1967) to govern the cellular slime mold populations in a South Dakota woodlot. There, as one might expect, *D. mucoroides* was most abundant, followed by *D. minutum, P. violaceum*, and *P. pallidum* in progressively smaller numbers. *D. lacteum, D. purpureum*, and a form designated *D. brevicaule* were encountered less frequently. Speaking of surface litter and moisture, they concluded: "Anything which influences these two factors influences greatly the population of Acrasiales present."

During the past decade Hiromitsu Hagiwara in Japan has published a series of papers, mostly under the general title "The Acrasiales of Japan," in which he has described several new species of *Dictyostelium*, three of *Polysphondylium*, and one of *Acytostelium* (considered in detail in Part II). Along with his descriptive texts Hagiwara has provided appreciable new information relative to the distribution of cellular slime molds in Japan (1971, 1972, 1973a,b, 1974, 1978, 1979) and elsewhere (1976a,b). From this list, the later publications are of particular interest from an ecological viewpoint, since in these Hagiwara traces the presence of dictyostelids in soils collected from different altitudes on Mount Margherita (Ruwenzori Mountains) in East Africa and on Mount Ishizuchi, Shikoku, Japan, respectively. Confirming the work of Cavender (1973) and more recently that of Frischknecht-Tobler, Traub, and Hohl (1979) and Traub, Hohl, and Cavender (1981b), Hagiwara found the number of species on Mount Margherita to be greatest at altitudes below 2000 meters and to disappear altogether between 3400 and 3900 meters. Species of *Dictyostelium*, numbered sp.-1 through sp.-6 but not named, were briefly characterized and accounted for most of the isolates, although *P. violaceum* and a few named Dictyostelia were recorded as well. It was suspected that many, but not all, of the isolates designated by number belonged to the so-called *D. mucoroides* complex. Turning to Mount Ishizuchi, Hagiwara (1976b) observed a definite correlation between altitude and the composition of the dictyostelid flora, which he summarized as follows:

> The distribution of the Dictyosteliaceae (cellular slime molds) in Mt. Ishizuchi (33°46′ N & 133°7′ E, 1,982 m alt.) was investigated. Ten species of *Dictyostelium* and two of *Polysphondylium* were isolated from seventy soil samples which were collected from the sites along the mountain route from Omogo-guchi (about 800 m alt.) to the summit of Mt. Ishizuchi in August of 1975. *Dictyostelium delicatum, D. firmibasis, D. mucoroides,* and *Polysphondylium violaceum* are distributed from the foot to the summit. *D. minutum, D. purpureum, P. candidum* and *P. pallidum* have each the distribution defined below 1,500 m in altitude. To the contrary, the distributions of *D. microsporum* and *Dictyostelium* sp. are defined above 1,400 m in altitude. The other three, *D. deminutivum, D. monochasioides, D. polycephalum,* seem to be occasional species, for these species were obtained from only one or two sites respectively.

Factors listed as influencing the occurrence and distribution of cellular slime molds at different altitudes included ''the degree of temperature and desiccation in the habitat'' and ''the change of vegetation.'' Changes in species diversity of the vegetation were considered especially important. Frequencies and densities were not given for any species, but *D. microsporum* and *Dictyostelium* sp. were reported as endemic for the cool-temperate forests. More recently K. Kawabe (1980) has reported on the occurrence of dictyostelids in the Southern Alps of Japan, noting the same type of altitudinal distribution previously shown by Cavender and Hagiwara, and recording *D. discoideum* among the species isolated. In Japan also, F. Kanda (1981) has studied the composition and density of dictyostelids in the Kushiro Moor, Hokkaido, and reported five species to be present, of which *P. pallidum* and *D. mucoroides* were especially common.

Not surprising, fewer cellular slime molds were found within the moor than in the surrounding forest.

Although more taxonomic than ecological in character, a paper by Y.-F. Lee (1971) also contained significant information about the occurrence of the genus *Dictyostelium* in Japan. He found *D. mucoroides* and *D. purpureum* could be isolated quite often from forest soils, whereas other species occurred less frequently: *D. discoideum* (four isolates), *D. minutum* (five isolates), *D. polycephalum* (four isolates), *D. deminutivum* (one isolate), and *D. rosarium* (one isolate, from wallaby dung).

Quite detailed investigations of the occurrence and distribution of Dictyostelidae in soil habitats of Switzerland have been reported recently by Frischknecht-Tobler et al. (1979) and by Traub et al. (1981b). Of these, the first investigation followed changes in the density and composition of the dictyostelid flora in the soil of an oak and hornbeam forest near Zurich during the course of a year. Attempts were made to correlate observed changes with environmental or nutritional parameters such as soil temperature, soil moisture, pH, soil carbonate, litter and cellulose decomposition, soil respiration, and soil microflora. Numbers of dictyostelids found in surface soils varied from approximately 700 to 1400/g dry soil with the greatest density occurring in spring and autumn, periods at which the bacteria also reached maximal numbers. The myxamoebae in the soil at any one time, however, seemed not to influence the number of bacteria substantially, thus playing the role more of indicators rather than regulators. Nine different species and types of dictyostelids were isolated, of which *D. mucoroides* was by far the most abundant, commonly representing as much as 70-85% of the total population. *P. violaceum* was next in importance and accounted generally for about 15-20%, while other species and types comprised the remainder. In one phase of the study, samples of soil were taken from 0, 5, 10, 20, and 25 cm deep, and the dictyostelid population ranged from 1375 clones/g at the surface to 43 at 20 cm and 0 at 25 cm. Of the different factors investigated, the slime mold population was found to follow most closely the amount of aerobic bacteria, which in turn was influenced primarily by available nutrients and soil temperature.

In the second study, Traub et al. (1981b) collected and analyzed soils from a wide variety of forests at 51 different sites in Switzerland. The types of forest and their altitudes follow: mixed deciduous forests (400-500 meters), oak forests (400-600 meters), beech forests (400-1100 meters), colline and montane conifer forests (450-1200 meters), and subalpine conifer forests (1400-1950 meters). The wealth of tabulated data concerning the occurrence and distribution of different species within these forests has been summarized in this way:

D. *mucoroides* was present at every site but Lac de Gruere, an acid bog. Average frequency and density are high in all forest types. Isolates designated as *D. mucoroides* are believed to represent a mixture of two or three species.

P. violaceum ranks second or third in all types of forest except subalpine conifer, where it is rare.

D. minutum is most abundant in mixed deciduous, oak, and some beech forests. Present in colline and montane conifer forests, it is rare in subalpine regions.

D. fasciculatum, a new species, is found in all forest types but is particularly abundant in the beech forest.

P. pallidum has low frequencies and relative densities throughout but is most abundant in beech forest.

P. filamentosum, a new species, also, shows a similar occurrence and distribution to *P. pallidum* but is less abundant at high altitudes.

D. sphaerocephalum is rare in mixed forest and absent from oak. It occurs in subalpine conifer forests and has been reported from tundra of Alaska and Iceland (Cavender, 1978).

D. giganteum is found infrequently in all types of forests.

D. aureo-stipes var. *Helvetium* prefers acid humus and is almost restricted to conifer forests; it is prevalent in the subalpine zone.

D. polycarpum, a third new species, is very rare and restricted to subalpine conifer forests.

D. polycephalum and *Acytostelium leptosomum* are found in beech forests but are very rare in Switzerland.

D. discoideum, *D. purpureum*, and *D. lacteum*, fairly common in the deciduous forests of North America (Cavender and Raper, 1965b and c; Cavender, 1972), were not found in Switzerland. Except for the absence of these three species and the presence of two new subalpine species, *D. polycarpum* and *P. filamentosum*, the dictyostelid populations of forested regions of Switzerland resemble those present in the northern United States and in Canada both in numbers and in relative abundance. In no other study, however, has the presence of particular dictyostelids been more convincingly identified with the soils that underlie particular plant communities.

Most of the studies relating to ecology of the cellular slime molds have centered upon the enumeration of natural populations, total and specific, and the isolation from among these of forms thought to be new or to possess special characteristics. And in most cases, several grams of soil have been collected and a portion of this, often 10 g, homogenized or shaken in water (to disperse spores, myxamoebae, or microcysts), substantially diluted, and plated as small fractions of the original—each then constituting a representative fraction of the whole. This could of course tell what species were present and in what numbers/g in a given area, but it would tell little about where the slime molds actually grew and fruited. Robert Eisenberg in Delaware (1976) attempted to answer these questions by a different and rather ingenious approach. Very small samples of soil were collected from close together and assayed for the presence of cellular slime molds. Three patterns of collection were used: in two cases he formed tight hexagonal bundles of 37 plastic drinking straws (6 mm diam) and with a flat board pushed these into the moist soil to a depth of 6-8 cm; in another he took similar samples at 5 cm intervals along a transect; and in still another he took random samples within a 0.25 m² area. Soils were later extruded, weighed, and diluted 1:50, and amounts equivalent to 0.02 g wet soil were removed and spread on each of two plates for the two sets of bundle samples and three plates for the transect and random samples. Cerophyll agar (Jones and Francis, 1972) was used as a base and pregrown *E. coli* was added as nutrient for dictyostelids. Results obtained from each set of samples clearly showed that the slime molds were in fact quite localized in their distribution, and that adjacent samples might contain high levels of one or more species or they might show different dictyostelids, or none at all. Overall the distribution was quite patchy (Fig. 3-4). A single sample contained four species, two others contained three, six contained two, several one, and many none. Of 101 total samples taken from a hardwood forest near Newark, Delaware, 59 contained one or more species of cellular slime molds, and of the five species present overall each was present in soils of all four sets. *Polysphondylium pallidum* was by far the most

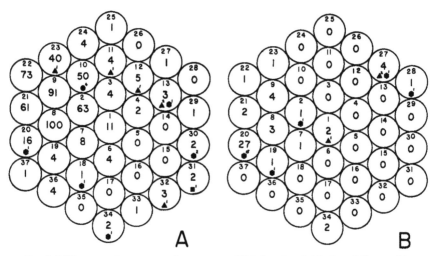

FIG. 3-4. Diagrammatic summary of survey to establish the microdistribution of dictyostelids in forest soils. *A*. Set 1, Bundles of small tubes (6 mm diam) were pressed into the soil, the contents extruded and then analyzed individually. Samples numbered concentrically at the top of each circle; numbers at center of circles indicate total isolates/sample. Triangles = *Dictyostelium discoideum*, hexagons = *D. mucoroides*, circles = *D. purpureum*, squares = *Polysphondylium violaceum*; where no symbol present, all = *P. pallidum*. Superscripts above the symbols indicate the number of isolates of that species. *B*. Set 2. Soil collections from a different site; legends unchanged. (From Eisenberg, 1976)

numerous, followed by *Dictyostelium mucoroides, D. discoideum, D. purpureum*, and *P. violaceum*.

Subsequent to Eisenberg's study, Kuserk (1980) examined the relationship between cellular slime molds and soil bacteria in the same forest. Several significant points emerged. Examining how the abundance and distribution of slime molds correlated with numbers of bacteria in the field, Kuserk concluded that the food supply was especially important in regulating species' numbers. *Dictyostelium mucoroides* was especially responsive to the bacteria in soil, whether they were indigenous or augmented. Dictyostelid populations peaked in spring and autumn when natural bacterial populations were greatest, at which seasons 51% and 24%, respectively, represented myxamoebae. In contrast, the myxamoebae present in winter and summer accounted for only 10-12% of the viable propagules, the remainder being spores, or spores and microcysts in species such as *Polysphondylium pallidum*. Whereas *P. violaceum* grew very rapidly in the laboratory, it failed to measure up to *D. mucoroides* when tested in the field, indicating that factors other than available food supply also influence community diversity under natural conditions.

In addition to the various types of population surveys, other investigations have centered upon the interrelationship among specific dictyostelids and their potential food sources. Such studies include those by Edward Horn (1971) of Princeton and Donald McQueen (1971a,b) of Vancouver.

Four cellular slime molds, *Dictyostelium mucoroides, D. purpureum, Polysphondylium violaceum*, and *P. pallidum*, and ten cultures of bacteria, designated by number but not otherwise identified, were isolated by Horn from soil of a "local area." Experiments of various types were then set up to analyze competition among the slime molds in the presence of a single (and sometimes limited) food source, and when offered a choice of different bacteria as food. Many variations on these themes were conceived, the experiments run, and the results analyzed. Unfortunately, they cannot be recorded in full, but a few general findings may be mentioned. Dictyostelids did not utilize all bacteria equally, thus confirming Raper's (1937) and Singh's (1947b) earlier studies. Dictyostelids did not have uniform growth and germination rates. Some dictyostelids could retard or inhibit the growth of other species. Given a choice, certain dictyostelids would invade and consume particular food sources and avoid others. Some dictyostelids succeeded better than others in competitive situations where two or more species were intermixed. Recognizing these variables and others, Horn concluded that a dispersed mixture of different types and sizes of bacterial colonies could account for the coexistence of the four species of *Dictyosteliaceae* in a small circumscribed area of surface soil or leaf litter.

McQueen's investigations (1971a,b) centered upon the interrelationship and competition between two structurally and physiologically contrasting dictyostelids. Spores of *Dictyostelium discoideum* (unbranched sorocarps and unable to fruit above 26-27°C) and of *Polysphondylium pallidum* (consistently branched sorocarps and able to fruit at 37°C) were intermixed and incubated in a manner that forced their myxamoebae to compete for the same food source, *Escherichia coli*, over a wide range of temperature. The first study was essentially exploratory in nature wherein cultural techniques were developed and standardized and the components in competition were identified and defined.

The second and more interesting report related to a series of experiments in which equal numbers of spores of *D. discoideum* and of *P. pallidum* were intermixed with a suspension of *E. coli*, spread over agar in a long narrow stainless steel culture dish (2.5 cm × 2.5 cm × 40 cm), and incubated on an aluminum block heated at one end and cooled at the other to provide a temperature range from 15°C to 30°C. Under these conditions *D. discoideum* could grow and fruit selectively at the cool end of the culture vessell whereas *P. pallidum* could do the same at the warm end. In between lay the jousting field. Numerous "cultural gradients" were run in which spores from sorocarps of one intermixed culture were used to inoculate the succeeding one, up to 12 in number, and in many cases the two grew and sometimes fruited separately in the early gradients but adapted to conjoint growth and fruiting as the series progressed. Significantly, in competitive mixtures *P. pallidum* made the adaptation and "learned" to fruit at a reduced temperature—the shift being from 24-37°C before growth in competitive cultures to 20-37°C following such intermixture. Additionally, before competition *D. discoideum* interfered with sorocarp formation in *P. pallidum*, an effect that was not observed after competition. There was some suggestion that the changes in *P. pallidum* were genetic and effected through parasexuality.

What emerges from all the above is, on the one hand, a very positive correlation between latitudinal and altitudinal geography and the numbers and types of cellular slime molds to be found in forest soils (Cavender, 1973); additionally this research provides a basis for predicting in a reasonable manner the composition of such populations. On the other hand, what it suggests but does not actually reveal are differences in the microenvironments (climatic and nutritional) within which the slime molds must grow, propagate, and fruit, and which in the final analysis determine whether or not the organisms can persist. For the present, criteria advanced several years ago still seem valid, albeit incomplete: "The requirements for optimal development of the cellular slime molds that seem to correlate best with their known distribution include the need for (1) moderate temperature, (2) a high oxygen tension, (3) a near saturated atmosphere, (4) sufficient soil moisture for cell movement, and (5) an adequate bacterial food supply" (Cavender and Raper, 1965c).

CHAPTER 4

Cultivation

Dictyostelids occur naturally in soil and in organic residues undergoing aerobic decomposition, where they grow and survive by feeding upon the bacteria that are present. This would suggest, and laboratory experiments have shown (Raper, 1937; Singh, 1947b), that myxamoebae of *Dictyostelium* can utilize a wide variety of bacteria as food. Experiments have shown also that some types of bacteria are consumed more rapidly and more completely than others. Additionally, the same bacterium on one substrate is often better nutrient for the myxamoebae than when it grows on a different substrate. Taken together, these observations clearly indicate that in the routine cultivation of the dictyostelids there is a continuing interplay between the slime mold, the bacteria upon which it feeds, and the substrate that nourishes the bacteria. Thus it behooves the investigator to establish conditions, cultural and nutritional, that are optimal for the growth and subsequent morphogenesis of the slime mold. Given a relatively low-nutrient substrate, such as hay or thin-hay agar (Raper 1937, 1951), one can, of course perpetuate cultures of many dictyostelia and polysphondylia by simply transferring myxamoebae, spores, or whole sorocarps to fresh substrates together with some of the associated bacteria. In doing so, however, the risk of enhancing some variation or perpetuating any mutation that may have occurred in either the dictyostelid or its bacterial host is increased. Since this is true, it is advisable at each serial transfer to select spores from sori of isolated, well-formed sorocarps and combine these with bacteria from a fresh stock culture, or introduce them into bacterial streaks preinoculated from a similar source. As dictyostelids are isolated from natural habitats, such as forest soil or dung, they are accompanied by bacteria, but seldom do these represent a single species, and if so, seldom is that species optimal for laboratory cultivation.

BACTERIAL ASSOCIATES

The choice of a bacterial associate should be governed in large measure by the nature of the investigation. If the principal objective is a study of the vegetative phase, with particular reference to the feeding habits of the myxamoebae, the slime mold can be cultivated advantageously on low-nutrient agar with *Bacillus megaterium* or some other large-celled bacillus (Raper, 1937). If a more general study is intended, or if the primary objective is to follow the fruiting process, particularly in its earlier stages, cultivation in association with some smaller-celled, non spore-forming species is recommended. This is true especially because *Bacillus* spores are not digested and may accumulate in sufficient quantities to obscure the normal movements of the myxamoebae during the period of cell orientation and the early stages of aggregation that characterize pseudoplasmodium formation.

Of the many bacteria employed by the writer (Raper, 1937, 1939) as hosts for *Dictyostelium discoideum* and other species of cellular slime molds, *Escherichia coli* was found most suitable. Two properties recommend this species. Particularly, (1) it is a small Gram-negative rod and on the majority of media it normally produces moist, nongummy colonies of a physical character especially favorable for feeding by the myxamoebae; and (2) it characteristically produces neither acid nor alkaline by-products in excessive quantities if available nitrogen and carbon sources in the medium are properly balanced. There are, of course, the further advantages that it is easily identified and is readily available to all investigators.

For many years *E. coli* strain 281 was generally used in this laboratory and in many others to which it was sent along with requested cultures of *D. discoideum* and other slime molds. More recently *E. coli* strain B/r has been used quite frequently by us, and even more generally in some other laboratories. Strain B/r was first introduced as a nutrient source for *D. discoideum* by G. Gerisch (1959) in his pioneering studies on myxamoebal growth in submerged culture. It offered the distinct advantage of not forming chains, hence was more readily ingested by the small amoeboid cells. Additionally, in our experience it showed less tendency than #281 to develop secondary areas of mucoid growth when cultures were permitted to remain for a few weeks at room temperature. For certain studies it has the disadvantage of growing more slowly than #281 at 22-23°C; hence the amount of slime mold growth is correspondingly reduced and sorocarp maturation may be somewhat delayed.

A closely related Gram-negative bacterium, *Aerobacter aerogenes* (now designated *Klebsiella pneumoniae*) was employed in 1951 as an alternative nutrient source by M. Sussman, and he and many others have used it quite successfully since that time. A comparable strain, identified as "*Aerobacter* 1912" (and possibly the same), had been used earlier by Singh for cultivating *D. mucoroides* and *D. giganteum* (1947a,b). Strains of *K. pneumoniae* compare very favorably with *E. coli* #281, and the two can be used interchangeably in many cases. Strain #900 (Singh's #1912) has been much used in our laboratory; and for cultivating some Acrasidae such as *Guttulinopsis nivea* (Raper, Worley, and Kessler, 1977) and *Fonticula alba* (Worley, Raper, and Hohl, 1979) it offers certain advantages over *E. coli*. *Dictyostelium discoideum* and some other species of this genus have been grown quite successfully with many other bacteria as well

(Raper, 1937), including a number of species pathogenic to plants or animals (Raper and Smith, 1939). Such growth is illustrated in Fig. 4-1 and suggests a positive role for dictyostelids in the microbiology of the soil.

For special experiments such as the grafting of pseudoplasmodial fractions (Raper, 1940a) or the intermixing of different species of slime molds (Raper and Thom, 1941), the strongly pigmented species *Serratia marcescens* affords exceptional advantages. In association with this red bacterium, the myxamoebae and pseudoplasmodia of *Dictyostelium discoideum, D. purpureum,* and *Polysphondylium violaceum* become colored by the undigested pigment, prodigiosin, which accumulates within vacuoles in the bodies of the myxamoebae. In contrast, the cells of *D. mucoroides,*

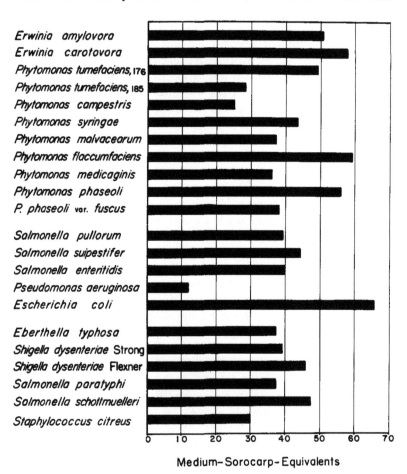

FIG. 4-1. Comparative growth of *Dictyostelium discoideum* in association with different bacteria pathogenic to plants, animals, and man (from Raper and Smith, 1939). Growth in a prescribed area is expressed in medium-sorocarp-equivalents as defined by Raper in 1937.

D. minutum, and *P. pallidum* digest the pigment, and hence remain colorless. When *Serratia* is used for staining pseudoplasmodia for grafting experiments it is essential to secure good pigmentation of the bacteria. However, this should not be too intense, for colonies so pigmented are apparently toxic to the myxamoebae and cell growth is limited or lacking. The desired goal can be achieved (1) through judicious formulation of the bacterial substrate, or (2) by growing the bacteria on or in a comparatively rich nutrient medium, harvesting the pigmented cells, and restreaking them on nonnutrient agar before introducing the slime mold.

Pseudomonas fluorescens, the bacterial species present in Nadson's (1899), Pinoy's (1903, 1907), and Skupienski's (1918, 1920) cultures, can provide excellent nutriment for the myxamoebae of *D. discoideum* and many other slime molds, but special precautions must be exercised to prevent the culture medium from becoming excessively alkaline as a result of its intense proteolytic activity. Here again, careful attention to the kind or composition of the substrate is essential. For example, *D. deminutivum* (Anderson, Fennell, and Raper, 1968) grows well and fruits optimally with this bacterium, but it does so only on a very dilute glucose-peptone agar. *D. discoideum* grows well and produces normal sorocarps with *Ps. fluorescens* on hay agar (a low-nutrient medium), and on balanced glucose-peptone agars as well, but only because it can ferment glucose to offset the alkaline by-products released from the peptone (Raper, 1939).

A shift from a two-membered culture with one host bacterial species to another can be readily accomplished as a rule. Spores from selected sorocarps are transferred directly to the ends of previously established streak cultures of the second host, and the slime mold is allowed to grow through the streak and fruit at the opposite end (Fig. 2-3). This procedure is then repeated one or more times as necessary. In the case of *D. discoideum*, this will be realized as a rule in the first transfer if spores are taken from isolated sorocarps formed from pseudoplasmodia that have migrated 1 cm or more beyond the limits of bacterial growth (Raper, 1937). In other large species, such as *D. giganteum*, *D. purpureum*, and *P. violaceum*, the same result may be attained on the first trial if spores are taken from sori borne on long sorophores and are removed 2-3 cm from the sites of sorophore origin. To move delicate species such as *D. minutum* or *D. lacteum* from one bacterial host to another, spores are taken from the cleanest possible area of the old colony and similarly implanted in preinoculated streaks of the recipient host. Several passages may be required.

With some bacterial hosts and on certain substrates there is a marked tendency for rough or slimy variants (or mutants) to develop after the bacterial colony has been visibly consumed. With *Serratia marcescens* #175 on hay infusion agar these often become very numerous within 2-3 weeks. They are quite rough, often less strongly pigmented, and are not consumed by the myxamoebae (Fig. 4-2). Development of such secondary

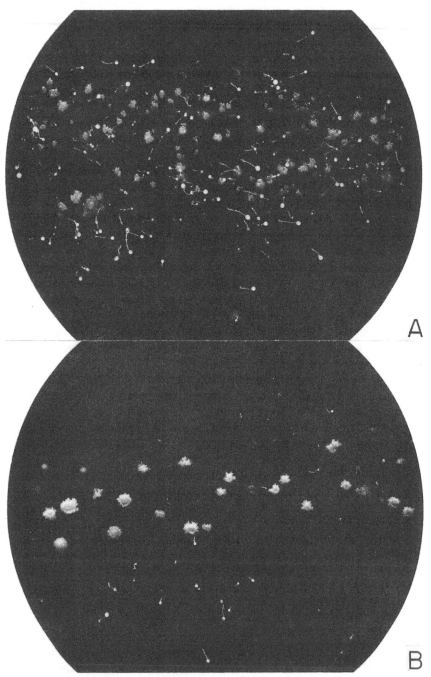

FIG. 4-2. Rough colonies of bacteria that have developed following visually complete consumption of the bacterial streaks by feeding myxamoebae of *Dictyostelium discoideum*. A. *Serratia marcescens*. B. *Micrococcus luteus*. × 3.

rough colonies is accentuated by the addition of 0.1% peptone and phosphate buffers. A comparable phenomenon has been observed in cultures of *Micrococcus luteus* #1018, in which case the secondary colonies are less numerous but equally striking. They are raised, compact, very rough, and, as in *Serratia*, they are not consumed by the myxamoebae. The appearance of rough colonies in plates where bacterial streaks seemingly had been completely consumed was first observed in cultures of *Dictyostelium discoideum*. The phenomenon is not unique to this slime mold, however, for it has been observed to occur with five other species of *Dictyostelium* and *Polysphondylium* after the myxamoebae have first consumed the visible bacterial growth. It is not known if the myxamoebae play some inductive role, or if they merely remove the predominantly "smooth" cells, thus allowing the occasional "rough" variants to grow unimpeded.

While quite different in appearance, a somewhat comparable phenomenon may occur in *Escherichia coli* and *Klebsiella pneumoniae*, where relatively large gummy secondary colonies often develop following myxamoebal clearance of the primary bacterial streaks. Although such gummy colonies will support some slime mold growth, it is to preclude their proliferation that we routinely take bacteria from fresh stocks when starting new cultures. As a rule, secondary bacterial growth of the type described is to be avoided insofar as possible, but it is an interesting fact that among macrocyst producing strains of *D. mucoroides* one sometimes sees these cysts in greatest numbers in just such situations—a reflection, we presume, of the especially moist environment the more mucoid colonies provide.

For maintaining stock cultures of *Escherichia coli* and other species of bacteria we have used a nutrient base commonly referred to as NTGY agar or broth and prepared as follows: tryptone (Difco) 5.0 g, yeast extract (Difco) 5.0 g, agar 20.0 g in 1000 ml distilled water; melt in autoclave and then add 1 g each of dextrose and KH_2PO_4; adjust to pH 7.0 prior to sterilization. The same nutrient solution with adequate aeration has been used to produce pregrown bacterial cells when these were needed.

Culture Methods

The conventional methods developed for the cultivation and systematic study of most species of *Dictyostelium* and *Polysphondylium* are based upon the growth of the host bacterium on a selected agar medium, coupled with the simultaneous or delayed introduction of the slime mold, which feeds upon the bacteria in situ. To investigate pseudoplasmodium formation, that is, the aggregation of myxamoebae, two drops of a suspension of slime mold spores and host bacteria may be spread evenly in a broad band (1-2 inches) upon the surface of a nutrient-poor agar plate, a bent glass rod being used to effect this distribution. Under these conditions the bacteria and the slime mold make an even but comparatively thin growth, and characteristic wheel-like pseudoplasmodia will begin to develop throughout the area within 40 to 46 hours when incubated at 22°-24°C. Such a culture is illustrated in Fig. 4-3. To observe the development of a succession of pseudoplasmodia, bacteria alone may be spread in the manner described, and spores of the slime mold may be subsequently added at the center of the plate. In this case, pseudoplasmodia will first appear at the site of inoculation within 2 days or less and will continue to develop for several days as the slime mold progresses toward the plate margin.

When large migrating pseudoplasmodia of *Dictyostelium discoideum* (or developing sorogens of other species) are desired (Fig. 4-4), richer substrates should be employed. Bacteria and slime mold may be inoculated simultaneously, or the introduction of the latter may be delayed until the bacteria have made a substantial growth under conditions which will preclude the development of an unfavorable pH or the production of toxic metabolites (Raper, 1939). In any case, the area of inoculation should be

FIG. 4-3. Pseudoplasmodia of *Dictyostelium discoideum* in early to middle stages of formation. Mixed spore and bacterial (*E. coli*) inoculum was spread on plate of 0.1% peptone agar, incubated at 23 ± °C, and photographed after 44 hours. Bacterial growth relatively sparse. × 6.

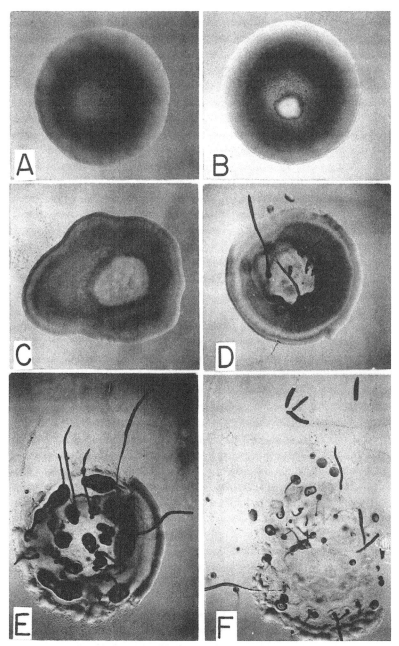

FIG. 4-4. Successive stages in the consumption of dense colonies of *E. coli* by *Dictyostelium discoideum* and the subsequent formation of migrating pseudoplasmodia. Note that pseudoplasmodia develop from blocs of myxamoebae rather than by a convergence of inflowing streams. × 3.

limited so that broad, bacteria-free expanses remain wherein the pseudo-plasmodia may migrate and the sorocarps develop. The migration of pseudoplasmodia in *D. discoideum* and the lengthening of sorophores in other species are greatly enhanced by incubation in one-sided light.

When semi-quantitative measurements of slime mold growth are contemplated, it is preferable to plant the bacterial associate 2-3 days prior to the introduction of the slime mold. Plates should be used in which the surface is sufficiently dry to prevent the bacteria from spreading beyond the intended area of the colonies thus established. For more preeise measurements of growth, liquid cultures should be used.

By adding fresh bacterial suspension to a dictyostelid culture at appropriate intervals, the onset of aggregation can be appreciably delayed, a phenomenon first demonstrated by Potts in 1902. By transferring vegetative myxamoebae to fresh colonies of bacteria at intervals of 36 hours or less, it would appear that cultures can be continued indefinitely in a vegetative state quite devoid of fruiting phenomena. Single myxamoebae when transplanted to favorable bacterial colonies reproduce the entire life cycle to yield wholly normal cultures of *Dictyostelium* or *Polysphondylium*—such cultures being marked only by a considerable delay in the appearance of the first pseudoplasmodia and sorocarps.

Although the myxamoebae of dictyostelids normally feed upon living bacteria, they can also be grown upon dead bacterial cells. As early as 1902, Potts demonstrated that *D. mucoroides* could grow upon cells of *Bacterium fimbriatum* killed by chloroform; and in 1937 Raper reported the successful cultivation and normal development of *D. discoideum* on cells of *Serratia marcescens* killed by exposure to flowing steam and on cells of *Achromobacter radiobacter* killed by exposure to ultraviolet light. Subsequently cells of *E. coli* were grown in liquid culture, harvested by centrifugation, autoclaved, and spread as a thin paste on lactose-peptone agar plates, into which bacteria-free spores of *D. discoideum* were introduced. Parallel cultures with live *E. coli* served as controls. The slime mold grew somewhat less rapidly with killed bacteria: migrating pseudoplasmodia appeared in 4 days rather than 2-3 days as in the controls. Consumption of the streaks of killed bacteria, however, was complete within 5-6 days, and the development of the slime mold was entirely characteristic. Migrating pseudoplasmodia were typical in form and size and not infrequently migrated 6-7 cm prior to the formation of normal sorocarps (Raper, 1951).

This type of recultivation was repeated every week for seven successive weeks, in each instance the spores being taken from sorocarps produced in the preceding series of cultures grown on killed bacteria. During the seven culture "generations" no evidence developed that the slime mold was becoming less vigorous in its growth or less characteristic in the pattern of its completed sorocarps. There seemed to be no reason for suspecting that the slime mold might not be grown indefinitely upon killed bacteria.

A decade later Kersten Gezelius (1962), Uppsala, Sweden, published a more comprehensive study of the growth of *D. discoideum* on autoclaved cells of *E. coli* that confirmed and expanded the earlier reports of Potts (1902) and Raper (1937, 1951). More specifically, Gezelius showed that if killed bacteria (10^{10}/ml) were suspended in Bonner's salt solution (1947) enriched with thiamine, riboflavin, glucose, and certain amino acids, and if the solution was set initially at pH 4.0-4.4 rather than pH 6.0-6.2, to which it rose during myxamoebal growth, the generation time was lowered to 3.5 hours and the lag phase to 10-12 hours. The final cell density reached 1.5×10^7 myxamoebae/ml, a very good yield by any standard. This newer method of cultivation in liquid media afforded special advantages for biochemical and physiological studies in the period prior to the emergence of strains capable of growing axenically.

CULTURE MEDIA

Just as the bacterial associate should be chosen with special regard to the nature of the study, so should the culture medium. In addition, it should be compounded with reference to the fermentative capabilities of the bacterial associate. These are generally not critical if nutrient-poor or well-buffered media are utilized, since neither an excess of acidity nor alkalinity is ordinarily built up. They may be quite critical, however, if inadequately buffered media containing nutrients such as peptones are employed. They will almost certainly be critical for nutrient-rich media unless special measures are taken to control the pH. This may be accomplished in many cases by properly balancing the carbon and nitrogen sources, or failing in this it can be attained generally by adequately buffering the medium with phosphate buffers (Raper, 1939). The favorable range for the Dictyostelidae as a group appears to be between pH 5.0 and 8.0, and for the species *Dictyostelium discoideum* between pH 5.0 nd 7.0, with an optimum about 6.0-6.2 when the bacteria and slime mold are grown concurrently.

Considered from an historical viewpoint, the substrates used by early investigators were usually based upon extracts of natural products, either dung (Brefeld, 1869, 1884; Olive, 1901, 1902; Harper, 1926, 1929; and others) or some plant material (van Tieghem, 1880, 1884; Potts, 1902; and Pinoy, 1907). Raper's early studies (1935, 1937) were based upon hay infusion agar made from an extract of weathered grass. Although now supplanted in large measure, this medium, with some modifications, is still a very useful substrate, particularly for primary isolations and for the maintenance of stock cultures. Some natural extracts continue to be used, but as the dependence of the myxamoebae on bacteria for food became fully recognized, there was a shift to substrates of more reproducible composition. Ideally, these were based upon the growth characteristics of the bacteria and of the slime mold(s) under investigation. When working with delicate species such as *Dictyostelium lacteum* or *D. minutum* a low-nutrient substrate is always indicated, and the same is generally advantageous when observing cell aggregation even in robust species such as *D. discoideum*, *D. mucoroides*, and *Polysphondylium violaceum*. For studying fructification in the last three species a richer substrate would be employed as a rule.

SOLID SUBSTRATES

The writers have found a number of solid media (1.5 or 2.0% agar) to be especially valuable in cultivating dictyostelids in association with *Escherichia coli*, and of these the following are particularly suitable for observing and studying the vegetative growth of myxamoebae and the process of pseudoplasmodium formation:

(1) Hay infusion agar (Raper, 1937, p. 293): 35 g of weathered grass infused in a liter of tap water for ½ hour at 110°C and 5-10 lbs pressure; infusion filtered, and filtrate made up to 1 liter; 0.2% K_2HPO_4 and 1.5% agar added; adjusted to pH of 6.0-6.2; and sterilized at 15 lbs pressure for 20 minutes.

(2) Hay infusion agar diluted to ½ or ⅓ strength with nonnutrient agar and designated thin-hay agar.

(3) Hay infusion agar enriched with 0.05% peptone, designated weak peptone-hay agar.

(4) Lactose-peptone or dextrose-peptone agar containing 0.1% of each ingredient together with 1.5 or 2.0% agar in distilled water. An agar medium containing 0.1% lactose and 0.1% peptone, and commonly referred to as 0.1L-P, or LP medium, is now recognized as something of a standard for comparing the growth and fructification of different dictyostelids. For more delicate species, 0.1L-P agar may be advantageously diluted to ½ strength with nonnutrient agar, being designated 0.1L-P/2. Alternatively, 0.1L-P agar made slightly acidic by the addition of KH_2PO_4, 2.05 g/l, and $NaHPO_4$, 0.33 g/l, designated 0.1L-P (pH 6.0), is generally more appropriate for the larger species.

(5) Carrot infusion agar prepared by boiling 300 g fresh carrots per liter of water, reconstituting the filtrate, and adding 2.0% agar to make a firm gel; the medium is mildly acidic.

Of the media listed, the first three are satisfactory for cultivating species of *Dictyostelium* in association with *Serratia marcescens*. They may be used also for cultivating the slime molds in association with *Bacillus megaterium*. Only the first two can be considered entirely satisfactory if *Pseudomonas fluorescens* is used as a bacterial associate, although combinations of dextrose and peptone can be used at relatively low temperatures.

A more luxuriant growth and development of robust dictyostelids may be obtained by cultivating them in association with *Escherichia coli, E. coli* B/r, or *Klebsiella pneumoniae* upon solid media of greater nutrient content. Of many such media evaluated, a few of those frequently employed or of special interest would include:

(6) Peptone-yeast extract-dextrose agar, wherein peptone and yeast extract are used in 0.2% concentration and dextrose 0.5%.

(7) Peptone-lactose and peptone-dextrose agars, wherein both nutrients are employed in balanced concentrations up to 1.0 or 1.2% (Raper, 1939).

(8) Enriched hay infusion agar, wherein 0.5% peptone is added and the medium is buffered with KH_2PO_4, 2.7 g/l and $Na_2HPO_4 \cdot 12$ H_2O, 7.16 g/l (Raper, 1951).

(9) Enriched carrot infusion agar, wherein peptone and phosphate buffers are added as in the preceding.

(10) V-8 agar: V-8 juice, 100 g (Campbell Soup Co.); $CaCO_3$, 1.0 g; agar, 15 g; distilled H_2O, 1000 ml (Traub and Hohl, 1976).

(11) Bonner's agar: Peptone, 10 g; dextrose, 10 g; $Na_2HPO_4 \cdot 12 \ H_2O$, 0.96 g; KH_2PO_4, 1.45 g; agar, 20 g; distilled H_2O, 1000 ml. This medium is compounded with a view to ensuring an initial pH of about 6.0 (Bonner, 1947).

(12) SM agar: Peptone, 10 g; dextrose, 10 g; $Na_2HPO_4 \cdot 12 \ H_2O$, 1 g; KH_2PO_4, 1.5 g; agar, 20 g; distilled H_2O, 1000 ml. Adjust to pH 6.5 before autoclaving (Sussman, 1951).

(13) Prepared agar media: Levine Eosin-Methylene Blue Agar (Difco), and to a lesser degree Endo Agar (Difco), constitutes a favorable substrate for the larger dictyostelids. The former is especially good when diluted to ½ to ¼ strength with nonnutrient agar. Cultures should be incubated at a temperature low enough to prevent the development of a metallic sheen on the *E. coli* colonies.

Media 8 and 9 normally support good growth and development of *Dictyostelium discoideum* in association with *Serratia marcescens* at 20-22°C, hence may be used for producing red pseudoplasmodia suitable for grafting experiments. None of the above are recommended for use with *Pseudomonas fluorescens* as the bacterial associate, albeit those containing dextrose (e.g., medium 7) generally give reasonably good results when incubated at 20-22°C. Medium 10 is useful at full strength for cultivating robust species of *Dictyostelium* and at half strength for most forms that Traub and Hohl (1976) regard as *Polysphondylium*-related dictyostelids, that is, produce spores with polar granules. Medium 11, recommended by Bonner in 1947, has been used extensively for more than 30 years in his laboratory, in ours, and elsewhere for the cultivation of *D. discoideum* and other large cellular slime molds. Medium 12, dubbed "SM Medium," presumably as an acronym of slime mold medium, has been used more widely perhaps than any other. Since this is true, its evolution may be of some interest.

The story began with this question: Since *E. coli* is a good bacterial associate for *D. discoideum*, and since Endo Agar and Eosin Methylene Blue Agar are used in diagnostic tests for its presence, would the slime mold grow and develop sorocarps in colonies of *E. coli* on these substrates? Trials were run and the answers were positive—in fact, myxamoebal growth was excellent and sorocarp formation was satisfactory in each case (Raper, 1939). Different components of the two media were then omitted singly, and it was found that the sodium sulfite and basic fuchsin of Endo Agar and the eosin and methylene blue of E.M.B.Agar could be removed without apparent reduction in the growth of *E. coli* or impairment of fructification in *D. discoideum*. Likewise the dipotassium phosphate of each medium could be omitted with minimal effect. Thus was left a medium consisting of 15 g agar and 10 g each of lactose and peptone. Media of such balanced composition could be used quite successfully, as could substrates where lactose and peptone were increased to 12 g each/liter (medium 7). It had been found (Raper, 1939) that mildly acidic substrates were optimal for the growth and development of *D. discoideum*. Acordingly, after Bonner took up the study of this slime mold, he substituted dextrose for lactose and included phosphate buffers (1947) to ensure a pH of approximately 6.0 (medium 11). Four years later (1951) Sussman published his version of the dextrose-peptone medium in which the dibasic sodium phosphate was rounded from 0.96 g to 1.0 g/l and the monobasic potassium phosphate was rounded from 1.45 g to 1.5 g/l with pH adjustment to 6.5—thus was born the generally favorable and since widely used SM medium (medium 12). A decade later Sussman (1961) cited the aforementioned formulation with the inclusion of 0.5 g/l MgSO$_4$ and 1 g/l yeast extract as "SM agar" (Sussman, 1951). Other investigators have reinstated lactose, and one suspects that the term may now have acquired a generic connotation to include almost any combination of dextrose or lactose and peptone.

In considering the media listed above, emphasis has been placed upon *Dictyostelium discoideum* because we have had the greatest experience with this species and because it has been and still is studied far more widely than any other cellular slime mold. Nonetheless, other large species such as *D. purpureum*, *D. giganteum*, and *Polysphondylium violaceum* can be cultivated equally well on these substrates with *E. coli* or *K. pneumoniae* as the bacterial hosts. Delicate species of *Dictyostelium* and other species of *Polysphondylium* can be cultivated satisfactorily only on media that are less rich in nutrients.

In the case of *Dictyostelium discoideum*, the suitability of existing culture conditions can be accurately gauged by the behavior and response of the slime mold itself. From our experience, and as recorded in 1951, six degrees of reponse may be cited, ranging from complete absence of growth to optimal growth and development, as follows:

(1) The spores fail to germinate.
(2) Limited growth of myxamoebae occurs, but no cell aggregates are formed (Fig. 4-5A).
(3) Fair growth of myxamoebae occurs, and few, generally atypical pseudoplasmodia form but seldom migrate.
(4) Good growth of myxamoebae occurs, typical aggregates develop, and some pseudoplasmodia migrate; but sorocarps are generally very atypical (Fig. 4-5B).
(5) Good growth of myxamoebae occurs, typical aggregates are formed, and extensive migration usually follows; but sorocarps are generally atypical with short and unevenly tapered stalks, ill-formed basal discs, and very irregular sori.
(6) Good growth of myxamoebae occurs, aggregation is wholly typical, migration is extensive, and mature sorocarps are characterized by erect and evenly tapered stalks, broad flattened basal discs, and citriform to nearly globose sori (Fig. 4-5C)

The character of the slime deposit left in the wake of the migrating pseudoplasmodium of *D. discoideum* likewise reflects to a considerable degree the suitability of a given culture. Regularly, a few myxamoebae are dislodged from the migrating body to remain in the collapsed slime deposit, where they often become vacuolate. Under conditions of too high temperature, too low humidity, or too great acidity or alkalinity, the number of myxamoebae remaining in the deposit increases tremendously, thus rapidly exhausting the migrating body and lending to the so-called slime sheath a conspicuously irregular cellular texture.

The same general relationship of slime mold to culture conditions can be observed in the other species of *Dictyostelium*. However, since few of these possess a stalk-free migrating stage and only one develops a basal disc, the degree of response is less striking and clean-cut. In *Polysphon-dylium violaceum* the extent and regularity of branching affords a fairly reliable index. As suggested by Singh (1947a), *Dictyostelium mucoroides* and *D. giganteum* appear to be less sensitive to pH than *D. discoideum*. The same is true of *D. purpureum*.

TEMPERATURE, HUMIDITY, AND LIGHT

In addition to the identity of the bacterial associate and the composition of the culture medium, other factors, including temperature, humidity, and

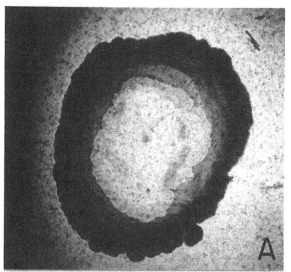

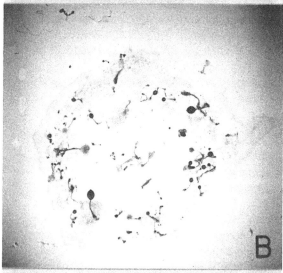

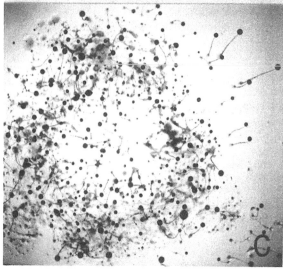

FIG. 4-5. Development of *Dictyo-stelium discoideum* in cultures of differing degrees of suitability. *A.* Very unfavorable culture: myxamoebae have made limited growth but no pseudo-plasmodia have formed. *B.* Unfavorable culture: bacterial colony has been consumed, pseudoplasmodia have formed, but sorocarps are few in number and atypical in pattern. *C.* Optimal culture: abundant sorocarps of typical pattern have developed. ×3. *A* and *B* with *Pseudomonas fluo-rescens*; *C* with *E. coli*. (From Raper, 1951)

light, markedly affect the growth and development of *Dictyostelium* and *Polysphondylium*. Optimum temperatures vary somewhat depending upon the species, but generally range between 20° and 25°C. However, some species such as *D. polycephalum* grow and fruit optimally at 30°C (Whittingham and Raper, 1957), and one natural isolate of *P. pallidum*, strain "Salvador," fruits at 37°C. Cultures of most species can be grown at lower temperatures, and developmental processes usually proceed in a normal manner; however, the rate of growth and development is somewhat retarded. *D. septentrionalis* Cavender (1978) represents a special case where growth is optimal at 15-17°C and normal fruiting does not occur above 20°C. It has been clearly demonstrated that the onset of the fruiting stage in *D. discoideum* can be hastened by raising the temperature of incubation (Raper, 1940a) and that migrating pseudoplasmodia will move from the cool (15-16°C) toward the warm side of a Petri plate culture. Bonner and coworkers (1950) have demonstrated this response in plate cultures where the gradient is as little as 0.05°C, and by calculation the difference in temperature at the two sides of a migrating pseudoplasmodium 0.1 mm wide is only 0.0005°C—an amazingly sensitive response, later confirmed independently by Poff and Skokut (1977). Bonner and Slifkin (1949) have shown that migration is prolonged at lowered temperatures, whereas pseudoplasmodia or young sorocarps exposed to an abrupt rise in temperature show a marked reduction in the percentage of stalk cells.

Maximum growth and optimum development generally occur in nearly saturated atmosphere; hence, cultures are usually grown in tubes with snug-fitting closures or in Petri dishes with close-fitting lids. An added advantage may be realized if cultures can be incubated in a chamber with controlled humidity. There are exceptions, however. Some isolates of *D. rosarium* (Raper and Cavender, 1968), including the type culture, fruit much better if glass plate covers are replaced by porous clay lids at the time of cell aggregation, and the same is true of *D. polycarpum* Traub, Hohl, and Cavender (1981a). *D. polycephalum* fruits best in a three-membered culture with *Klebsiella* and *Dematium nigrum* (Raper, 1956a) or in plates held at 98% relative humidity (Whittingham and Raper, 1957).

We know of no evidence to indicate that light either enhances or retards the rate of myxamoebal growth. In many species, however, light exerts a marked influence on the fruiting process. Konijn and Raper have reported in some detail the influence of light on the time (1965) of cell aggregation and on the size (1966) of aggregations formed in *D. discoideum*. Raper (1940a) showed that pseudoplasmodia developed from 2 to 4 hours earlier and were smaller and more numerous in cultures grown in diffuse daylight than in similar cultures incubated in total darkness. Potts (1902) and Harper (1932) had reported previously that sorocarps of *D. mucoroides* and *Polysphondylium violaceum*, respectively, were larger when formed in darkness than in light. Subjected to one-sided illumination, the responses of the larger species to light are quite striking and no less sensitive than those to temperature. For example, Bonner (1959a) has reported positive responses to very weak sources, including cultures of fluorescent bacteria and phosphorescent paint. In *D. discoideum* there is a consistent and pronounced migration of pseudoplasmodia toward light prior to sorocarp formation (Fig. 4-6A). Fruiting structures produced under these circumstances may be formed at an angle of 30° or more from the vertical, whereas in darkness or in uniform light sorocarps are characteristically built perpendicular to the surface where they originate. Sorocarps of *D. mucoroides*, *D. purpureum*, and *P. violaceum* formed in one-sided illumination are regularly constructed in the direction of the light source (Fig. 4-6B,C), and appear to be characterized by longer sorophores than are typical of sorocarps produced in the dark; they are unquestionably longer than those produced in evenly illuminated cultures. In contrast, light seems to exert little effect upon the orientation or dimensions of fruiting structures in such delicate species as *D. minutum* and *D. polycephalum* (Raper, 1941b, 1956a).

As outlined above, the successful cultivation of a dictyostelid normally depends upon the use of a suitable nutrient source, such as *E. coli* or *K. pneumoniae*, grown on a favorable substrate at an appropriate temperature. But finding this combination may be very time-consuming, and often one must settle for something short of ideal. Under these circumstances it is frequently possible to secure improved if not optimal fructification by adding activated charcoal to the culture plate at the time cell aggregation should begin (Bonner and Hoffman, 1963), that is, when the lawn of bacteria is virtually consumed and the agar surface is covered with an even and nearly continuous layer of myxamoebae, or when the aggregation process is just getting underway. Later additions are less effective but often helpful, nonetheless. While a fine powdered charcoal such as Norite can be used effectively, we have found a granulated type of activated coconut charcoal (6-14 mesh) to be more convenient and far less messy. If inoculations have been made in streaks, chips of charcoal may be liberally distributed along the intervening spaces, or if the culture can be incubated upside down they may be placed in the lid of the Petri dish. However inserted, such additions of charcoal are very beneficial for sorocarp formation in many dictyostelids, particularly the more delicate species. Whether charcoal exerts its salutary effect soley via the absorption of some gaseous compound inimicable to sorocarp formation (see Bonner and Dodd, 1962b), or in part by some other means, has not been firmly established. It should be noted that in species with fruiting structures of intermediate size, such as *D. aureum*, there is a marked tendency for the sorocarps to build in the direction of the charcoal, and not infrequently to collide with it (Fig. 4-7).

A similar favorable effect can often be achieved by replacing the glass lid of a Petri dish with an unglazed porcelain cover (Coors), or if too rapid

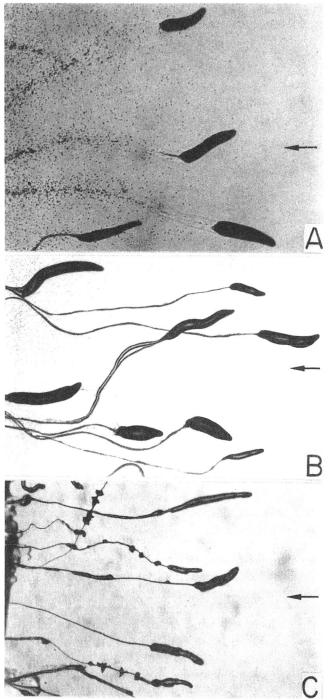

Fig. 4-6. Response to light in developing dictyostelids. *A*. Migrating pseudoplasmodia of *Dictyostelium discoideum*. *B*. Developing sorocarps of *D. purpureum*. *C*. The same of *Polysphondylium violaceum*. Direction of light indicated by arrows. × 7. (From Raper, 1951)

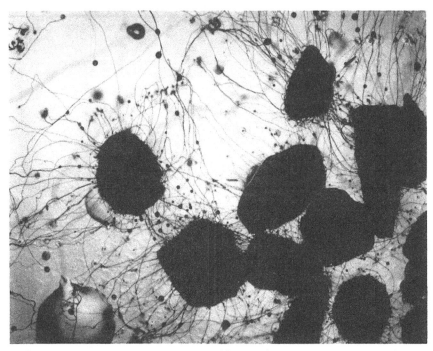

FIG. 4-7. Response of developing sorocarps of *Dictyostelium aureum* var. *aureum* to chips of activated charcoal. During fructification sorogens of this and many other species build toward the chips and often collapse against their sides. × 12.

evaporation creates a problem, lids that are unglazed below but glazed above may be used. In this case we suspect that the effect is largely one of regulating the relative humidity at the plane of sorocarp construction, although allowing the escape of some gaseous inhibitor cannot be ruled out. The beneficial effect of the poreclain covers is nowhere more dramatic than in cultures of *D. polycarpum* (Traub et al., 1981a) as we have cultivated the species with *E. coli* B/r on 0.1L-P agar in this laboratory. Wherever used, the porecelain covers have one serious disadvantage—you cannot observe their beneficial effect without removing them, whereupon many of the delicate sorocarps collapse.

Finally, concerning the relative merits of glass versus clear plastic Petri dishes, we have found that while each affords certain advantages, the balance in our experience favors glass. Cultures in such dishes dry out less rapidly, yield superior photographs when using transmitted light, collect fewer water droplets on the underside of the cover, and, perhaps most important of all, do not upon handling build up sufficient static electricity to lure onto the dish lid sori from the tallest sorocarps.

SUBMERGED CULTURE

Surface cultures grown on agar plates are most useful for comparative taxonomic studies of the type emphasized in this volume. For many other investigations, however, some type of submerged culture is highly advantageous. Recognizing this need, Gunther Gerisch in 1959 described a system wherein pregrown cells of *E. coli* B/r were harvested from nutrient broth by centrifugation, washed in phosphate buffer pH 6.0, and resuspended in the same buffer solution at a concentration of 1×10^{10} bacteria/ml as food for the slime mold. Spores of *D. discoideum* were then added at 5×10^3/ml, and air was blown through the culture to provide aeration and prevent clumping of the growing myxamoebae. Incubation was at $23 \pm 0.5°C$, and consumption of the bacteria was followed nephelometrically. Cells agglutinated spontaneously upon cessation of strong movement of the culture liquid. When removed from the solution and placed on a cover glass, such agglutinates began immediately to form sorocarps.

Further details were given in a longer paper the following year (Gerisch, 1960). *E. coli* B/r was used rather than *E. coli*, as usually recommended, and cultures were grown in strongly aerated Jena culture flasks (ca. 150 ml). The generation time was reported to be 3.3 hours, which compares very favorably with surface cultures on agar. To facilitate and better control cell agglutination, small aliquants of the submerged culture were transferred to roller tubes (as used in tissue cultures). Using agglutinates thus formed, he made many important observations concerning pseudoplasmodial polarity and induction of the fruiting process. For example, transformation of isodiametric agglutinates into cone-shaped structures occurred only when these were in contact with an air/liquid interface[1] and when both sides of the agglutinate were in contact with air, as on a vertical silk screen, twin fruiting structures developed pointing in opposite directions! These papers, together with others published by Gerisch in 1961 and 1962, had a profound effect in stimulating researches on the cellular slime molds and in shaping the course of future investigations.

During the same period, and working independently, M. Sussman (1961) introduced a method for growing myxamoebae of *D. discoideum* and a host bacterium, *Aerobacter aerogenes*, concurrently in a liquid medium of the following composition, expressed in g/l: yeast extract (Bacto), 0.5; peptone (Bacto), 5.0; glucose, 5.0; KH_2PO_4, 2.25; $K_2HPO_4 \cdot 12 H_2O$, 1.5; $MgSO_4 \cdot 7 H_2O$, 0.5; to final pH 6.0-6.3. The medium, designated "Liquid

---

[1] Using different techniques, other investigators have demonstrated limited anterio-posterior polarity and divergent cellular differentiation (spores and stalk-like cells) within aggregates formed in submerged culture, processes that were enhanced by incubating pregrown myxamoebae in roller tubes under an atmosphere of oxygen (Sternfeld and Bonner, 1977).

Medium A,'' was dispensed in 18 ml amounts in 125 ml Erlenmeyer flasks and autoclaved at 121°C. Flasks were inoculated with $1\text{-}2 \times 10^5$ slime mold spores and 1 ml of a 48 hour broth culture of *Aerobacter* and incubated at 22°C on a reciprocating shaker. Cell growth was monitored directly via haemocytometer counts and indirectly via clonal platings. Two diploid strains, RA and H-1, and a haploid derivative of RA were compared. The generation time for RA was calculated to be 3.6 hours, for the others 3.0 hours. The lag period varied with the inoculum, this being longer with older spores, shorter with young spores, and lacking when logarithmic phase myxamoebae were used. In a separate experiment, one liter of Liquid Medium A contained in a large Fernbach flask, and aerated by a sintered glass sparger, was inoculated with *Aerobacter* and 10^4 myxamoebae/ml. The generation time attained was 3-4 hours and the yield of myxamoebae 10^7 ml, thus demonstrating that substantial quantities of cells could be produced when needed.

AXENIC CULTURE

Having solved the problem of producing quantities of myxamoebae in submerged culture, either with pregrown bacteria as nutrient (Gerish, 1959 et seq.) or with bacteria concurrently grown in the same liquid medium (Sussman, 1961), investigators directed their attention toward an even more important objective: growing the cellular slime molds in the total absence of bacteria. Prior to this, Bradley and Sussman (1952) and Sussman and Bradley (1954) had reported the isolation of a ''protein growth factor'' from cells of three Gram-negative species (*E. coli, A. aerogenes,* and *Pseudomonas putida*) that supported limited growth of *D. discoideum* myxamoebae when the material was spread as a paste over the surface of a nutrient-poor agar. Incorporation of the bacterial factor (fraction) was thought to be at least in part via ingestion of the substance in particulate form. Whereas limited myxamoebal growth was reported, preparation of the factor was tedious, reproducibility poor, and cell growth only 10-20% of that with live bacteria. The approach was soon discontinued.

Beginning about 1960 Hohl and Raper took up the challenge of developing a procedure for the axenic cultivation of cellular slime molds, and a preliminary report was presented at the Second Annual Meeting of the American Society for Cell biology, San Francisco (1962). *Dictyostelium purpureum, Polysphondylium violaceum,* and *P. pallidum* were investigated along with *D. discoideum*; and of the four species *P. pallidum* proved most responsive. An early hint of its potential came in the rapid and substantial growth of the myxamoebae on killed bacteria (Hohl and Raper, 1963a). Fortunately, this intimation was verified when spores or myxamoebae were implanted in a thoughtfully formulated substrate devoid of any bacteria, living or dead (Hohl and Raper, 1963b). The formula of the liquid medium is given in Table 4-1.

Of many formulations tested, one was found to be superior. The medium was compounded from the following refrigerated stock solutions: Difco TC Bovine embryo extract, 100%; Difco Bovine serum albumin, 5%; Difco TC Vitamins (Eagle, 1955), 100×; modified Hank's salt solution (Paul, 1960); tryptose, 1%; and glucose, 1%. The desiccated embryo extract was reconstituted by adding salt solution and was used within 10 days.

To make 10 ml of medium, 100 mg of tryptose and 100 mg of dextrose were dissolved in 1 ml of salt solution (10×) and 0.6 ml of distilled water. After autoclaving for 20 minutes at 121°C, 0.2 ml of embryo extract (100%), 8 ml of serum albumin (5%), and 0.2 ml of Eagle's vitamin solution (100×) were added aseptically prior to inoculation. The tubes were incubated at 25°C on a rotary shaker at 250 rpm. Larger quantities were grown in aerated tissue culture bottles incubated on magnetic stirrers with agitation via teflon covered magnets.

If the medium was to be solidified, agar was added prior to autoclaving to give a final concentration of 1%. The other components were added when the autoclaved solution was still warm but not hot enough to precipitate the albumin. The mixture was poured into Petri dishes as a thin

TABLE 4-1. Composition of complex medium developed for growing myxamoebae of *Polysphondylium pallidum* in axenic culture (Hohl and Raper, 1963b)

Constituent	Amt. (mg/100 ml)	
Bovine serum albumin	4000.0	
Bovine embryo extract (100%)	2.0	(ml)
Tryptose	1000.0	
Dextrose	1000.0	
D-Biotin	0.048	
Folic acid	0.088	
Niacinamide	0.024	
Ca-pantothenate	0.048	
Pyridoxal-HCl (USP)	0.040	
Thiamine-HCl	0.068	
Riboflavine	0.008	
Choline chloride	0.028	
NaCl	80.0	
KCl	40.0	
$CaCl_2 \cdot 2\ H_2O$	10.0	
$MgSO_4 \cdot 7\ H_2O$	10.0	
$MgCl_2 \cdot 6\ H_2O$	10.0	
$Na_2HPO_4 \cdot 7\ H_2O$	6.0	
KH_2PO_4	6.0	

(2 mm) layer. Usually, 2% rather than 4% serum albumin was used for the solidified medium.

Among the ingredients of the complex medium, embryo extract was judged to be the most important, although none could be omitted without some loss in growth. The medium with embryo extract supported growth through 125 generations without a decrease in growth vigor or the ability to form normal sorocarps when postlogarithmic phase cells were harvested and transferred to agar or some other solid surface. By way of comparison, the generation time of *P. pallidum* myxamoebae was 2.5 hours when grown on living bacteria, 3.5 hours on dead bacteria (Hohl and Raper, 1963a), and 5-6 hours when grown axenically in the complex medium listed in Table 4-1.

Of the different cellular slime molds investigated only *P. pallidum* grew well in the complex medium, and of this species different isolates varied in their ability to grow and subsequently fruit. Strain Pan-17 gave the maximum response and FR-47 the least, whereas WS-320, the most studied strain of the species, gave an intermediate response, the relative values being about 1.1×10^7, 4.2×10^6, and 6.9×10^6/ml myxamoebae, respectively, at stationary phase. For *P. violaceum* and several strains of *Dictyostelium*, including *D. discoideum, D. purpureum* and *D. muco-roides*, growth of the myxamoebae was minimal, representing increases of six- to eight-fold at most and in some strains none.

Further attempts were made to simplify the complex medium and, if possible, to construct a truly defined liquid substrate wherein the identity of all constituents was known (Hohl and Raper, 1963c). Toward the first objective we made limited progress, and it was possible to devise a minimal medium for strains Pan-17 and FR-47. We did not succeed, however, in finding a formulation that would support continuous serial growth of WS-320, and we were unable to eliminate completely the serum albumin from the media used in growing the other test strains. In a simplified but still complex medium, column 1 of Table 4-2, cell densities up to 10^7 myx-amoebae/ml were obtained in Pan-17, and by supplementing this with amino acids, vitamins, nucleic acids, and glucose, cell yields up to 3×10^7 myxamoebae/ml were realized with a generation time of 4.5 hours (column 2, table 4-2).

Parallel with the above studies in Wisconsin, M. Sussman (1963) was investigating also the possibilities for axenic cultivation of *Polysphondy-lium pallidum*. The culture used for his work was designated PP-1 and said to be "derived" from WS-320 (cited above). Sussman's medium (1963) consisted of lipid-free milk powder (Starlac), 0.5-2%; proteose peptone, 1%; 0.05 M phosphate buffer at pH 6.5; and 400-800 μg/ml of lecithin (Glidden Products). Sterilization was by autoclaving for 25-30 minutes at 120°C. Aliquots of 5 ml were transferred to foil-covered Erlenmeyer flasks (125 ml), inoculated with 10^4 myxamoebae or spores, and incubated on a

TABLE 4-2. Media developed for the cultivation of strains of *Polysphondylium pallidum* with details of their composition (Hohl and Raper, 1963c)

Complex medium (simplified)		Complete medium		Minimal medium	
Component	Amt.	Component	Amt.	Component	Amt.
Tryptose	2.0%	Amino acid mixture 1		Amino acid mixture 2	
Serum albumin	4.0	Vitamin mixture		Riboflavine*	
Inorganic salts		Dextrose	1.0%	Dextrose	1.0%
		Bases		Bases	
		Trace elements		Trace elements	
		Serum albumin	1.0	Serum albumin	1.0
		Inorganic salts		Inorganic salts	

Amino acid mixture 1:		Amino acid mixture 2:		Vitamin mixture:	
Glycine	75†	Glycine	300†	D-Biotin	0.35†
DL-Isoleucine	131	DL-Isoleucine	520	Choline chloride	2.10
L-Lysine	182	L-Lysine	730	Folic acid	0.20
DL-Methionine	149	DL-Methionine	595	Niacin	1.00
L-Tryptophan	204	L-Tryptophan	204	Ca-pantothenate	1.10
L-Tyrosine	36	L-Tyrosine	11	Pyridoxal·HCl	2.20
L-Arginine	210	*Bases:*		Thiamine·HCl	0.70
L-Cysteine·HCl	175	Guanine·HCl	25†	Riboflavine	1.10
L-Histidine·HCl	191	Adenine sulfate	25	Pyridoxine·HCl	2.20
L-Leucine	131	Thymine	25	Inositol	2.80
DL-Valine	117	Uracil	25	p-Aminobenzoic acid	0.20
L-Alanine	89	Cytosine	25	B_{12}	0.003
L-Asparagine	150	Xanthine	25	Ascorbic acid	0.30
L-Aspartic acid	133	Hypoxanthine	25	*Trace elements:*	
L-Proline	115	*Inorganic salts:*		$FeSO_4 \cdot 7 H_2O$	10†
L-Glutamic acid	147	NaCl	800†	$FeCl_3$	10
DL-Phenylalanine	165	KCl	400	$MnCl_2 \cdot 4 H_2O$	1
DL-Serine	105	$CaCl_2 \cdot 2 H_2O$	100	H_3BO_3	1
DL-Threonine	119	$MgSO4 \cdot 7 H_2O$	100	$(NH_4)_2Al_2 (SO_4)_4 \cdot 24 H_2O$	1
L-Cystine	48	$MgCl_2 \cdot 6 H_2O$	100	$ZnSO_4 \cdot 7 H_2O$	1
DL-Norleucine	131	$Na_2HPO_4 \cdot 7 H_2O$	60	$MnSO_4 \cdot H_2O$	1
		KH_2PO_4	60	$CuSO_4$	1
				$CoCl_2$	1

* Riboflavine: 0.1 mg/l.
† All amounts in the mixtures are expressed as mg/l of the final medium.

shaker. Generation time was 3.7 hours with a yield of 1.6×10^7 myx-amoebae/ml. The milk fraction was found to be nonessential, albeit the generation time was lengthened to 4.5 hours and the yield of cells reduced to 1×10^7 when milk was omitted. Lecithin and at least one protein were essential. *P. pallidum* was maintained through seven serial passages representing about 100 generations without decrease in rate of growth or yield.

Neither *D. discoideum* nor *D. mucoroides* could be cultivated in the medium just considered.

H. S. Hutner and coworkers at the Haskins Laboratories in New York took up the search for a defined medium for the dictyostelids about 1962, for which then current studies with *P. pallidum* served as a starting point. A preliminary report (abstract) on the growth of this species in "defined media" was published by Allen, Hutner, Goldstone, Lee, and Sussman in 1963 citing moderate growth through at least two serial transfers. A year later, and using a modified medium, growth through subcultures of vegetative myxamoebae at a level of 7.4 × 10⁵ cells/ml was reported by Banerjee, Allen, Goldstone, Hutner, Lee, and Diamond (abstract). Subsequent to this, a formal paper (1966) was published under the authorship of Goldstone, Banerjee, Allen, Lee, Hutner, Bacchi, and Melville. This contained substantial information. A defined medium with all nutrient constituents known and controlled was reported, as was appreciable vegetative growth of myxamoebae in *P. pallidum* strain WS-320. The composition of the defined medium is reproduced in Table 4-3.

As listed, the "defined medium supported growth of 2.0 × 10⁶ cells/ml through 20 serial transfers," while supplementation with "complete supplement No. 15 (2) raised yields to 5-7 × 10⁶ cells and decreased the generation time from 42 to 26 hrs." Complete supplement No. 15 is not

TABLE 4-3. Defined medium for vegetative growth of *Polysphondylium* (Goldstone et al., 1966)

	g/100 ml		g/100 ml
DL-Lactate	0.05	L-Glutamic acid	0.15
K₃ citrate · H₂O	0.04	L-Histidine (base)	0.12
Na acetate · H₂O	0.04	DL-Leucine	0.002
Glycerol	0.1	L-Lysine HCl	0.06
Sucrose	0.1	DL-Methionine	0.002
KH₂PO₄	0.1	DL-Phenylalanine	0.008
MgSO₄ · 3 H₂O	0.002	L-Proline	0.004
Ca (as Cl⁻)*	0.002	DL-Serine	0.01
Metals No. 59†	0.004	DL-Threonine	0.006
DL-Alanine	0.04	DL-Tryptophan	0.003
DL-Aspartic acid	0.05	L-Tyrosine	0.005
L-Arginine HCl	0.09	DL-Valine	0.008
L-Cystine	0.002	Vitamins mix	0.5
diethylester HCl		No. 15‡ pH 6.5	

* "Prepared by suspending 62.5 g CaCO₃ in 250 ml H₂O and adding conc HCl dropwise until dissolved; volume then brought to 500 ml; slight amount of Tris (to neutralize excess acid) and preservative added. Stored at room temperature."

† "Metals No. 59" yields the following (mg/100 ml final medium) when used at the indicated level: Fe 0.15, Mn 0.12, Zn 0.12, Mo 0.015, Cu 0.0075, V 0.0015, Co 0.0015, B 0.0015."

‡ "Vitamins 15 is nearly identical with a previous mixture (2)."

identified as such in this paper or in that cited as "(2)" in the table legend; however, the writers note that, in the substrate-free medium containing a complete amino acid mixture, glycerol or acetate increased growth by 100% and sucrose, maltose, or glucose improved growth by 50% when added individually. Difficult to understand are generation times of 42 or 26 hours; cultures incubated in near static rather than shaken or stirred vessels; a lack of sorocarp development from cells adhering to the sides of culture tubes; and especially, the misbelief that higher concentrations of nutrients might favor fructification when starvation was generally recognized as an important trigger for initiating this process. In this connection it is perhaps significant to recall that when Toama and Raper (1967a,b) wanted to produce large quantities of myxamoebae of *P. pallidum* (strains WS-320 and Pan-17) for their studies of microcyst formation and analysis, they chose to use the older methods of Hohl and Raper (1963b,c).

While a fully defined medium useful for most investigations still waited in the wings, an approach to the same objective slowly materialized from a different direction—namely, the emergence of variants or mutants capable of growing in media of known composition in the total absence of bacteria. Progress in this regard has been most dramatic with *Dictyostelium discoideum*. Using a medium designated CF[3], R. and M. Sussman in 1967 succeeded in adapting strain NC-4, or a derivative thereof, to grow in the absence of bacteria. The medium was one that Chandler Fulton had used for the axenic growth of the amoebo-flagellate *Naegleria gruberi* and communicated to the Sussmans in 1967 (published 1974), which in turn was a modification of a medium previously recommended by W. Balamuth for cultivating the same protozoan (1964). Composition of the medium used by the Sussmans follows.

I	KH_2PO_4	3.4 g	in 100 ml H_2O (autoclaved)
	$Na_2HPO_4 \cdot 7 H_2O$	6.7 g	
II	Glucose	13.5 g in 50 ml H_2O (autoclaved)	
III	proteose peptone	10.0 g	in 650 ml H_2O (autoclaved)
	yeast extract	5.0 g	

The above are mixed in the ratio 2:2:65

To 3.5 ml of the mixture are added 1.0 ml of a 5% solution of Wilson's Liver Concentrate (NF), autoclaved, cleared by centrifugation and reautoclaved, and 0.5 ml of sterile fetal calf serum (Grand Island Biologicals Co.).

To develop the axenic strain, spores from carefully selected sori were suspended in nutrient broth overnight to facilitate germination and to verify the absence of bacterial contamination. Small aliquots were inoculated into 5 ml volumes of CF[3] medium in 125 Erlenmeyer flasks and incubated with gentle shaking at 22°C. The rate of growth and yield of cells were very poor initially, but these improved steadily in serial subcultures from one passage to the next. In the ninth passage the myxamoebae grew exponentially and without lag. The generation time was 12 hours and the final cell yield 1×10^7 myxamoebae/ml, or, except for elapsed time, approximately equal to that using bacteria as nutrient. Cells that were harvested and dispensed on Millipore filter pads in the usual manner fruited normally. By common usage the subculture so derived came to be known as Ax-1. The gradual improvements in growth rate and cell yield realized over some months are shown in Table 4-4, copied from Sussman and Sussman (1967).

Few developments have had a greater impact on the nature and amount of research devoted to the cellular slime molds, especially *Dictyostelium discoideum*. A culture that could grow axenically opened the way for many meaningful investigations in areas of biology previously closed by the presence of bacteria, even when dead and fragmented. During the ensuing years substrates have been modified and simplified to meet specific needs in different laboratories. We shall not attempt to catalogue all of these; instead, we shall consider a few that seem to merit special attention. All are intended for the growth of myxamoebae in shaken liquid cultures.

Medium HL-5, introduced by Coccuci and Sussman (1970) for cultivating strain Ax-1 of *D. discoideum*, contains the following ingredients: glucose, 16 g/l; proteose peptone, 14 g/l; yeast extract, 7 g/l; $Na_2HPO_4 \cdot 7$ H_2O, 0.95 g/l; and KH_2PO_4, 0.5 g/l. Subsequently, Schwab and Roth (1970) modified this by omitting the glucose and reducing the proteose peptone to 10 grams/l.

In 1970 also, Watts and Ashworth announced the isolation from Ax-1 of a new strain, Ax-2, that was selected for its ability to grow on a medium of the following composition: Oxoid bacteriological peptone, 14.3 g/l; Oxoid yeast extract, 7.15 g/l; D-glucose, 15.4 g/l; $Na_2HPO_4 \cdot 12 H_2O$. 1.28 g/l; and KH_2PO_4, 0.486 g/l, with final pH 6.7. Optimum temperature was 22-23°C and the generation time 8-9 hours in shaken culture. With some modifications this is still being used.

Using a medium differing from that of Sussman and Sussman (1967)

TABLE 4-4. Record of serial passages on CF medium (Sussman and Sussman, 1967)

Passage	Days incubated	Inoculum (No./ml)	Stationary phase yield (No./ml)
1	20	5×10^3	8×10^4
2	20	1×10^4	1.2×10^6
3	13	2×10^4	5×10^6
6	7	2×10^4	9.5×10^6
9	6	1×10^5	1.1×10^7

primarily in the reduced amount of glucose (10 instead of 13.5 g/l), Loomis in 1971 isolated a haploid strain from his culture of NC-4 that grew axenically and, as reported, rapidly. The subculture, designated A3 (sometimes cited A-3 or AX3), has been used extensively in Loomis's laboratory and elsewhere for a variety of biochemical, physiological, and genetic investigations (Dimond et al., 1973; Firtel and Bonner, 1972; Rossamondo and Sussman, 1973; and others). A formulation known as TM Medium, reported by Free and Loomis (1974) for cultivating this strain, contained the following nutrients: trypticase (tryptic digest of casein) in lieu of proteose peptone, 10 g; yeast extract, 5 g; and glucose, 10 g/l of 2 mM sodium phosphate buffer at pH 6.5. The doubling time was 10 hours.

Substantial progress has been made toward the development of defined media for growing myxamoebae of *D. discoideum* strain Ax-2 by Watts (1977) and for strain AX3 by Franke and Kessin (1977). Using such defined media, these researchers succeeded in isolating a number of auxotrophic mutants that broaden substantially the possibilities for genetic analyses in this species (Franke and Kessin, 1978). By substituting glycogen for glucose in the defined medium, Borts and Dimond (personal communication) have used it to test the physiological role of the acidic glucosidase in glycogen utilization by selected mutants of *D. discoideum*.

Whereas some of the aforementioned axenic media may be stiffened with agar and used as solid substrates for cloning experiments and other purposes, they are generally employed as liquids, with cell growth being monitored optically or by direct cell counts. To follow the fruiting stage, which cannot develop in shaken culture, the myxamoebae are harvested by centrifugation at stationary phase, or earlier if the experiment demands, washed in Bonner's salt solution (1947) or phosphate buffer, and dispensed on the surface of nonnutrient agar, or, as now seems more commonplace, on the surface of Millipore filters resting upon filter pads saturated with an appropriate buffer solution. Such an LPS (lower pad solution) contains the following ingredients per liter distilled water: KCl, 1.5 g; MgSO$_4$, 0.5 g; K$_2$HPO$_4$, 1.6 g; KH$_2$PO$_4$, 1.8 g; streptomycin sulfate, 0.5 g (Sussman, 1966).

By working with myxamoebae pregrown in stirred or shaken culture, one has the advantage of essentially synchronous growth, and the opportunity to harvest the cells at any stage in the life cycle. Upon resuspension at predetermined densities these can be dispensed upon nonnutrient agar and their further behavior followed by continual microscopic observation. If cells are spread evenly and thinly the observer can follow in detail all phases in cell aggregation as centers arise and inflowing streams of myxamoebae converge to produce in ordered sequence an upright column, a migrating pseudoplasmodium (or sorogen), and subsequently the terminal fructification, or sorocarp. When cells are dispensed on filters, only the latter stages of this developmental cycle can be seen clearly. Even so, for many types of study the filters afford compensating advantages. The researcher can interrupt the developmental cycle at any stage by lifting the filter off the supporting pad, and then by gently shaking it in cold buffer remove the cells, or developing fructifications, for whatever type of assay or analysis befits his purposes.

CHAPTER 5

Culture Maintenance

Insofar as is known, all species of the Dictyostelidae can be maintained indefinitely in laboratory culture by periodic transfer onto appropriate agar media with suitable bacterial associates. Such stocks, whether in plates or tubes, should be renewed every 4-6 months, and following growth and fructification should be stored at 3-5°C to reduce evaporation and extend viability. Low-nutrient substrates are generally preferable, and media such as numbers 1, 3, and 4 listed in Chapter 4 are recommended. In our experience hay infusion agar, with or without 0.05 or 0.1% added peptone, has been used most commonly. In recultivating dictyostelid stocks one should bear in mind that the culture is two-membered. Failure of the slime mold to grow following direct transfer may result either from its demise or that of the associated bacteria, as happens occasionally; hence the precaution of adding fresh bacteria at the time of transfer.

Cultures of cellular slime molds grown with *Escherichia coli* or *Klebsiella pneumoniae* on agar slants and stored at 3-5°C commonly remain viable for 8 months or more. Even so, if valuable stocks are to be maintained, prudence requires that they be checked more frequently. For example in one series of tests nine strains of *Dictyostelium* and *Polysphondylium* were examined after storage for one year. Five were still viable, but four were not; so why take the risk?

In some laboratories such cultures have been covered with sterilized heavy mineral oil, as is sometimes done to conserve filamentous fungi (Buell and Weston, 1947). Results have been equivocal. Experience in Bonner's laboratory (personal communication) may be regarded as representative: one culture of *D. discoideum* was viable at 33 months, while others were dead at 24. One culture of *D. purpureum* grew after 2 years; another of the same age failed to do so. Cultures of *P. violaceum* and *P. pallidum* failed to grow when tested at 2 years. Results were sufficiently variable overall to suggest that environmental conditions (pH, metabolic products, etc.) existing in the tubes when oil was applied might have determined longevity. At best the method affords only limited reprieve, and Bonner, like many others, turned to lyophilization.

LYOPHILIZATION

The lyophil process, as used for the preservation of molds (Raper and Alexander, 1945; Fennell et al., 1950) and yeasts (Wickerham and Flickinger, 1946), has proved especially useful for conserving cultures of *Dictyostelium* and *Polysphondylium*. Whereas attempts to preserve myxamoebae or vegetative cells by this tecnhique have been unsuccessful, spores, or microcysts if formed, can be preserved superbly. Some lyophil preparations for each of 17 selected strains were prepared in May and June 1941, and viability tests have been made from time to time since that date. A rather detailed summary was compiled and published by Raper more than a quarter century ago (1951) wherein he reported favorable viabilities after 15-16, 31-32, and 105-106 months. The tests of nearly 9 years were especially encouraging, as have been the results when other preparations were opened as needed during the intervening years. However, no purposeful reevaluation had been made, and since we still had two or three tubes of most of the cultures processed in 1941, we decided in December 1979 to expend one of each to obtain a reading at 38 + years. Although we do not know the effect of storage temperature on the viability of lyophilized cultures of cellular slime molds, we have taken the precaution of holding them in a cold room at 3-5°C throughout the storage period. Tubes were checked for vacuum before being opened, and, with a single exception (where the thin tip of the sealed tube had broken off), all registered quite well. The vacuum-less tube (#V-1a of 5/29/41) of course yielded no growth, either of slime mold or bacteria, and a more recent preparation was substituted. In all other cases growth of myxamoebae was evident within 2-2½ days. In several cases pseudoplasmodia began to appear within 42 hours following rejuvenation; and some mature sorocarps were present at 60 hours or before. To make the viability test, the content of a tube was dissolved in sterile water (ca. 0.5 ml) and streaked via 5 mm loop on plates of Hay, T-Hay, 0.1L-P, and 0.1L-P (pH 6.0) agars; then fresh *E. coli* was added to the remaining suspension and loopfuls of this were streaked on the opposite side of the same plates. Thus we could determine if both, one, or neither of the two organisms remained viable. The results of the tests are summarized in Table 5-1.

Unfortunately, for most of the strains listed we have only one, two, or no preparations left from the 1941 processing, and so we cannot look forward to extended future tests. We think it is quite remarkable, however, that the cultures survived so long in a living but virtually inactive state, and that they could be brought back to active growth and fruiting in a period of time but little if any longer than required for freshly matured spores.

The ability to conserve slime molds in a desiccated state proved especially valuable during the period from 1941 to 1949, attention then being directed wholly toward the search for better penicillin-producing molds and other problems. It is probable that valuable stocks, upon which much work had already been done, would have been lost had it proved necessary to transfer these cultures every few months. Although we cannot estimate the probable viability of properly processed lyophil preparations, we can now report that spores thus conserved for nearly 4 decades still showed neither apparent loss of viability nor reduction in vigor.

Whereas no systematic search for slime molds was made during the period 1941-1949, many soils from different stations in this country and abroad were examined for antibiotic-producing molds. From time to time slime molds were observed, isolated, and lyophilized, and a considerable backlog of strains was built up. By continuing to lyophilize new strains as they were isolated or received from collaborators, it has been possible to build up a large collection of species and strains numbering in the hundreds upon which to draw for the current monographic study.

The technique of culture lyophilization is possibly familiar to most readers, and various types of processing equipment are now available commercially. Nevertheless, because of the excellent results obtained over many years and the simplicity of the apparatus employed, we feel justified in summarizing briefly the methods that have served well. The apparatus as constructed in local shops (Fig. 5-1) consists of these principal parts: a vertically adjustable manifold designed to accept 30 small homemade Pyrex tubes (6 mm O.D × 100 mm L.) connected to a Welch Duoseal vacuum pump (½ HP); an insulated freezing bath containing dry ice and methyl-cellusolve for quickly freezing spore suspensions; a water vapor trap between the manifold and pump; a reliable thermometer to read to $-50°C$; a Bruner-type vacuum gauge to monitor the progress of desiccation; a gas-oxygen torch to seal the evacuated tubes; and a Tesla coil for confirming evacuation of the apparatus and of the sealed preparations. A vacuum of 1 mm of mercury in the system when tubes are sealed is probably satis-factory, but 200-500 μm is preferred.

Spores with associated bacteria for processing may be suspended in sterile beef serum or in a dried fat-free milk reconstituted at $2 \times$ normal strength to yield solid pellets upon drying—otherwise the suspension tends to adhere to the sides of the tube and removal is difficult. Spores from well-isolated sori (if possible) are collected in sterile water on a 3 mm platinum-iridium loop until the drop is faintly clouded, then added to about 0.5 ml of sterile reconstituted milk, together with a loopful of freshly grown *Escherichia coli* or other bacterial associate, and thoroughly inter-mixed. Using a Pasteur pipette, aliquants of about 0.05 ml are added to sterilized lyophil tubes. When the tubes are firmly seated in the manifold sleeves, the tips containing the spores are immersed in a bath of dry ice and methylcellusolve at about $-30°C$ for instant freezing; evacuation is begun, and the temperature is permitted to rise gradually to $-5°C$, where it is held until the suspensions appear thoroughly dry. The manifold is then raised to ambient temperature and evacuation continued for ½ hour or more. The lower portions of the tubes (4.5-5.0 cm) are finally removed by hermetic sealing with a Hoke gas-oxygen torch. Tubes containing the

TABLE 5-1. Viability of lyophilized spores of *Dictyostelium* and *Polysphondylium* produced and tested in association with *Escherichia coli* (test 4)

Genus & species	Strain	Date lyophilized	Age yr:mo	E. coli	Slime mold	Remarks
D. discoideum	NC-4	6-20-41	38:6	+++	++++	Aggregation at 44 hr, sorocarps at 60 hr
	V-12[a]	11-25-48	31:1	+++	++++	Migrating pseudo-plasmodia at 42 hr, sorocarps at 60 hr
	V-11[b]	4-20-50	29:7	++	++++	Young sorocarps at 44 hr
D. mucoroides complex	S-2	6-20-41	38:6	++	++	Myxamoebae at 42 hr
	V-10	6-10-41	38:6	++	+++	Sorogens at 44 hr, sorocarps at 60 hr
	Hall	6-14-41	38:6	+++	++	Myxamoebae at 44 hr, sorocarps at 60 hr
	Thom	6-19-41	38:6	++	++++	Young sorocarps at 42 hr
D. purpureum	V-15	6-19-41	38:6	++	+++	Aggregation at 48 hr, sorocarps at 72 hr
	D-6	6-4-41	38:6	++	+++	Aggregation at 48 hr,

	V-1a[a]	11-3-48	31:1	+++	+++	Aggregation at 44 hr, sorocarps at 66 hr
	G-4	5-18-50	29:7	—[c]	+++	Aggregation at 48 hr, sorocarps at 72 hr
P. violaceum	P-6	6-15-41	38:6	+++	++++	Aggregation at 42 hr, sorocarps at 66 hr
	V-6	6-19-41	38:6	++	++++	Aggregation at 42 hr, sorocarps at 66 hr
P. pallidum	V-16[a]	11-17-45	34:1	++	++++	Aggregation at 42 hr, sorocarps at 66 hr
	Martin[a]	5-26-47	32:6	++	+++	Aggregation at 42 hr, sorocarps at 66 hr
D. minutum	V-3	5-29-41	38:6	—[c]	++	Few amoebae at 60 hr, aggregation at 72 hr, sorocarps at 84 hr

KEY: ++++ = excellent viability   +++ = good viability   ++ = fair viability   + = poor viability   — = no growth

NOTE: For results of tests 1, 2, and 3 see Raper, 1951, p. 186. Storage at 3-5°C.
[a] No lyophil tubes from 1941 remained, hence next oldest tested.
[b] Not included in 1951 tests.
[c] Bacteria did not grow from lyophil tube; recorded growth was in streaks with freshly added *E. coli*.

Fig. 5-1. Lyophil apparatus used for conserving cellular slime molds: (1) manifold; (b) vacuum gauge; (c) thermometer; (d) lyophil preparations in final stage of processing; (e) Dewar flask containing water vapor trap immersed in dry ice and methyl cellosolve; (f) vacuum pump; (g) insulated bath for quick freezing of preparations and maintenance of frozen state during desiccation; (h) Tesla-type vacuum tester; (i) Hoke oxygen-gas torch for sealing preparations under vacuum. (From Raper and Alexander, 1945)

desiccated cultures are checked for vacuum after 2-3 days, one is opened for a viability test, and the remaining four, five, or more are stored in labeled vials or bottles at 3-5°C. A full account of the procedures developed, and in the main still being used, is given in the paper "Preservation of Molds by the Lyophil Process" published by Raper and Alexander in *Mycologia* 37 (1945): 499-525.

To test preparations for viability, tubes are marked with a file scratch and surface sterilized, broken open, and the pellets allowed to dissolve in about 0.5 ml sterile water. The resulting suspension is then streaked on one or more appropriate agar substrates and incubated at a temperature favorable for growth. Whereas bacteria are routinely included with spores at the time of processing, it is prudent to add fresh cells at the time of recultivation to ensure their presence in a living state when the myxamoebae emerge.

SILICA GEL

Silica gel granules have been used successfully for conserving spores of *Dictyostelium discoideum* for periods up to 11 years (W. F. Loomis and R. L. Dimond, personal communication). The method is credited to Joshua Lederberg, University of Wisconsin, sometime prior to 1957, when it was cited by Freeman A. Weiss in a chapter on culture maintenance written for the then current *Manual of Microbiological Methods* (Soc. Amer. Bacteriologists). A rather detailed account of its use as applied to stock cultures of *Neurospora* was published a few years later by Perkins (1962) wherein he reported good viability and retention of mutant characteristics in most cases (80-100%) over a period up to 5-6 years. Four years later, Reinhardt (1966) reported using the same procedure to preserve *Dictyostelium purpureum* with 100% recovery when tested after 31 months, and *Acrasis rosea*, a nondictyostelid cellular slime mold, for 13 months. The latter becomes more impressive when one recalls that this acrasid cannot be lyophilized, at least by conventional means, and storage in refrigerators is not wholly satisfactory (Olive and Stoianovitch, 1960). Since many people have used silica gel, at least for short-term storage, and since most seem to have been guided by Perkins, we shall summarize very briefly the salient features of this procedure:

Test tubes 13 × 100 mm are half filled with silica gel (Davidson Chemical Corp., Baltimore, MD; 6-12 mesh, grade 40, desiccant, activated, without humidity indicator dye that may be toxic). Tubes are plugged with cotton, dry sterilized at 180°C for 1½ hours, and stored in tightly sealed containers. Spores or microcysts to be preserved are suspended with bacteria in reconstituted and sterilized nonfat dry milk, and 0.5 to 1.0 ml of the suspension is carefully deposited dropwise on granules of the silica gel with a Pasteur pipette. (Precooling the gel in ice may be advantageous to reduce heat generation.) The tube is tapped gently to loosen granules and further distribute the inoculum. Viability is checked at 1 week by transfer of a few granules to an appropriate substratum, and tubes are closed with Parafilm (Marathon Corp., Menasha, WI) and stored at 5°C in a sealed container.

Whereas inquiries concerning the use of silica gel for culture preservation have not been uniformly enthusiastic, we believe it to represent a generally useful method for preserving over considerable periods of time dictyostelids that form either spores or microcysts. Its convenience cannot be questioned, for granules can be removed repeatedly from the same tube to reinitiate cultures as needed. Additionally, no specialized apparatus is required.

PRESERVATION OF MYXAMOEBAE

Methods of culture maintenance considered above have focused upon strains that sporulate and in some cases form microcysts as well. In recent years, and with greater emphasis on genetic, physiological, and developmental studies, an acute need has developed for some means of preserving aggregateless and other nonsporulating mutants. This has been attempted in several ways with varying success.

As early as 1954 Maurice Sussman showed that some "morphogenetically deficient" mutants when grown together would respond synergistically and produce fruiting structures that contained spores of the two parent types, although neither could form sorocarps and spores alone. Capitalizing upon this phenomenon, he has over the years cultivated nonsporulating strains of *D. discoideum* synergistically with a normal culture such as wild type NC-4 and preserved the spores by lyophilization. Recovery of the mutant is by cloning at the time of culture revitalization. Another method, mentioned by Sussman (1966) and investigated in some other laboratories as well (Laine et al., 1975), is similar to that developed for preserving animal cells. Myxamoebae are suspended at high density in 5% (Sussman, 1966) or better still 10% glycerol (R. L. Dimond, personal communication), after which small aliquants are sealed in foil-wrapped ampules, then frozen and retained at liquid nitrogen temperature.

The use of dimethyl sulfoxide (DMSO) for preserving nonsporulating cultures has gained wide acceptance within recent years. Probably first used by Frank Rothman, but unreported, a detailed account of the method was published by Laine, Roxby, and Coukell in 1975. Several variations have been investigated but, insofar as we have been able to determine, the essential features remain as described originally except that the DMSO is heated to 80°C for 15 minutes before being added to the storage solution. The method of Laine et al. follows:

Non-axenic strains of cellular slime molds are grown on Petri plates in association with *Aerobacter aerogenes* (or other bacterial associate), as previously described (Sussman, 1966). When the bacterial lawns begin to clear, the amoebae are harvested and washed 3 times by centrifugation (500 × g for 3 minutes) in standard salts solution (SS) (Bonner, 1947). Axenic strains are grown on plates as described above or in liquid HL-5 medium (Cocucci and Sussman, 1970). In the latter case, the cells are harvested at a cell density of ~5 × 10^6 cells/ml and washed once in SS. After washing, the cells are resuspended in HL-5 + 10% dimethyl sulfoxide (DMSO) to give a concentration of ~10^7 cells/ml. (The DMSO, not sterilized, is added to the HL-5 medium just before the cells are frozen.) Exactly 0.5 ml of this suspension is placed in an 8-mm Taab capsule (Marivac Services, Halifax, Nova Scotia); the capsule is placed in an aluminum rack (similar to the one received with the capsules), frozen immediately at $-73°C$ and stored at this temperature.

To thaw the cells, the vials are shaken in a water bath (usually at 45°C) until thawed. The cells are drawn up in a sterile Pasteur pipette, transferred to a sterile test tube, and 0.5 ml of fresh HL-5 medium is added. The cell suspension is left at room temperature for 30 min, shaken gently, and left for another 30 min. The cells are then plated out with *A. aerogenes* and incubated at 22°C for 4-6 days. To ensure single clones, the original suspension should be diluted an additional 5- or 10-fold in SS before plating. Usually four vials of each strain are frozen and one is thawed and plated out 1-5 days later to ensure that the cells have survived the freezing.

The percentage of the amoebae which survived the freezing process was calculated from the plating efficiency (E.O.P. = viable cells/total cells) of the cells before and after freezing (E.O.P. after freezing/E.O.P. before freezing $\times$ 100).

Best survival rates are obtained with densities of 10^7 myxamoebae/ml. Of the freezing regimens tested, fast freezing at $-73°C$ and storage at the same temperature gave the most satisfactory results; and of the three methods of thawing investigated, rapid thawing at about 45°C was most efficaceous. Viable cell yields sometimes reached 10% and often exceeded 3-4%. Successful preservation up to 15 months was reported by Laine et al. in 1975, but correspondence with current investigators now confirms viabilities up to 6 years (Barrie Coukell, personal communication). Peter Newell is reported to store his myxamoebae not at $-73°C$ but in liquid nitrogen (fide Coukell). Cells are suspended in sterile serum or HL-5 medium plus 5-10% DMSO, dispensed in plastic bull sperm straws, placed at $-20°C$ for 2 hours and then at $-73°C$ for 2 hours before being moved to liquid nitrogen. The percentage survival is said to be improved. Storage in DMSO would seem to be withal an increasingly satisfactory method of cell conservation that offers great promise for the future.

Although strains of *Dictyostelium* and *Polysphondylium* can in theory be continued indefinitely by periodic transfer from one agar plate or tube to the next, it behooves the investigator to preserve his cultures as soon as possible by lyophilization if spores or microcysts are produced. If the culture cannot be lyophilized, then some other method for long-time storage should be used. This is recommended not only as a measure of insurance against outright loss, but because many isolates, particularly the more delicate forms, tend to become less vigorous in growth and less consistent in form when kept in continuous culture. If one were to apply the type of nomenclature employed by phytopathologists for certain nonspecific plant diseases, this condition might be appropriately termed "dictyostelid decline."

Another hazard of which to be mindful is contamination, either by adulteration or substitution of the intended bacterial associate. Even worse is the invasion of a culture by a foreign amoeboid organism. Fortunately, the latter is rare, but it can occur—not as a rule by another cellular slime mold, but rather by dust-borne, cyst-forming, free-living amoebae such as species of the genus *Hartmanella* (for descriptions see Singh and Das, 1970). A case in point is that of the reputed "PC behaviorial mutant of *D. discoideum*" described by Frantz in 1980.

CHAPTER 6

Vegetative Stage

The vegetative phase of all cellular slime molds consists of amoeboid cells that feed upon microorganisms smaller than themselves, grow, and by repeated division build up large populations of free-living cells prior to fructification. Among these myxamoebae are cells of different patterns, and in substantial part it is upon such differences that the three subclasses of the Acrasiomycetes are recognized, namely: the Protostelidae, wherein the myxamoebe (and small plasmodia) have nuclei with compact, centrally positioned nucleoli and filose pseudopodia; the Acrasidae, with nuclei of similar pattern but with broad, lobose pseudopodia; and the Dictyostelidae, characterized by nuclei with peripherally lobed nucleoli and filose pseudopodia (Raper, 1973).

Myxamoebae of the Dictyostelidae lack structural features that an observer might employ to distinguish between vegetative cells of different species, or even different genera, when using conventional light or phase microscopy. Although limited differences in average cell size do exist between certain species of *Dictyostelium*, and between some species of *Dictyostelium* and *Polysphondylium*, one would be hard pressed to identify the myxamoebae of any dictyostelid if they were intermixed with those of another species (Fig. 6-1). Size measurements of actively moving myxamoebae are not very meaningful, as any experienced observer knows. Still it may be said that the apparent dimensions of the trophic stage when moving on agar may measure up to 25 μm in maximum diameter in many species of *Dictyostelium*, while those of *Polysphondylium* and *Acytostelium* rarely exceed 20 μm (Raper, 1941a; Raper and Quinlan, 1958).

SPORE GERMINATION

Growth begins with the germination of spores, which in all but a few dictyostelids are narrowly elliptical, or capsule-shaped, and represent the principal dormant stage in the life cycle. As if to ensure the widest possible dispersal of the emergent myxamoebae, nature has in most species, but not all, provided autoinhibitors which preclude the germination of spores en masse but permit them to germinate when spread apart. Working with *Dictyostelium mucoroides*, Russell and Bonner (1960) were the first to investigate this phenomenon, and in a few carefully conducted experiments showed a definite correlation between spore density and the percentage of germination when bacteria-free spores were deposited on an agar surface.

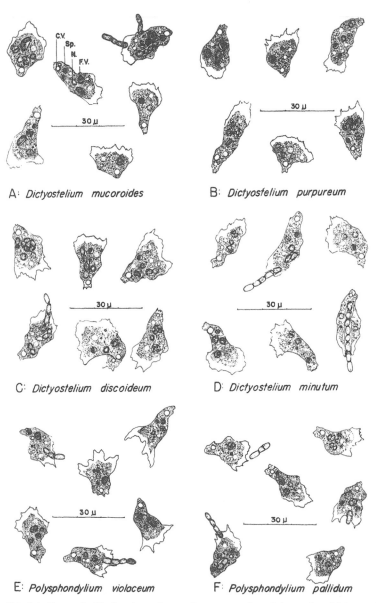

A: *Dictyostelium mucoroides*

B: *Dictyostelium purpureum*

C: *Dictyostelium discoideum*

D: *Dictyostelium minutum*

E: *Polysphondylium violaceum*

F: *Polysphondylium pallidum*

FIG. 6-1. Camera lucida drawings of vegetative myxamoebae of six species of *Dictyostelium* and *Polysphondylium*. Cultures grown in association with *Bacillus megaterium* upon hay infusion agar at $22 \pm °C$. All figures drawn to scale as shown. F.V., food vacuole; N., nucleus; Sp., undigested bacterial spore; CV, contractile vacuole. (From Raper, 1941a)

Germination was almost 100% on nutrient agar in 4-5 hours when the concentration was less than 33 spores/mm^2, while it failed to reach 25% in 5-6 hours at concentrations above 400 spores/mm^2. On nonnutrient agar germination percentages were significantly lower. No marked differences in germination were noted among spores from 2 to 15 days in age.

Stimulated in part by the aforementioned study, and perhaps even more by the work then in progress on spore germination in bacilli, David Cotter undertook an in-depth study of spore germination in *Dictyostelium discoideum*. Now sixteen years and more than twenty papers later the work continues. Much has been learned—so much, in fact, that space does not permit an adequate digest here. Instead the interested reader is referred to his original papers beginning in 1966, and to comprehensive reviews by Hans Hohl (1976) and Cotter (1981) that summarize their work along with that of several other investigators.

There are certain behavioral manifestations that merit special attention, quite aside from the underlying biochemistry and physiology of spore germination, which has been studied intensively (Cotter and associates, 1966 et seq.; Gezelius, 1959; Snyder and Ceccarini, 1966; Ceccarini and Cohen, 1967; Hohl and Hamamoto, 1969; Hashimoto and Yanagisawa, 1970; George et al., 1972; Sussman and Douthit, 1973; Bacon and Sussman, 1973; Bacon, Sussman, and Paul, 1973; Hashimoto, Tanaka, and Yamada, 1976; Dahlberg and Cotter, 1978; Dowbenko and Ennis, 1980). In *D. discoideum*, as in *D. mucoroides* cited above, germination of newly formed spores is strongly depressed as the spore concentration per unit area is increased, which together with attendant autoinhibitors accounts in large measure for the virtual absence of germination among contiguous spores of a collapsed sorus. Spores of *D. discoideum* that are harvested and washed twice in buffer solution germinate at very low levels, or not at all, when deposited on nonnutrient agar buffered with potassium phosphate and adjusted to pH 6.5, or upon buffered agar enriched with 1% glucose. In contrast, if similar spores are heat-shocked for 30 minutes at 45°C and then deposited on buffered agar, they germinate at percentages up to 80% or more within 4-6 hours. Also, if spores are deposited on buffered agar containing 1.0% peptone, germination approaches 100% within 6-7 hours. Going a step further, it has been shown that, of the amino acids present in peptone, tryptophan, phenylalanine, and methionine are especially important in promoting germination. Thus it was seen that spores require some type of activation to break their dormancy, and this could be either physical or nutritional in character (Cotter and Raper, 1966). Other investigators have found that activation may be effected by gamma irradiation (Khoury et al., 1970; Hashimoto and Yanagisawa, 1970) and by judicious exposure to urea (Cotter and O'Connell, 1976), dimethyl sulfoxide (DMSO) (Cotter et al., 1976), or an unidentified substance present in culture broths of *Aerobacter aerogenes* (Hashimoto et al., 1976).

Cotter and Dahlberg (1977) have shown that some mutant strains require no activation, and that the spores of wild type strain NC-4 will autoactivate upon aging for 2 weeks in sori.

In studying spore germination in *Dictyostelium discoideum*, Cotter and Raper (1966, 1968a) considered the process to consist of three phases: activation, swelling, and emergence (Fig. 6-2). *Activation* referred to the stage between the time spores were subjected to an activating event, or substance(s), and the first sign of swelling. *Swelling* involved the enlargement of a lateral bulge until the entire spore appeared swollen. This was accompanied by a loss of refractility, the appearance of granules and contractile vacuoles, and lastly the longitudinal splitting of the outer two walls of the three-layered spore case. *Emergence*, which was generally rapid, involved breaking the innermost wall and the escape of the nascent myxamoeba from the spore case, which was left behind as an empty shell. The term was considered to be equivalent to "outgrowth" of the bacterial endospore or "germtube formation" in the fungi. Once free, and in the presence of bacteria, the myxamoebae began feeding, growing, and dividing.

Whereas all natural isolates of *D. discoideum* tested behaved in the manner described above, spores of other species and strains showed different germination responses. For example heat shock clearly enhanced spore germination in *D. purpureum* but the increase was less dramatic than in *D. discoideum*, and in *D. rosarium* enhanced germination after heat shock was evident only on media containing peptone. Strains of *D. mucoroides* gave varied results. In fact *D. mucoroides* var. *stoloniferum* (Fig. 12-2) is apparently devoid of any autoinhibitor; spores in a collapsed sorus germinate within 2-4 hours, reaggregate immediately, and form multiple small sorocarps, all within 8 to 10 hours (Cavender and Raper, 1968).

TROPHIC STATE

As seen on an agar surface, the vegetative myxamoebae vary in shape from nearly spherical when immersed in a bacterial colony to very irregular and of constantly changing form when feeding near the colony margin. In the latter case the myxamoebae are somewhat flattened and quite often appear more or less triangular with the advancing edge relatively broad and bearing few to several protoplasmic extensions in the form of short pseudopodia. A sharp demarcation is generally evident between the outer, clear ectoplasm that is especially prominent at the advancing front and the inner, granular endoplasm that contains the cell's complement of organelles (Fig. 6-3). As seen in liquid cultures, where cells can move in any direction, the myxamoebae are even more irregular, and prominent filopodia may be extended from any part of the constantly changing surface. Myxamoebae may be observed in different ways, but in our work two methods have

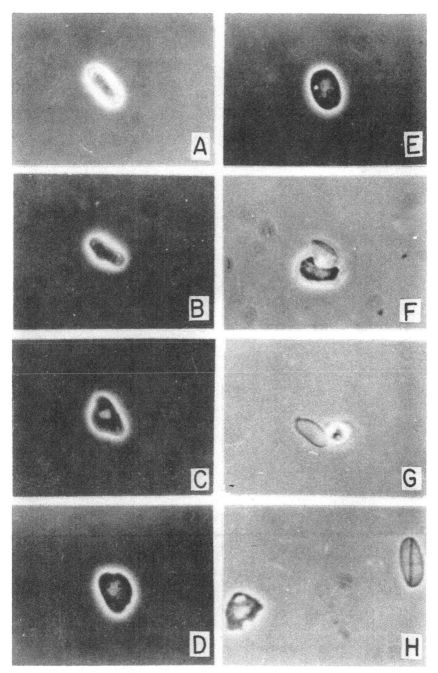

FIG. 6-2. Spore germination in *Dictyostelium discoideum*. *A*. Dormant spore, haploid. *B-E*. Early, intermediate, and late swelling, respectively. *F,G*. Protoplast emergence. *H*. Post-emergence. The myxamoeba has moved away from the empty spore case. Phase microscopy. × 1000. (Rearranged from Cotter and Raper, 1966)

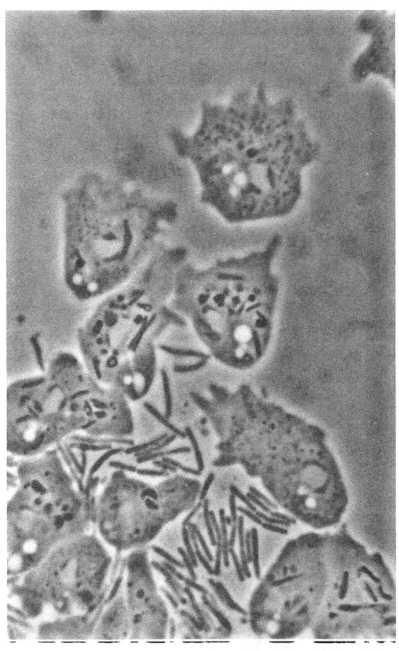

Fig. 6-3. Vegetative myxamoebae at the margin of a bacterial colony, showing nongranular ectoplasm at the front of the moving cells and the finely granular endoplasm that comprises most of the cells' volume and contains their organelles: food vacuoles (containing bacteria), clear contractile vacuoles, single vesicular nuclei with two or three peripheral nucleolar lobes (dark) and possible mitochondria. *Dictyostelium aureum* var. *luteolum* cultivated in association with *Escherichia coli* on 0.1% lactose-peptone agar at 23°C. Myxamoebae slightly compressed. × 1750. Phase microscopy.

proved especially rewarding: (1) growth of the cells on thin layers of low-nutrient agar, or (2) growth in low-nutrient broth as hanging drop preparations. If the first method is used, the coverglass should be supported so that the cells are not seriously compressed; and if the second is employed, the slides should be inverted for an hour or more before examination to permit substantial numbers of myxamoebae to adhere to the underside of the coverglass.

Vegetative myxamoebe are attracted to bacterial colonies over considerable distances by positive chemotaxis, as Samuel (1961) and Konijn (1969) have clearly shown, possibly in response to folic acid excreted by the bacteria (Pan et al., 1975). Once having reached the food source they begin to feed immediately. Whereas bacteria are generally ingested along the anterior surface of a moving myxamoeba, engulfment can occur at almost any place on the cell surface. This may be effected as the myxamoeba advances by simply "rolling" over one or more bacterial cells broadside and enclosing them in a food vacuole. Such an engulfment of E. coli B/r cells by the myxamoebae of *Dictyostelium purpureum* is beautifully recorded in a movie directed by G. Gerisch and available as Film #E 629 from the Institut für den Wissenschaftlichen Film, Göttingen, Germany. Alternatively, the myxamoeba may approach one or more bacterial cells, and upon contact put out pseudopodia on either side of the bacteria and so entrap them in a vacuole. This process is particularly striking when the bacteria are quite large, as with *B. megaterium*, and several cells still adhere in a chain (Fig. 6-4). In this case engulfment of the cells is usually endwise and, because of their size, separate vacuoles are formed for each two, three, or rarely more bacteria. The resulting food vacuoles are at first quite elongate with the bacteria lying end to end, but these soon contract and as they become more rounded the bacteria are shifted so that they lie side by side, or one is reoriented perpendicular to the other(s) (Fig. 6-4, D1-6). Once this occurs, the bacterial cells gradually weaken at midpoint (presumably along lines of future fission) and break into smaller, rounded fragments as the process of digestion continues (Raper, 1937). This process of progressive fragmentation is not seen in the digestion of small Gram-negative rods like E. coli (Hohl, 1965).

Working with submerged cultures of *Dictyostelium discoideum* cultivated in association with pregrown cells of *Escherichia coli* B/r, Gerish (1960) calculated the generation time to be 3.3 hours, and estimated that one myxamoeba consumed about 1100 bacteria prior to division. This figures out to be about 5 bacteria consumed per minute. Although we have no quantitative data to refute or even contest this number, it seems quite high based upon our visual observations of feeding myxamoebae. Using bacteria labeled by overnight growth on media containing ^{14}C-lysine, Glynn (1981) carefully studied the phagocytosis of E. coli by myxamoebae of D. discoideum. At high bacteria: myxamoeba ratios ingestion was ap-

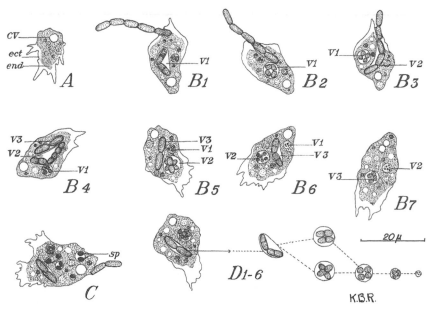

FIG. 6-4. Camera lucida drawings of myxamoebae of *Dictyostelium discoideum* feeding on cells of *Bacillus megaterium*. A. Young myxamoeba; CV, contractile vacuole; ect, hyaline ectoplasm; end, granular endoplasm. *B1-B7*. A feeding myxamoeba drawn at intervals of approximately 10 minutes during a period of 65 minutes; V1, V2, and V3, food vacuoles. C. Myxamoeba containing undigestible spores (sp). *D1-D6*. Food vacuoles showing bacilli in progressive stages of digestion. (From Raper, 1937)

proximately linear during the first half of the time course, then continued at a steadily declining rate and finally leveled off at approximately 60 minutes, the average number of bacteria ingested being about 67. Other approaches have been used as well. Vogel et al. (1980) investigated phagocytosis by wild type and mutant strains of *D. discoideum* using latex beads (protein coated and uncoated) and erythrocytes in addition to bacteria. Some years prior to this Githens and Karnovsky (1973) had studied the phagocytosis of polystyrene beads by the myxamoebae of *Polysphondylium pallidum* and found beads of about 1.0 μm diameter to be optimal for ingestion. When compared with *Acanthamoeba* and guinea pig leucocytes, the rate of ingestion was somewhat slower than the former and appreciably slower than the latter.

The ingestion and phagocytosis of one dictyostelid myxamoeba by another of the same species is not an uncommon occurrence, as was recorded by Huffman and Olive in 1969. Quite recently an especially interesting example of carnivorous behavior has been reported by David Waddell of Princeton University (1982). Among cellular slime molds isolated from bat guano taken from an Arkansas cave was a delicate *Dictyostelium*, designated *D. caveatum* but not described,[1] of which the myxamoebae feed upon the cells, vegetative and sorogenic, of other dictyostelids, and in so doing quite effectively arrest their growth and subsequently preclude sorocarp formation by the prey species.

In a different vein, and somewhat surprising, myamoebae are less sensitive to ultraviolet radiation than are the spores, particularly cells exposed during logarithmic growth (Hashimoto and Wada, 1980). The same is not true, however, of γ radiation (Gillies, Hari-Ratnojothi, and Ong, 1976).

DIVIDING CELLS

Cell division in vegetative myxamoebae may be observed in any of the dictyostelids by preparing cultures especially for this purpose. The bottom of a scratch-free Petri plate is covered with a thin layer (1-2 mm) of clear, low-nutrient agar (e.g., 0.025% dextrose, 0.025% peptone in 1.5% agar) and inoculated with a dilute suspension of bacteria (*E. coli*) and slime mold spores by carefully spreading a single drop with a bent glass rod. Growth of the bacteria should be discontinuous but consist of minute colonies closely spaced, each of which should contain a few slime mold spores. Myxamoebae should be evident in some colonies within about 24-30 hours and the cultures should be optimal for examination about 6-8 hours later. Observations may be made directly with a good water-immersion lens, but clearer images are possible with phase optics. In this case the growth should be covered with a thin coverglass (#0 thickness) supported here and there by chips of broken coverglass to preclude undue compression of the myxamoebae. A good quality oil-immersion lens and a phase condenser with a front element of long focal length (6.0 mm) are essential, while a mechanical plate carrier is a great convenience.

One begins the search by scanning the plate and selecting an area where myxamoebae are present in an interrupted monolayer of bacteria. The positions of a few larger cells are mentally recorded and these are carefully watched for the earliest signs of approaching division: a rounding up of a myxamoeba and the appearance of a slightly crenulate surface. Mitosis progresses rapidly (Fig. 6-5). The nucleus disappears as an identifiable organelle and soon thereafter one detects two darker objects (clusters of chromosomes) that gradually move apart. Concurrently, the cell becomes somewhat elongate in the direction of such movement, the membrane becomes indented near midpoint, and the cytoplasm begins to constrict. As the mitotic process continues the chromosome clusters come to occupy opposite poles, the cell separates into two parts, and identifiable nuclei can soon be seen in each of the two daughter cells as they go their separate

---

[1] A description has been prepared recently and is included in Chapter 12 as *Dictyoste-* *lium caveatum* Waddell, Raper, and Rahn sp. nov.

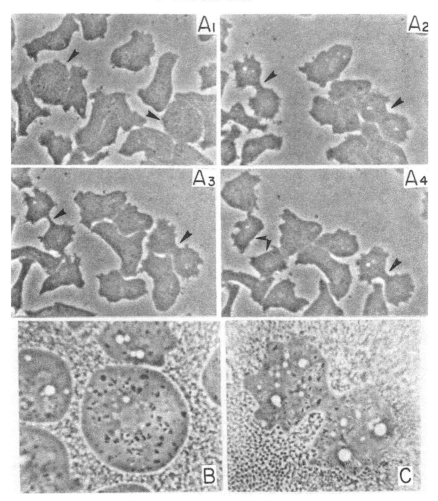

FIG. 6-5. Cell division in vegetative myxamoebae of *Dictyostelium discoideum*. A_1. Myxamoebae, of which two (arrows) have rounded up preparatory to division. A_2, A_3, and A_4. The same after 3.5, 5.0, and 6.0 minutes, respectively—division is completed in one case and well advanced in the other. × 600. *B*. Single myxamoeba approaching division; note vesicular nucleus at center and absence of contractile vacuole(s). *C*. Myxamoeba in process of division; nuclei not apparent, but contractile vacuoles have reappeared. *B,C*, × 1350. Phase microscopy. (*B* and *C* from Ian Ross, 1960)

ways. The whole process may require less than 10 minutes. Cell division is beautifully illustrated in the Gerisch film of *D. purpureum* cited above.

An alternative method of studying cell division in vegetative myxamoebae was described by Ian Ross in his study of *D. discoideum* (1960). Thin rings 0.5-1.0 mm thick were cut from polyethylene tubing (ca. 2.0 cm diameter), cemented to slides with paraffin, partially filled with a drop of 2% nonnutrient agar, and covered with a #0 coverglass bearing on its underside a thin smear of *E. coli* and a few slime mold spores. The coverglass was fixed to the ring with petrolatum. The myxamoebae grew well and small aggregates developed between the agar and the coverglass. The preparations were excellent for phase microscopy, including use of an oil immersion lens (Fig. 6-5).

Between mitoses the nuclei of dictyostelid myxamoebae appear as relatively clear, generally circular areas with well-defined deformable membranes and show one or more (commonly two or three) peripheral nucleolar lobes. Other organelles identifiable by phase microscopy (Fig. 6-3) include food vacuoles containing bacteria in various stages of digestion, one or more contractile vacuoles, and sometimes mitochondria. The contractile, or pulsating, vacuoles are especially prominent and are normally located in the posterior region of moving cells. They often arise as multiple small vacuoles and commonly but not always merge before discharge. No residual structures are apparent after discharge, although new contractile vacuoles seem to arise in the same position as the old.

The haploid number of chromosomes in *Dictyostelium discoideum*, the most widely studied species, is seven (Bonner and Frascella, 1952; Wilson, 1952, 1953). This is also true of *D. mucoroides* (Filosa and Dengler, 1972), the oldest species and the one most widely distributed in nature. In *Polysphondylium violaceum*, another cosmopolitan species, the haploid number was first reported to be 8 or 9 (Wilson and Ross, 1957). More recently, Williams has reported (1980) that the number appears to be 11 or 12, as it is in *P. pallidum*, which he has studied quite carefully.

It is an interesting and sometimes confusing fact that all strains of *D. discoideum* when first isolated from forest soils have spores in the general range of 6.5-8.0 × 2.5-3.5 μm, indicative of the haploid state, whereas the same strains after months (or years) of laboratory cultivation often show a preponderance of larger, commonly reniform to banana-shaped spores 11-12 × 3.5-4.5 μm (Fig. 6-6), characteristic of the diploid state (Sussman and Sussman, 1962, 1963). In other species as well, strains with small and/or large spores are sometimes encountered, and these too are interpreted as having been produced by haploid or diploid cells, respectively. A striking example of such difference in spore size was noted and illustrated by Raper and Fennell in their description of *D. rhizopodium* (1967).

For reasons unknown investigators have often experienced difficulty in staining the nuclei of vegetative myxamoebae (Bonner and Frascella, 1952; LaBudde, 1956). Of several methods reported, those of R. Sussman (1961) and Brody and Williams (1974) seem to have been most successful. Sussman used Carnoy's solution for fixation and stained with 2% Gurr's acetoorcein in 45% acetic acid following hydration through an alcohol series and hydrolysis in 1N HCl. Brody and Williams fixed cells in methanol/glacial acetic acid (3:1, v/v) and stained with 10% Gurr's Giemsa stain

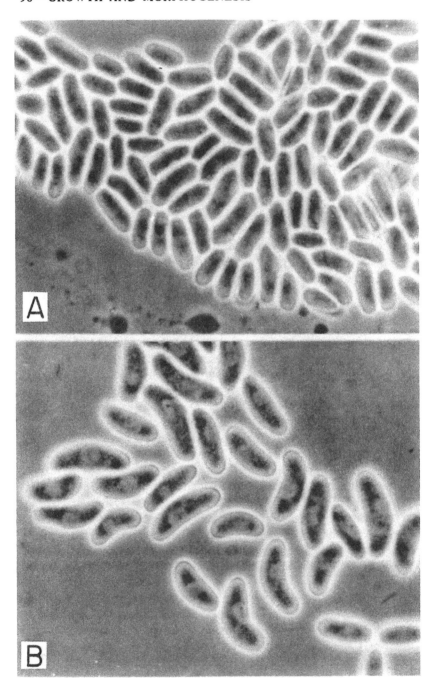

FIG. 6-6. Spores of *Dictyostelium discoideum*. *A*. Haploid. *B*. diploid. Note differences in form and dimensions. × 950. Phase microscopy.

(Improved R66) in M/15 Sorensen's phosphate buffer (pH 6.8). Preparations could be examined immediately or dried and preserved with Euparal. Good nuclear division figures were obtained in haploid and diploid cells, and in aneuploids as well. Among the latter, nuclei were found with all possible chromosome numbers from 14 to 7, indicating that haploidization in the parasexual cycle of dictyostelids follows a pathway similar to that of *Aspergillus nidulans*. Since the strains examined were primarily axenic ones, the problem of ingested bacteria staining as chromosomes was not encountered.

ULTRASTRUCTURE

The first ultrastructural study of a dictyostelid, *D. discoideum*, was published by Kurt Mühlethaler in 1956. The major emphasis was given to an overview of the fruiting phase and little was said about the trophic myxamoebae. A year later, Kirsten Gezelius initiated her studies on the fine structure of the dictyostelids, and in the first of these (with B. G. Ranby, 1957) paid special attention to the "framework substance," or cellulose, in the spores, stalk cells, and stalk sheath of *D. discoideum*. In a second study, which included *Acytostelium* and other species of *Dictyostelium* as well as *D. discoideum*, Gezelius (1959) reported in detail on the chemical nature of the cellulose and its location and fibrillar orientation within the walls of spores, stalk cells, and the extracellular sheath that enveloped the stalk. Additionally she called attention to the tubular content of the mitochondria, and was intrigued by the prominent concentric lamellae that were often present in old food vacuoles following the engulfment and digestion of bacteria. Special attention was given to this lamellar system by Hohl in a beautifully illustrated paper published in 1965. In his view the multilayered, concentric lamellae represented a condensation of undigested material, probably lipoprotein in nature, deposited successively as unit membranes to conform with the external wall of the vacuole or with the surface of the engulfed bacteria. The whole membrane system is subsequently ejected (as waste), which Hohl believes indicative of a fairly inefficient digestive process.

Subsequent to this Hohl and coworkers devoted special attention to the formation and ultrastructure of dictyostelid spores, tracing the transition from prespore myxamoebae to mature spores both via phase and electron microscopy. In the first of such studies, based upon *D. discoideum*, Hohl and Hamamoto (1969) identified a special type of vacuole, termed prespore vacuole (PV), that was present in all postvegetative cells destined to produce spores. During the process of spore formation the PVs moved to the periphery of the cell, emptied their content outside, and took up positions as parts of the initial spore coat. Once this was formed, two more layers constituting the bulk of the spore case were soon produced (Hohl, 1976).

Of these the heavier middle layer consisted essentially of cellulose, oriented parallel with the spore axis on the outside but inwardly unoriented. Composition of the third and innermost layer remained in doubt, although Hemmes, Kojima-Buddenhagen, and Hohl (1972) found it to be sensitive to cellulase and pronase. Another investigation centered upon *Acytostelium* (Hohl, Hamamoto, and Hemmes, 1968), including the spores which differ from those of *D. discoideum* in shape and seem to be structurally similar to the microcysts. Meriting attention also is an investigation of ultrastructural changes that accompany spore germination in *D. discoideum* (Cotter, Miura-Santo, and Hohl, 1969).

Information about the ultrastructural aspects of nuclear division in *Polysphondylium violaceum* has been provided by U.-P. Roos, now of the University of Zurich (1975a,b); and because of the apparent similarity among the trophic stages of different species and genera it would seem probable that this may be representative of other Dictyostelidae as well. The mitotic process in *P. violaceum* as summarized by Roos follows:

> Mitosis is characterized by a persistent nuclear envelope, ring-shaped extranuclear spindle pole bodies (SPBs), a central spindle spatially separated from the chromosomal microtubules, well-differentiated kinetochores, and dispersion of the nucleoli. SPBs originate from the division, during prophase, of an electron-opaque body associated with the interphase nucleus. The nuclear envelope becomes fenestrated in their vicinity, allowing the build-up of the intranuclear, central spindle and chromosomal microtubules as the SPBs migrate to opposite poles. At metaphase the chromosomes are in amphitelic orientation, each sister chromatid being directly connected to the corresponding SPB by a single microtubule. During ana- and telophase the central spindle elongates, the daughter chromosomes approach the SPBs, and the nucleus constricts in the equatorial region. The cytoplasm cleaves by furrowing in late telophase, which is in other respects characterized by a reestablishment of the interphase condition.

In commenting about possible taxonomic relationships, Roos (1975a) notes that mitosis in *P. violaceum* is unlike that in the plasmodial myxomycetes and in the protists, notably the amoeboid protozoa. Continuing, he states (1975a) that there is a striking similarity with certain fungi. He notes that spindle pole bodies and a tightly organized central spindle in the late stages of division have been documented for zygomycetes, ascomycetes, basidiomycetous yeasts, and an imperfect fungus, and concludes with this summary: "Only a systematic study within a group of closely related organisms can reveal the degree of variation or constancy, and therefore their value in taxonomy, of ultrastructural features of mitosis, but the ensemble of similarities pointed out above indicates that cellular slime molds may be more closely related to extant fungi than to protozoa."

Roos has recently published additional papers (1980; Roos and Ca-menzind, 1981) with special emphasis on the formation and dynamics of the spindle during mitosis, the second of which is centered upon *D. dis-coideum* and is especially noteworthy for its excellent illustrations, in-cluding both phase-contrast and electron micrographs (Fig. 6-7).

MICROCYSTS

Following vegetative growth the myxamoebae typically aggregate in large numbers to form in turn pseudoplasmodia, sorogens, and sorocarps. However, in some species and strains, but not in others, many myxamoebae never progress through this cycle. Instead they round up and encyst as individual cells, forming microcysts, and in so doing enter a transient resting stage (Fig. 6-8). In certain isolates, and under some conditions, more cells actually differentiate in this fashion than as cellular elements, spores or stalk cells, of completed sorocarps. Long ago E. W. Olive concluded (1902) that such encystment, first mentioned by Cienkowski in 1873, represented a response to suboptimal culture conditions, and there is much evidence pointing to such a causal relationship. There is at the

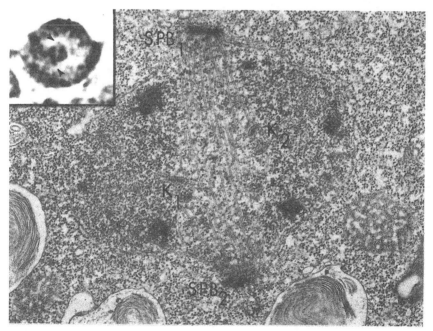

Fig. 6-7. Electron micrograph of a section through a metaphase nucleus in *Dictyostelium discoideum*. SPB$_1$ and SPB$_2$, spindle pole bodies; K$_1$ and K$_2$, pairs of kinetochores of two chromosomes. × 23,000. Inset: Phase contrast micrograph of same cell embedded in plastic. Arrows mark spindle poles. × 1300. Courtesy of U.-P. Roos and R. Camerzind (1981)

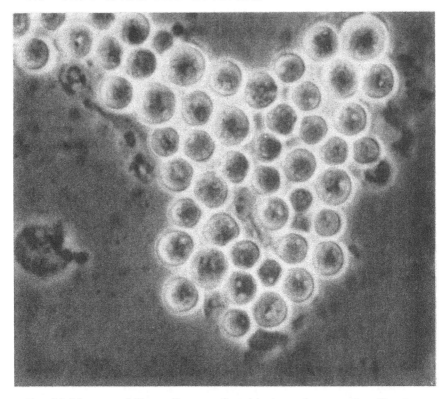

Fig. 6-8. Microcysts of *Dictyostelium tenue* formed in situ on the agar surface. Note the granular character of the cyst contents, and the presence of undifferentiated myxamoebae in the area adjacent. × 1750. Phase microscopy.

same time, we believe, a genetic component of equal or greater importance, for true microcysts have not been recorded for such robust species as *Dictyostelium discoideum, D. purpureum, D. giganteum*, or *Polysphondylium violaceum*. In other and more delicate species such as *D. mexicanum, D. lavandulum, D. polycephalum, Polysphondylium pallidum*, and *Acytostelium leptosomum* the microcysts may at times dominate the cultures. It is probably not without significance that Traub and Hohl (1976) cite the formation of microcysts, along with the presence of polar spore granules and the lack of an aggregative response to cAMP, as a chief diagnostic characteristic of a cluster of dictyostelids for which Traub (1977) proposed the creation of a new genus, *Heterosphondylium*.

One cannot question the influence of the culture environment in the production of microcysts by those species and strains that possess this capability; and of several variables that have been implicated, increased osmolarity is well documented. Working with a strain of *Polysphondylium pallidum*, Pan-17, previously used in nutritional studies by Hohl and Raper

(1963c), and using different concentrations of potassium chloride, Toama and Raper (1967a,b) showed that the level of salt required for maximum microcyst formation (ca. 80%) by pregrown myxamoebae in a shaken nonnutrient solution was 0.08 M KCl, whereas that for similar cells deposited on nutrient-free agar was 0.12 M KCl. NaCl was ineffective when used alone, but if an iso-osmotic solution of the salt was supplemented with 2 mM Ca^{++}, Mg^{++}, or K^+, microcysts were formed by 50-75% of the test myxamoebae. A second strain, WS-320 (Raper, 1960; Hohl and Raper, 1963b; Hohl, Miura-Santo, and Cotter, 1970), behaved somewhat differently in that 0.06 M KCl was sufficient to elicit 96% encystment in liquid. On agar the required level was twice this amount and the cultures needed to be incubated in darkness to preclude aggregation. The optimum pH for encystment was 6.0 (Toama and Raper, 1967a). More recently Lonski (1976) has shown that ammonia profoundly influences microcyst formation in *Polysphondylium pallidum* when added to cultures either in a gaseous form or as an ammonium salt. In the case of NH_4Cl, virtually 100% of the myxamoebae formed microcysts when the concentration of ammonia in the agar reached 0.08-0.09 M, while less than $\frac{1}{10}$ this concentration of gaseous NH_3 produced an almost equal result. A year later, Lonski and Pesut (1977) reported ethionine and putrescine to be stimulatory as well, while glycerol at 0.4-0.5 M was also found to be an effective inducer of microcysts.

Microcysts are generally globose or nearly so and the cytoplasm is somewhat denser than that of the vegetative myxamoebae, but less so than that of mature spores. The onset of encystment is marked by a progressive rounding of the cell content followed by the deposition of a loose fibrillar layer which becomes firmer in texture and, as determined by electron microscopy, appears to consist of a dense inner and a loose outer layer (Hohl et al., 1970) in contrast to the three-layered wall of the mature spore. The nuclei are generally obscured, vacuoles are small if present at all, and the cyst wall contains substantial amounts of cellulose along with glycogen-like materials, lipids, and proteins (Toama and Raper, 1967b). Consistent with its transient nature the microcyst does not require activation prior to germination, but seemingly rehydrates and swells; and instead of splitting the cyst case the protoplast seems rather to dissolve a limited part of the wall, through which it then escapes (Blaskovics and Raper, 1957).

Superficially, the dictyostelid microcysts closely resemble the cysts of many small, free-living soil amoebae, and in the absence of any sorocarps they could easily be mistaken for such protozoa. In fact, this similarity could lend strength to the not unreasonable argument that the Dictyostelidae may have arisen from previously free-living amoebae that acquired a penchant for social contact.

CHAPTER 7

Cell Aggregation

Of the several features that distinguish the Acrasiomycetes none is more singular or more striking than the phenomenon of cell aggregation that marks the transition from a vegetative stage of independent myxamoebal growth to a fruiting stage of cellular interdependence and divergent differentiation. It is a phenomenon that fascinated Brefeld (1869) despite his early misinterpretation of its significance, and it is the phenomenon that captivated van Tieghem and led him (1880) to select the name *Acrasis*[1] for the assemblage of collaborating cells in his culture vessels. That the cells retained their individuality but still built multicellular fructifications of predictable pattern quite clearly amazed him, as it has all subsequent students of his "Acrasiée." Today, a full century after van Tieghem's discovery, researchers in scores of laboratories throughout the world are still seeking answers. How and why do they effect this transition from a unicellular to a multicellular mode?

Various theories have been advanced to account for the aggregation of previously independent myxamoebae, some based upon purely physical factors, others upon nutritional changes, and still others upon autoinduction. Prominent among the latter have been suggestions of some type of self-generated chemotaxis. At the beginning of this century, E. W. Olive attempted unsuccessfully to induce aggregation in *Dictyostelium* and *Polysphondylium* by exposing myxamoebae to dilute solutions or organic acids and sugars in the manner Pfeffer (1900) had done to attract fern antherozoids. Following his seeming failure, however, he observed something of even greater significance, to wit:

> No evidence whatever was gained as to any directive or repellent influence of these substances on the myxamoebae. At any rate, the substance which causes the contagious stimulus and exerts an attracting influence on the myxamoebae must evidently vary in different forms, since two well marked species of *Dictyostelium*, one for example, with white spores and another with dark, sown in the same spot of a nutrient agar tube, will result in fructifications showing the two distinct forms growing side by side without any trace of intermixture (Olive, 1902)

In the same year, and working independently, George Potts (1902) recognized nutrient exhaustion as a primary stimulus for fructification in

[1] Derivation: Greek, indicating "without fusion."

Dictyostelium mucoroides, and relative to its initial stage, cell aggregation, he wrote: "The ray-like pattern of the inflowing arms around the midpoint appears almost to indicate a chemotactic influence."

In describing *Dictyostelium discoideum*, Raper (1935) had this to say about the aggregative process (Fig. 7-1):

> The first evidence of aggregation is the appearance of certain areas of closely packed myxamoebae. The stimuli involved in this aggregation have never been determined, but these strategically placed individuals apparently exert some stimulus which reaches myxamoebae in the surrounding area and causes them to become elongated and oriented in the direction of these centers and to flow toward them. Gradually the orientation proceeds outward and the colony of aggregating myxamoebae takes the form of definite streams converging toward definite centers.

In his comprehensive account of development in *Dictyostelium mucoroides*, Arndt (1937) gave special attention to cell aggregation and was the first to employ time-lapse cinematography. He was able in this way to reveal a dramatic pattern of wave movement in converging streams of myxamoebae. He concluded that the stimulus for aggregation had its origin within the congregating cells. He did not, however, suggest the nature of the attractant or how it might have produced the inflowing waves he observed.

Clues to these questions began to appear a few years later. The first of these was a series of simple but highly revealing experiments reported by Ernest Runyon in a seminar at the Marine Biological Laboratory (Woods Hole, Mass.) in August 1942, and subsequently published (1942) as an abstract in the laboratory's house organ, *The Collecting Net*. Since the circulation of this publication is somewhat limited, parts of Runyon's succinct but informative account are included here:

> Study of the aggregation phase was found to be facilitated by the use of non-nutrient agar over which are thinly spread the amoebae, previously washed, concentrated and roughly separated from the associated mass of bacteria by centrifuging. Size, pattern and rate of development of aggregates are much affected by the thickness of the film or layer of water in which the amoebae are dispersed. If the culture is relatively dry (the aqueous film very thin) collecting points (centers) are many and aggregation progresses rapidly. In thicker films aggregation is slower, centers fewer and the amoebal strands leading to the centers longer and less compact. . . . Potts in 1902 recorded fruition of *Dictyostelium* under oil. The pattern and rate of aggregation under a layer of paraffin oil is much the same as without the oil: the collecting points are many and aggregation progresses rapidly. Aggregation with center determination occurs under a glass cover slip or dialyzing membrane (Visking) placed

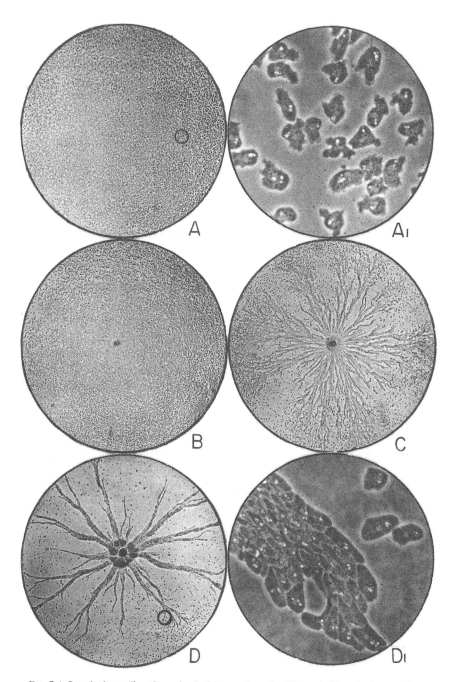

Fig. 7-1. Pseudoplasmodium formation in *Dictyostelium discoideum*. *A*. Myxamoebae evenly distributed on an agar surface just prior to the beginning of cell aggregation. *A1*. Small area of the same enlarged, showing irregular form of individual myxamoebae and absence of cell orientation. *B,C,D*. Progressive stages in aggregation. *B*. Emergence of an aggregation center. *C*. Formation of definite streams by inflowing myxamoebae. *D*. Later stage in stream convergence. *D1*. Distal end of small stream enlarged, showing elongate form and uniform orientation of aggregating cells. *A,B,C,D* × 10. *A1,D1*, phase microscopy × 275. (From Raper, 1962)

over the amoebae on agar. *Amoebae which are on top of such a dialyzing membrane become oriented corresponding to patterns of aggregation below the membrane*[2]; the amoebae on top accumulate along strands and centers below the membrane. Thus it seems likely that the determination of centers of aggregation depends upon the diffusion of a substance that is water- but not oil-soluble, active in thin films and of a molecular character such that it can pass through a dialyzing membrane.

ACRASINS

Few discoveries have influenced dictyostelid research more. In one simple experiment Runyon showed that the substance guiding cell aggregation in *Dictyostelium* was of myxamoebal origin, that it was readily diffusible, and that it was of relatively low molecular weight. This was during the early months of World War II, and we know not whether the work might have been continued under different circumstances. We know only that it was not, and that five years elapsed before publication of John Bonner's classic paper on the formation of cell aggregates in *D. discoideum* by chemotaxis (1947).

Although cognizant of Runyon's experiments, Bonner chose to investigate a number of potential stimuli, physical and otherwise, when investigating cell aggregation in *Dictyostelium*. Suffice it to say that of many experiments performed and of many hypotheses explored, the only one that gave positive results and could account in all particulars for the behavior of the aggregating myxamoebae was the production by the slime mold of a diffusible substance of unknown properties. Foremost among his ingenious experiments were the underwater tests: myxamoebae were deposited evenly over the surface of a coverglass which was then immersed in saline (Bonner, 1947), and when the cells were about to aggregate, water was caused to flow slowly over the field. The resulting aggregation patterns were "atypical" in that myxamoebae joined the emerging centers only from downstream (Fig. 7-2). In another experiment, when preaggregative cells were spread on the underside of a coverglass and an aggregation center placed on its upper surface, myxamoebae from the underside formed streams, turned the corner, and joined the transplanted center above (Fig. 7-3A). In a third test, when two coverglasses, one bearing a preformed center and the other unaggregated cells, was suspended close together but not touching, the scattered cells aggregated at the open gap nearest the preformed center (Fig. 7-3B). If the distance was not more that 20-30 μm (two cell lengths) the myxamoebae formed a bridge across the gap and joined the established center. Experiments were also conducted to determine the distance over which a center could attract myxamoebae in the environs. Strong attraction was evident up to about 200 μm, or about 13

[2] Raper's emphasis.

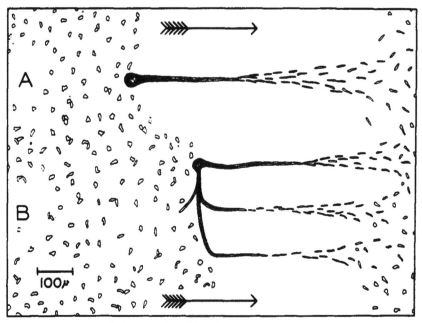

FIG. 7-2. Diagrammatic representation of the effect of moving water on aggregation patterns in *Dictyostelium discoideum*. *A*. Aggregation formed in moving water. *B*. Aggregation formed in still water, then subjected to a stream of moving water. Arrows indicate direction of water movement. (From Bonner, 1947)

cell diameters, within 3-5 minutes, whereas a weak response occurred at 350 μm, and even at 800 μm under special circumstances. Bonner (1947) summarized the work in this way:

> We have deduced from this flowing water experiment that during the aggregation of *Dictyostelium* there is some type of chemical substance (which is not necessarily homogeneous but might consist of a group of compounds) produced continuously or at frequent intervals by the center, which freely diffuses, and the myxamoebae move in the resulting gradient of this substance toward the point of its highest concentration. The final proof of the existence of the substance (and an important problem for future research) must be its isolation *in vitro*. But considering the present weight of evidence, it seems fitting to propose tentatively a name for the substance. The term *acrasin* is suggested, and it can be defined for the moment as a type of substance consisting either of one or numerous compounds which is responsible for stimulating and directing aggregation in certain members of the Acrasiales. It also may perform other duties in the development of these organisms but such considerations are not within the scope of this paper.

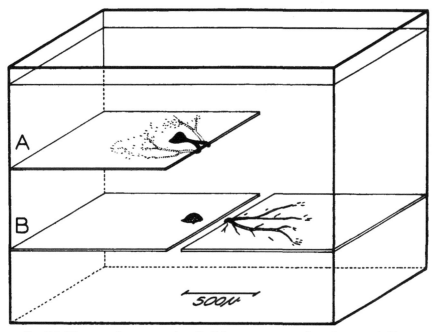

Fig. 7-3. Diagrammatic representation of cellular responses to preformed centers. *A*. Myx-amoebae of *Dictyostelium discoideum* spread on the underside of a coverslip turn a corner to join the aggregation center on the upper surface. *B*. Despite an open space between two coverslips, myxamoebae on one stream toward the point nearest the preexisting center on the other. (From Bonner, 1947)

At about the same time (1950) Rosemarie Pfützner-Eckert in Alfred Kuhn's laboratory in Tübingen independently reported evidence of a dif-fusible chemotactic substance of dictyostelid origin. Working with *D. mucoroides* she showed that when the center of an aggregation (conus) was moved aside a short distance, the myxamoebae comprising the now truncated streams gradually reoriented, changed their direction of move-ment, and rejoined the displaced center (Fig. 7-4). If the distance separating the proximal ends of the streams from the displaced center was too great, the myxamoebae that remained then formed a new center and further development proceeded separately at the two sites. Agar blocks from which centers had been removed were reported to act as foci for cell attraction also, but this could not be confirmed by either Shaffer (1962) or Bonner (personal communication).

Although the presence of a diffusible chemotactic substance in *D. dis-coideum* had been demonstrated convincingly and named appropriately, it remained for Brian Shaffer of Cambridge University to provide some insight into its chemical characteristics. Beginning in 1953 and for a dozen

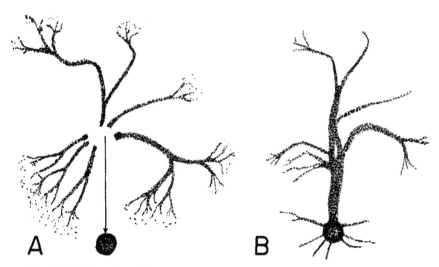

Fig. 7-4. Evidence for a diffusible attractant in *Dictyostelium mucoroides*. *A*. The center of a developing pseudoplasmodium is moved a short distance from its original site. *B*. Myxamoebae comprising the truncated streams reorient as necessary to rejoin the displaced center. (From Pfützner-Eckert, 1951)

years thereafter, Shaffer published a series of papers wherein he painstakingly reported ingeniously conceived and expertly performed experiments. At the outset he collected washings from aggregation centers and observed their effect upon preaggregative cells sandwiched between thin blocks of agar (Fig. 7-5) and glass slides (1953, 1956a, 1957a). The washings were quite stimulatory at 1 minute, weakly so at 5 minutes, and inactive at 15 minutes. Fresh washings if frozen within a few seconds retained their activity indefinitely, but lost it rapidly upon thawing. Remembering Runyon's experiment, he then confirmed that acrasin would pass through a dialyzing membrane and subsequently used this principle effectively in separating it from another extracellular product that failed to do so. The cell-free acrasin solution so obtained was stable, but only in the absence of a nondialyzable, heat labile substance that he judged to be an enzyme and was later shown to be a phosphodiesterase (Chang, 1968). A method for concentrating acrasin was soon developed based upon methanol extraction, vacuum drying, and reextraction of the residue to yield a dry, stable product. In aqueous solution this possesed high acrasin activity and would withstand boiling for more than one hour and exposure to 0.1 N HCl or 0.01 N NaOH (Shaffer, 1956b). The acrasin so recovered not only oriented sensitive myxamoebae, but it caused these cells to secrete acrasin in turn, thus initiating a chain reaction that greatly extended the distance over which the initial source (or aggregation center) could operate via an

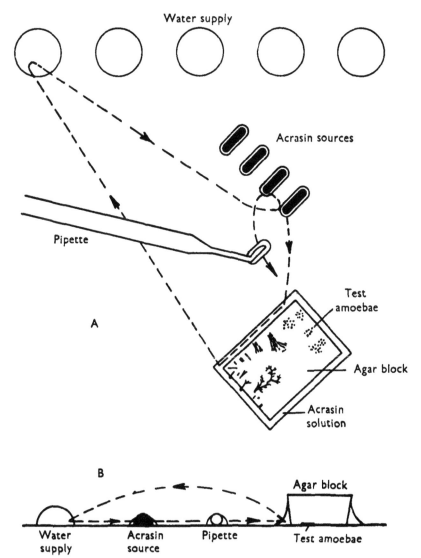

FIG. 7-5. Demonstration of acrasin and its role in cell aggregation. Preaggregative (test) myxamoebae are sandwiched between a thin agar block and a glass slide, the block being shaped to form a meniscus at its border. Acrasin sources (migrating pseudoplasmodia of *Dictyostelium discoideum*) are bathed with water, which is subsequently moved by pipette to the edge of the agar block. The acrasin contained in this water causes the cells beneath the agar to orient uniformly and to move outward toward it. Water from the test area is periodically removed and replaced by fresh acrasin solution. *A*. Diagram of the experimental layout. *B*. Elevation of the same. Myxamoebae under one side of the agar block are represented as unoriented and on the other as having responded to acrasin. (From Shaffer, 1956a)

acrasin relay—a relay system that is reflected in the wave-like convergence of myxamoebae first seen by Arndt and now much studied by current investigators. Other papers followed, and in these many aspects of cell aggregation (and later of fructification) were considered, including acrasin production and its inactivation as complementary phenomena essential for cell aggregation in dictyostelids (Shaffer, 1957a); the origin and significance of intercellular adhesivity in the aggregative process (Shaffer, 1957b); interspecific and intergeneric variability in aggregative behavior of cellular slime molds, together with effects of manipulative alterations in patterns of preformed aggregations (Shaffer, 1957c); and contrasting patterns of cellular integration and disintegration in aggregates of *Dictyostelium discoideum* and *Polysphondylium violaceum* (Shaffer, 1958). No other investigator has so carefully observed or meticulously analyzed the aggregative process of dictyostelids as has Brian Shaffer.

The net result of Shaffer's researches, and those of Sussman and coworkers (Sussman and Lee, 1955; Sussman, Lee, and Kerr, 1956), was that cell aggregation in the dictyostelids could be explained by the production of an attractant, acrasin, and the concurrent production of an enzyme capable of destroying it. Because of the latter, the accumulation of acrasin in amounts sufficient to negate a directed chemotactic stimulus due to background "noise" was avoided.

But what is acrasin? The answer is neither simple nor complete, but much information has accumulated during recent years concerning the nature and activity of an acrasin that is operative in *Dictyostelium discoideum* and in at least three other large species of the genus, *D. mucoroides* and *D. purpureum* (Konijn et al., 1968) and *D. rosarium* (Konijn, 1972; Bonner, Hal, Noller, Oleson, and Roberts, 1972). The first clue to its identity stemmed from an observation by Theo M. Konijn that preaggregative myxamoebae of *D. discoideum* when placed near a colony of *Escherichia coli* would move toward it. In the course of studies on the influence of various factors such as temperature, light, and cell density upon aggregation, Konijn (1965, 1968; Konijn and Raper, 1965, 1966) had developed a technique for confining myxamoebae within small droplets on specially prepared agar *unless* they were lured outside by a diffusible external attractant. The bacteria produced such a substance that apparently enabled the myxamoebae to sense the nearness of food and to orient their movements in its direction. Prior to this Maxman and Sutherland (1965) had reported that *E. coli* synthesized adenosine-3', 5'-monophosphate (cyclic AMP or cAMP), and it occurred to David Barkley that the attractant in *E. coli* could be cyclic AMP and, further, that this just might be the chemotactic substance in *Dictyostelium discoideum*. Supporting evidence was soon obtained by Konijn with his cAMP assay and then amply confirmed through cooperative studies between Konijn and coworkers of the Hubrecht Laboratory in Utrecht and Bonner and his group at Princeton

University. First, they clearly showed that crystalline cyclic AMP possessed acrasin activity (Konijn et al., 1967). Additional studies followed in which they identified adenosine-3', 5'-monophosphate as the bacterial attractant for myxamoebae of *D. discoideum* (Konijn, van de Meene, Chang, Barkley, and Bonner, 1969), and then demonstrated the synthesis of cyclic AMP in *D. discoideum* and *Polysphondylium pallidum* (Konijn, Chang, and Bonner, 1969).

Using low-rigidity agar with a hydrophobic surface perfected in his earlier experiments on cell aggregation, Konijn was able to devise a quantitative assay for cyclic AMP (1970). This was used for monitoring the isolation and identification of the compound and also for determining the levels at which cyclic AMP was active in promoting cell aggregation in *D. discoideum* and other slime molds. Using the Konijn assay, Bonner and coworkers (1969) had shown quite significantly that there was a hundred-fold increase in cAMP production during normal aggregation in *D. discoideum* and, additionally, that this was accompanied by a hundred-fold increase in the cells' response to cAMP. Meanwhile, Chang (1968) had characterized the extracellular enzyme in culture broths of *D. discoideum*, and Konijn et al. (1968) had shown that this acrasin-degrading enzyme, first reported by Shaffer (1956a), was similar to the mammalian phosphodiesterase described by Butcher and Sutherland (1962). In fact, confirmation of this similarity was based upon phosphodiesterase supplied by Butcher.

It is presumed that the chemotactic substance indigenous to large species of *Dictyostelium* such as *D. mucoroides, D. purpureum,* and *D. rosarium* is also cyclic AMP, since myxamoebae of these slime molds aggregate in response to it in the same general manner as those of *D. discoideum* but with appreciable differences in the timing of responses. The acrasins for the genus *Polysphondylium* and some of the more delicate cellular slime molds currently assigned to *Dictyostelium* are not cyclic AMP, for their myxamoebae do not aggregrate in response to it. This becomes all the more interesting when we recall that Konijn, Chang, and Bonner (1969) used a strain of *P. pallidum*, along with *E. coli*, as their source for the laboratory synthesis of cyclic AMP. That it cannot function as a cell attractant in this species, although abundantly produced, may result from the limited number or lack of appropriate cAMP receptors on the surface of the myxamoebae to generate a chemotactic response (Mato and Konijn, 1975). More probable, however, was the existence of a different kind of acrasin, as was indicated earlier by Bonner et al. (1972) and as now confirmed (see below). It is of further interest, as emphasized by Traub and Hohl (1976) and Traub (1977), that the cellular slime molds now classified in the genus *Dictyostelium* that are unable to respond to cyclic AMP actually share important characteristics with species of *Polysphondylium*. Among these are certain similarities in their aggregation patterns, the presence of polar granules in mature spores, and the frequent production of microcysts by unaggregated myxamoebae.

Appreciable attention has been accorded the "acrasin(s)" of *Polysphondylium violaceum* and *P. pallidum*. In his first paper Shaffer reported (1953) that washings from young centers of *P. violaceum* attracted only myxamoebae of *Polyspondylium*, but that similar washings from older *P.* centers attracted *Dictyostelium* myxamoebae as well. Hence he was led to propose that *P.* and *D.* acrasins possibly represent modifications of the same basic acrasin molecule. In parallel studies Sussman and coworkers suggested that acrasin consisted of two (Sussman et al., 1956) or three (Sussman, 1958) fractions and that these needed to be present in the proper combination and amounts in order for chemotactic activity to be expressed.

More recently the identity of *Polysphondylium* acrasin(s) has received serious attention in the laboratories at Princeton. Beginning in 1972 Pan, Hall, and Bonner reported a second bacterial factor with the properties of folic acid to be an effective chemotactic substance for a number of cellular slime molds, including five species of *Dictyostelium* and two of *Polysphondylium*. Responses varied with the different species, and the concentrations of acid required to elicit chemotactic responses were appreciably higher than those of cyclic AMP in *D. discoideum*. Amoebal responses were assayed by the "cellophane square" method of Bonner, Kelso, and Gillmor (1966), and of special interest was the attraction by folic acid of *Polysphondylium* myxamoebae (Pan, Hall, and Bonner, 1972). Following further studies and three years later, however, these investigators (1975) concluded: "Despite the fact that slime mold amoebae secrete small amounts of folic acid-related compounds, there is no evidence that folates are acrasins; rather it is postulated that attraction to folates may be a food-seeking device for the amoebae which prey upon folate-secreting bacteria in the soil."

The following year, hopes were again raised that a *Polysphondylium* acrasin had been found when Wurster, Pan, Tyan, and Bonner (1976) reported the preliminary characterization of a chemotactic substance from *Polysphondylium violaceum* that specifically attracted the myxamoebae of *P. violaceum* and *P. pallidum* and was degraded by a specific "acrasinase." The putative acrasin was isolated and purified and some of its chemical properties determined. It was thought to be a small molecule of less than 1500 daltons, and as stated: "One possibility is that it might be a peptide." The quotation proved to be prophetic. From recent work in the same laboratory, Shimomura, Suthers, and Bonner (1982) have now been able to isolate pure *P. violaceum* acrasin and have determined its structure. It turns out to be an unusual dipeptide made up of glutamic acid and ornithine which the authors are calling "glorin" (N-propionyl-γ-L glutamyl-L-ornithine-δ-lactam ethyl ester). The substance has been synthesized, and the synthetic compound shows the same high chemotactic activity at very low concentrations as the natural, purified product. Both the purified natural compound and the synthetic one seem to have maximal chemotactic activity at 10^{-8} M, which is roughly tenfold less concentrated than the 10^{-7} maximal for cyclic AMP in D. $discoideum$, according to their chemotaxis tests.

Parallel with the isolation and identification of $Polysphondylium$ acrasin in Princeton, researchers in Konijn's laboratory in Leiden (Netherlands) were isolating and characterizing a derivative of pterin as the acrasin of $Dictyostelium$ $lacteum$. Whereas this acrasin awaits complete identification, substantial information concerning its chemical properties and biological activity has been amassed by van Haastert, DeWit, Grijpma, and Konijn (1982). Identification as a pterin derivative is based upon (1) the UV spectrum, (2) the inhibition of enzymatic degradation by 6-methylpterin, (3) the antagonistic effect of 6-amino-pterin on chemotaxis toward both pterin and D. $lacteum$ acrasin, and (4) the degradation of the acrasin to pterin. The acrasin is species specific and attracts cells at a very low concentration (10^{-7} to 10^{-8}), a property shared by several naturally occurring stereoisomers of 6-polyhydroxyalkylpterin. A deaminase which converts pterin into 2-deamino-2-hydroxypterin (lumazin) has been identified as the acrasinase in D. $lacteum$. It is very interesting, and to us surprising, that three compounds as different as cyclic AMP, an unusual dipeptide, and a derivative of pterin should all perform similar functions among organisms as closely related as the dictyostelids appear to be.

Shifting from the chemical identity of acrasins to one of demonstrable presence or absence, Shaffer (1961a) had earlier noted that the presence of neither D. $discoideum$ nor P. $violaceum$ influenced the development of $Acytostelium$ $leptosomum$, nor did aggregating cells of $Acytostelium$ influence the myxamoebae of the other two species. Shaffer also observed (personal communication) that cells of $Acytostelium$ would aggregate only in the presence of air, thus differing from those of D. $discoideum$. While working in our laboratory in 1959, he encouraged a graduate student, Marietta Hostak, to perform a series of experiments that have not been reported elsewhere. They should be of interest to future students of the dictyostelids, particularly those interested in cell aggregation and sorocarp construction in $Acytostelium$. Using specially assembled vessels, Hostak (1960) prepared a type of culture sandwich wherein a small growing culture of $Acytostelium$ was covered with a wider but very thin sheet (150-250 μm) of nonnutrient agar. The myxamoebae beneath this thin agar could not aggregate unless induced by some external stimulus. Such stimulus might come from a preformed center of $Acytostelium$ myxamoebae implanted on the surface or at the margin of the film, or it might come from some foreign material or object placed on the film. During the period of these studies steroids as possible attractants were commanding attention (Wright, 1958; Heftman et al., 1960), and of a limited number tested estrone, 16-dihydro-pregneneolone, 11-α-hydroxyprogesterone, and some others induced aggregation. Of even greater interest was the effect of alkaloids applied to the surface of the covering agar. Of 14 alkaloids tested, seven stimulated cell aggregation within 6-9 hours and two within 1.5 hours. Among the most active were yohimbine tartrate, 3-Epiyohimbine, and methyl reserpate. It was not then nor is it now presumed that any of these compounds represent a true "acrasin" of *Acytostelium*; rather it was felt that they might have supplied cofactors necessary for acrasin production, or that they might have in some physiochemical way merely unleashed the latent but natural phenomenon of autoinduced cell aggregation.

INITIATION

When and how does cell aggregation in the dictyostelids begin? This question has engaged the attention of all early and many later students of the group; and if a consensus has emerged, it would be that aggregation begins following exhaustion of a favorable food supply. This is generally accompanied by a brief period of accelerated and undirected movement of the cell population, for which the term *interphase* is often employed. Actually, the *when* is not quite this simple, for a number of environmental factors also influence the onset of aggregation, most of which, it is suspected, exert their influence through alterations in the physiology and metabolism of the as yet free but newly inducible myxamoebae.

Following the work of Runyon (1942) and Bonner (1947) it was quite clear that in *Dictyostelium discoideum*, the most studied species, the myxamoebae aggregated in response to some self-generated chemotactic stimulus. But a puzzling question still remained: How is the aggregative process initiated? Beginning in 1952, M. Sussman and coworkers published a series of papers purporting to show that initiation resulted from the presence of special initiator cells (I-cells) which occurred in their populations in rather specific heritable ratios: that for *Dictyostelium discoideum* wild type being about 1:2100, and that for *D. purpureum* about 1:300. In the former species, such cells were first postulated upon the bases of the aggregative capacities of washed cells dispensed at various densities on washed agar (Sussman and Noel, 1952), and in later papers they were identified as morphologically different, substantially larger, and more active than the remainder of the population (Sussman, 1958; Ennis and Sussman, 1958; Sussman and Ennis, 1959). A high positive correlation was said to exist between the presence of such an I-cell and the ability of a small population to form an aggregate. This relationship was not confirmed in other laboratories. Exhaustive studies performed in our laboratory (Konijn and Raper, 1961), together with those of Gerisch in Germany (1961), Shaffer in England (1961a), and others, clearly showed that populations of 100-200 myxamoebae could aggregate in virtually every case when tests were conducted under conditions that were optimal for aggregation. No correlation was observed between such aggregation and the presence of any special cell that we or they could identify (Fig. 7-6). The myxamoebae quite normally entered, dispersed, and reentered loose and transitory cell associations, commonly referred to as "clouds," before establishing definite centers of aggregation, further suggesting that the initiation of aggregation was a community response generated by a group of intercommunicating cells.

It is our belief that in *D. discoideum* a small population of cells grown or dispensed within reasonable proximity of each other, such as 200 or more myxamoebae/mm² (Sussman and Noel, 1952), approach in time a physiological state where they as a community of myxamoebae are capable of joining together to complete their developmental cycle. The stimulus to aggregate must start somewhere, and we do not say that such potential may not vary among individual cells at any given moment, or that aggregation may not be triggered by a single cell that reaches a crucial acrasin-emitting state a little earlier than its neighbors. We do say that no visually unique or specially endowed cell is required or need be present to initiate aggregation. Of interest here is a recent study by Glazer and Newell (1981), who found that up to 1 in 5 wild type cells could initiate aggregation when intermixed with cells of a special aggregateless mutant at ratios of 1 in 500 and 1 in 10,000. While the relevance of this finding to normal aggregation in unaltered strains may prove difficult to assess, the work clearly suggests that initiation may depend upon the interplay between cells of differing potential.

Based upon the limited information available, it would appear that cell aggregation in the other larger species of *Dictyostelium* begins in the same manner. It is regarded as highly significant that myxamoebae of *D. discoideum* and of *D. mucoroides* when occupying a common frontier may enter the same wheel-shaped aggregate even though they draw apart before forming sorocarps; under similar circumstances myxamoebae of *D. discoideum* and of *Polysphondylium violaceum* aggregate separately and inflowing streams may at times overlap (Fig. 7-10), as shown by Raper (1940b) and Raper and Thom (1941).

Similar initiation is clearly not the case in *Polysphondylium violaceum*, for Shaffer (1961, 1962) showed that in this species a single cell, termed the *founder*, does in fact serve as the focus for aggregation. Such a cell (Fig. 7-7) becomes identifiable only when it rounds up and neighboring myxamoebae first orient themselves in its direction and then crowd in against it. Its inductiveness normally persists for some time, for if the preparation is disturbed so that the cells become dissociated, they will again reorient and reconverge upon it in a matter of seconds, a performance that can be repeated several times before the founder loses its appeal.

Founder cells also initiate aggregation in *P. pallidum* (Francis, 1965) and in *D. minutum* (Gerisch, 1964a, 1966, 1968), and upon the bases of general aggregation patterns one might expect them to be found in other

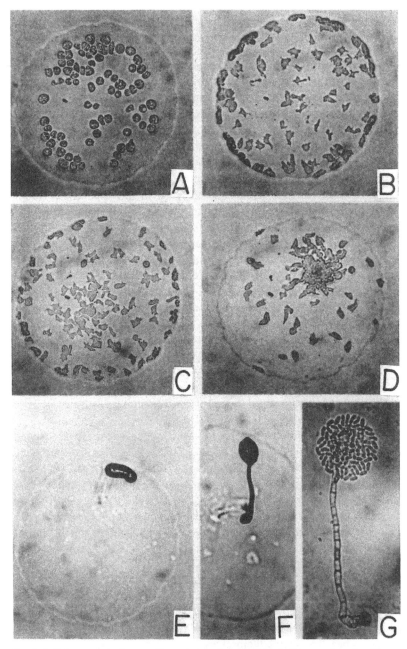

FIG. 7-6. Cell aggregation by a small population of myxamoebae in *Dictyostelium discoideum*. Spreading of the cells was precluded by implantation on hydrophobic agar. *A*. A few minutes after implantation, the 77 myxamoebae are becoming flattened preparatory to resuming amoeboid movement. *B*. Four hours after deposition, the cells are moving actively and have tended to collect at the boundary of the drop. *C*. At 12.5 hours the myxamoebae are beginning to concentrate in a subcentral area. *D*. At 14 hours a definite and fixed center has developed and outlying myxamoebae are converging toward it. *E*. At 25 hours every myxamoeba has joined the pseudoplasmodium and it is now migrating. *F*. A small sorocarp formed from a different but similar population. *G*. A mature sorocarp developed in a third and similar population that is compressed under a coverglass to reveal structure of the sorophore and spore complement of the collapsed sorus. *A-F* × 150; *G* × 350. (From Konijn and Raper, 1961)

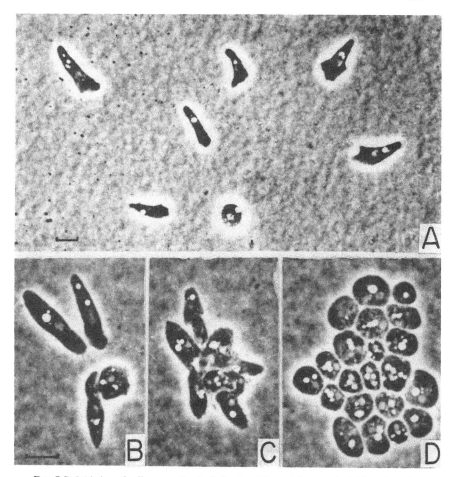

FIG. 7-7. Initiation of cell aggregation in *Polysphondylium violaceum*. *A*. A "founder cell" (lower center) toward which other myxamoebae are converging. *B*. Another newly active founder cell. One myxamoeba has begun to move around it; two others are about to arrive. *C*. The same 10 minutes later; the founder is virtually obscured. *D*. The same soon thereafter and immediately after tapping the coverglass. All cells are rounded and the founder is not obvious. Scales 10 μm. (From Shaffer, 1961b)

species as well, particularly those that produce elliptical spores with polar granules. It is a census worth taking.

There has been a tendency in some quarters to equate the founder cell of *P. violaceum* with the reputed I-cell of *D. discoideum*, but the two have little in common apart from cell individuality (Shaffer, 1962). The former cannot be identified prior to assuming its inductive role; its function is performed at the point of aggregation both in time and place; and it retains the capacity to reinitiate aggregation following spatial disarrangement of the responding myxamoebae. The I-cell, on the other hand, is morphologically different, larger and very active; more importantly, it exercises its inductive function several hours before any evidence of aggregation can be detected, and it may even remain outside the pseudoplasmodium when this is formed.

Investigations to date suggest that the process of cell aggregation in other large species of *Dictyostelium* is basically similar to that in *D. discoideum*. As early as 1937, Arndt discussed at some length the aggregative process in *D. mucoroides*, noting that the myxamoebae formed and reformed temporary cell associations, termed "cumuli," prior to the emergence of definitive centers. The same has been reported by Bonner, Shaffer, and others of more recent date. Particularly interesting is the fact that pregrown myxamoebae of *D. mucoroides*, and of *D. purpureum* as well, will form aggregates within a matter of 3-4 hours after being washed by centrifugation and deposited at appropriate densities on an agar surface. On the other hand, myxamoebae of *D. discoideum* similarly manipulated require up to 10-12 hours to attain a comparable stage in development.

Another phenomenon reported by Arndt (1937) and reemphasized in papers by Bonner and his associates (Bonner and Dodd, 1962a; Bonner and Hoffman, 1963) and by Shaffer (1962) is the spacing of aggregations in a field of uniformly disposed preaggregative myxamoebae. Arndt observed that the magnitude of developing pseudoplasmodia increased as the density of the cell population, but that the number of aggregates per unit area remained relatively constant, resulting in larger fructifications in dense populations and smaller ones where the myxamoebae were less numerous. Of several species of *Dictyostelium* and *Polysphondylium* studied by Bonner and Dodd (1962a), each was found to have its characteristic aggregation territory when tested at several cell densities under controlled environmental conditions. Bonner and Hoffman (1963) advanced the hypothesis that the substance responsible for establishing aggregation territories is gaseous in nature, just as it is for sorocarp orientation (Bonner and Dodd, 1962b), of which we shall say more later. Support was drawn from the fact that numbers of aggregates increased dramatically in the presence of activated charcoal or heavy mineral oil, substances that effectively removed the center-inhibiting substance (see also Kahn, 1968, regarding spacing of aggregates in *P. pallidum*). While the main premise relative to the control of aggregation territories by inhibitory spacing substances is not questioned, the conditions of the tests must be carefully defined and the reader must appreciate that the results apply specifically to those conditions. To illustrate, in very dense fields of myxamoebae such as develop in narrow streaks of *E. coli* grown on a rich medium such as SM agar, the aggregations occupy substantially less surface area than on a nutrient-poor substrate such as 0.1L-P agar—in fact, they are virtually streamless in the former case and arise more by blocking out masses of cells than by obvious cell convergence (cf. Figs. 4-3 and 4-4). Furthermore, the observer must differentiate in the latter case between primary centers of aggregation and secondary foci of cell convergence that result from the disruption of nascent and existing streams when relatively minor changes occur in the physical environment.

Although the exception is perhaps as common as the rule, it may be said that in the larger species such as *D. discoideum, D. mucoroides, D. purpureum*, and *Polysphondylium violaceum* the myxamoebae entering an aggregate typically collect into a single pseudoplasmodium and subsequently form a single fructification. Many other species differ in this regard, and only in very thin cultures does one aggregate normally give rise to a single sorocarp. It is not a consistent pattern in *P. pallidum*; it seldom happens in *D. lacteum* or *D. minutum*; and it almost never occurs in *D. polycephalum* or *Acytostelium leptosomum*.

The discussions thus far, and the examples cited, have all related to aggregative behavior of natural isolates, or wild types, of the several species considered. No less interesting is the behavior of mutant strains produced in the laboratory and studied either as aberrations of normal development or as analytical tools for clarifying genetic or physiological problems. There is, of course, a wide variety of such mutants, ranging from nearly normal strains with minor lesions to strains, more seriously impaired, that cannot aggregate. Still, some of the aggregateless mutants can join with aggregation-competent strains and synergistically enter into the aggregative process and subsequently fruit, as M. Sussman and coworkers showed many years ago (Sussman, 1954), and in doing so they form sorocarps with viable spores. In fact, and as previously noted, such conjoint cultivation has been used to secure spores for conserving mutants via lyophilization.

A more dramatic example of synergistic or inductive aggregation and development can sometimes be obtained by pairing two aggregatelesss mutants. Experiments by A. T. Weber may be considered representative (Weber and Raper, 1971). Having a substantial number of aggregateless mutants, Weber planted these in confluent lawns of *E. coli* on agar plates in such fashion that myxamoebae of the mutants became intermixed at common frontiers. In most cases no aggregation occurred, but a limited number of strains, when so commingled, formed quite normal-appearing aggregations and subsequently sorocarps. Whereas no attempt was made to identify the deficiency in either mutant, it would seem apparent that insofar as aggregation and sorocarp construction were concerned each mutant was able to supply what the other lacked in order for its myxamoebae to aggregate and fruit.

Some interesting examples of unusual aggregative behavior may be cited from studies conducted in this laboratory by Hans R. Hohl (Hohl and Raper, 1964). In one case the mutant was capable of forming only very small aggregates and very small sorocarps when allowed to develop unmolested on an agar surface, and when much larger masses of cells were brought together artificially, the sorocarps were still petite—that is, further development was not abetted by imposed cell aggregation. In a second case only small aggregates and sorocarps formed under natural culture conditions; however, when preaggregative cells were pushed together in small cup-like depressions, sorocarps of normal form and dimension were formed. Apparently, aggregative capacity was present but ineffective except over short distances. When the cells were "aggregated" artificially, sorocarps of normal form and dimensions developed. In the third case the mutant was aggregateless; but when postvegetative cells were pushed together, normal sorocarps were formed. This mutant seemingly possessed no inherent aggregative capacity, but when its cells were compacted it fruited normally. This would seem to represent a mutation not so much of the myxamoebae per se but of their capacity to communicate unless intercellular contacts were established artificially, thus representing a kind of societal mutation.

INTERCELLULAR COMMUNICATION

Once aggregation has begun there remains an even larger question: How do the cells communicate to establish a basically radial aggregation and subsequently to draw together in orderly fashion myxamoebae from the surrounding area? No other phase of the life cycle has received greater attention, for reasons clearly apparent. Where else can one witness within so few hours a comparable shift, or conversion, of a large population of free-living individuals into a single functional community of highly integrated cells, which in its further morphogenesis behaves as a multicellular organism? Whereas there are many dictyostelids, most of the research and published reports relating to cell aggregation have centered upon *Dictyostelium discoideum*, as does the discussion that follows. Scores of relevant papers have issued from many different laboratories during recent years—so many, in fact, that we cannot possibly discuss them here. Fortunately, there have been a number of good reviews (Gerisch, 1977 and 1979; Loomis, 1975 and 1979; Newell, 1977a and 1978; and, of course, Shaffer's earlier papers) from which to draw in presenting an abbreviated account of the intercellular communication that underlies the aggregation process.

Following initiation an aggregation center begins to emit pulses of an attractant, or acrasin, which in the case of *Dictyostelium discoideum* and other large species of the genus is the nucleotide cyclic AMP. As the acrasin is released myxamoebae near the center reorient, become elongate in the direction of the center, and move toward it. These activated cells are now stimulated to secrete acrasin; so they in turn attract myxamoebae outward from themselves. In this way a relay is built up causing cells in a widening area first to respond and then to produce acrasin (Shaffer, 1957a). All the while the center retains its dominance by the periodic emission of pulses of cAMP and their outward propagation via the relay system. Thus the influence of the center is extended far beyond the range of attraction of which it alone is capable. As aggregation proceeds the myxamoebae collect into fewer and larger inflowing streams that continue to converge, forming in due course a compact mass or column preparatory to sorocarp formation.

Much has been learned from the use of time-lapse cinematography, first employed by Arndt (1937) and later by Bonner (1944, 1958b), Gerisch (1963a-d, 1964b), Konijn (1972), Robertson et al. (1972), Alcantara and Monk (1974), and others. Not only do motion pictures show the wave-like character of cell aggregation, they also permit accurate timing of certain critical steps in the process. From an analytical viewpoint the films and resulting paper of Alcantara and Monk (1974) are especially helpful. Cultures of *D. discoideum* were grown with *Aerobacter aerogenes* as a mono-layer of myxamoebae on specially prepared agar under controlled humidity and photographed by time lapse with dark-field illumination. Under these conditions the *early* aggregations appear as alternating concentric light and dark bands (Fig. 7-8). The narrower, light bands contain elongate, inward moving cells whilst the wider, dark bands contain cells that are more rounded and essentially stationary. The overall gradient of cyclic AMP within the aggregation is maintained, as previously noted, by the prompt destruction of excess acrasin by extracellular phosphodiesterase following a brief period of excitation and inward movement of the stimulated cells. A resting period then follows while the myxamoebae await the next relayed impulse. The difference in refractility between alternatively moving and resting cells accounts for the light and dark bands seen with darkfield illumination, a phenomenon first illustrated nearly a decade earlier by Gerisch (1965). With reference to the bands and what they reveal, Alcantara and Monk (1974) wrote as follows:

> The width of bands of moving cells represents the distance the signal travels in the time that cells remain elongated after stimulation; it should therefore not vary with signal frequency. The interband distance, on the other hand, *should* depend on signal frequency since it reflects the distance that waves are propagated in the time elapsing between successive signals. This prediction can be tested since we observed that during the course of aggregation the period of signalling within a given territory decreases from 10 min to 3 min or less, by which time the cells are beginning to form streams and it becomes hard to discern individual movement steps.

Further analysis of the films revealed that the interval between the time a cell received a signal (cAMP) and the time it began to emit cAMP was approximately 12 seconds. Additionally, the distance over which this dif-

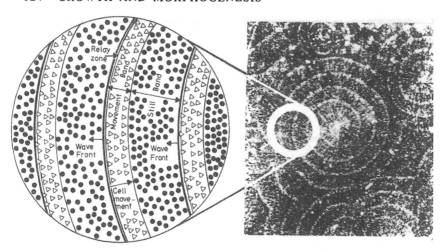

Fɪɢ. 7-8. Diagrammatic interpretation of the light and dark bands (moving and stationary myxamoebae, respectively) seen with dark-field optics of a dense population of myxamoebae on an agar surface. The circular field shown on the left is approximately 5 mm in diameter. (Reprinted from Newell, 1978; original photograph by M. Peacey and J. Gross)

fused before being destroyed by phosphodiesterase was about 60 μm. Of special significance for the success of cell aggregation, therefore, is the repeated amplification (tenfold or more) of the cAMP signal as the relay moves progressively outward (Roos, Nanjundiah, Malchow, and Gerisch, 1975).

As first reported by Gerisch and Hess (1974), pulsatile signal generation occurs also in stirred oxygenated suspensions of starved myxamoebae. When examined by spectrophotometer such cell suspensions showed autonomous oscillations in light absorbance that matched quite well the periodicity of the chemotactic waves that spread outward from the centers of aggregation on agar plates. Since changes in cell shape are synchronous throughout a cuvette and since these are accompanied by changes in light absorbance, Gerisch and coworkers were able to determine concentrations of cAMP and of the cAMP-forming enzyme, adenyl cyclase, at precise times during the oscillation cycle (Gerisch and Wick, 1975; Roos, Scheidegger, and Gerisch, 1977; see also Shaffer, 1975). When intra- and extracellular cAMP concentrations were measured, oscillations in optical density were found to correspond to pulses of cAMP synthesis and its release by the myxamoebae.

Several of the larger species of *Dictyostelium* show comparable responses to cyclic AMP but differ in their timing. For example, aggregating cells of *D. purpureum* respond much as do those of *D. discoideum* and show pulsations at approximately 5-minute intervals (Konijn, 1972). In *D. mu-*

coroides the time between pulses varies with the strain and may range from 5 to 12 minutes, whereas in *D. rosarium* the rhythmic waves appear at approximately 20-minute intervals, and the cells may actually disperse before the next pulse brings them together again. In *D. aureum* the interval between pulses is said to be long also (Konijn, 1972, and personal correspondence). Timing may vary also with the developmental stage and with the temperature (Nanjundiah et al., 1976).

Interesting also are the effects of induced pulses of cAMP (Robertson and Drage, 1975). If cells of *D. discoideum* in early preaggregative phase are stimulated by repeated pulses of cAMP, they start to oscillate autonomously earlier than control cells and induce pulsatile signaling much as in normal aggregation. Additionally, some aggregateless mutants can be stimulated to initiate aggregation if pulsed with cAMP, and once so induced they may continue through normal development to sorocarp formation. Or, as summarized by Gerisch (1979): "The elements of the cAMP-signal system—cAMP-receptors, adenyl cyclase and cell-surface phosphodiesterase—are connected such that they form a network that promotes the transformation of nonsignaling, weakly responding cells into cells with a fully developed communication system."

Signal reception has been carefully studied by several groups of investigators, particularly Gerisch and coworkers (Beug, Katz, and Gerisch, 1973; Gerisch, Beug, Malchow, Schwarz, and Stein, 1974; Gerisch, Hüsler, Malchow, and Wick, 1975; Roos and Gerisch, 1976; Roos et al., 1977; Müller and Gerisch, 1978; and Gerisch, 1977, 1979, and 1980). Upon the bases of their work, and that of others, signal reception is attributed to the presence on the cell surface of cAMP receptors that appear during starvation and herald the onset of aggregation competence leading to morphogenesis. It is thought that they may function in a manner comparable to certain hormone receptors in mammalian tissues (Mullens and Newell, 1978). Present also are contact sites of two types, designated A and B. Contact sites A can be purified and functionally characterized by immunological means. They are undetectable on vegetative myxamoebae, begin to appear on preaggregative cells in response to pulses of cAMP, and reach their maximum number when cells become aggregation-competent. They are EDTA stable, and evidence suggests that they are primarily responsible for end to end adhesion of streaming cells. Contact sites B are present throughout the developmental cycle, are not responsive to cAMP, are EDTA sensitive, and may facilitate side-to-side cell adhesion in stream formation. Just how the cells perceive the cAMP signal is still an open question. Two possibilities have been proposed. One suggests that the myxamoebae measure simultaneously the difference in concentration of cAMP from front to back of the cell, thereby sensing the presence of a gradient (Bonner, 1947, 1977, and Mato, Losada, Nanjundiah, and Konijn,

1975). The other suggests that the myxamoebae monitor the change in cAMP concentration with respect to time (Gerisch et al., 1975), much as do chemotactic bacteria (Koshland, 1977). However activated, the cells respond within a very few seconds by moving toward the center and releasing their own cAMP pulses before entering quiescent periods that decrease as aggregation progresses.

For the relay system to operate there must be a way to remove the signal once it has been discharged, otherwise it would accumulate to a continuous and ineffective level. Such removal does in fact occur, and about this Newell (1977b) concludes: "To avoid the amoebae becoming swamped by the cAMP signals, some sort of signal 'sink' must obviously be employed. Such a sink is provided by the phosphodiesterase enzymes. There are at least two types of these enzymes, one being soluble and excreted by the amoebae and the other being bound to the outside of their plasma membranes. These two enzymes ensure that the signal does not build up in concentration and are responsible for destroying it before it can diffuse far from its site of emission."

CHEMOTAXIS

The payoff from signal production and reception comes, of course, in the chemotactic movement of cells, that is, the directed convergence of myxamoebae to form a multicellular community.[3] Not much is known about the regulation of myxamoebal movement, but much has been reported concerning the observable behavior of aggregating cells. We know that the response of an aggregation-competent cell to an impulse of acrasin is very rapid, about 5 seconds, fide Gerisch (1979); and when filming a wave of cells we know that the initial response appears to be virtually instantaneous throughout the viewing field. Viewed as individual cells, one sees the emergence of one, rarely more, pseudopods from a portion of the cell surface and the progressive movement of the cell content into the enlarging projection. As a consequence the whole body moves forward. One hypothesis is that cAMP may cause the local release of calcium ions, and that these in turn activate the contractile actomyosin-like proteins shown to be present in the cells (Spudich, 1974) and thus bring about cell movement. The extent of movement in response to a single impulse is about two-cell lengths and may extend over a period of 50-100 seconds, the rate gradually slowing until the cell assumes a more rounded form and enters a period of inactivity prior to the next impulse.

The cumulative result of such individual cell movements is a general

---

[3] See also Bonner, 1977: "Some aspects of chemotaxis using the cellular slime molds as an example." Mycologia 59: 443-459.

progression of the whole population toward the aggregation center, and as the myxamoebae advance they collect into inflowing streams. Typically this progressioin continues to be controlled by the center through the relay system, but as the streams increase in size they too become strong acrasin emitters and dominance by the center may be lost, particularly when a break occurs in the continuity of a stream. The probability of such disruption is lessened but not obviated by the adhesive character of the inflowing cells, first reported by Shaffer (1957a) as "stickiness," and by the tendency of cells and cell clusters to continue moving in the same direction, at least for a short distance.

The typical aggregation in all but a few dictyostelids may be described as radiate in pattern, and where plates of favorable media are inoculated by spreading mixed suspensions of bacteria and slime mold spores across the surface, the first aggregations to appear are quite generally of this form. Furthermore, in the larger species of *Dictyostelium* it can be said that the cells will, as a rule, converge and form a single sorogen or, in the case of *D. discoideum*, a migrating pseudoplasmodium to yield a single sorocarp. This is rarely true in many species and never so in some. For example, in *Polysphondylium pallidum* the initial aggregation may be beautifully symmetrical, but with the emergence of sorogens (usually more than one) streams that are still converging seem to break apart, freeing the constituent cells which then disperse and may or may not reaggregate. In *D. polycephalum*, aggregations that are plane and radial in pattern at an early stage gradually become more compact and nodular and invariably give rise to multiple thin and tenuous migrating pseudoplasmodia. In *D. minutum* primary aggregations are formed by the convergence of myxamoebae into small fruiting mounds without evident stream formation (Gerisch, 1964a). Subsequent to this, and following limited sorocarp formation, residual myxamoebae may form sheet-like streams in relatively thin cultures that are poorly focused and often fail to fruit (Raper, 1941b). A fourth category includes some of the smaller species of *Dictyostelium* where the initial aggregations are radially symmetrical with relatively short, well-defined inflowing streams; but with the emergence of sorogens (commonly multiple) the streams surrounding the young fructifications seemingly dissolve whilst in a circle or arc some distance beyond the primary site scores of small, nearly parallel streams appear and all are oriented toward the original center (Raper and Fennell, 1967) (Fig. 7-9A). Insofar as we are aware, no explanation for this phenomenon, tentatively referred to as a "halo" pattern, has been offered, nor has anyone attempted to isolate or characterize the acrasin(s) responsible.

In addition to the foregoing diagnostically useful aggregation patterns, it is not unusual to see fixed centers replaced by spiral or helical rings into which outlying cells flow for varying periods of time before a definitive

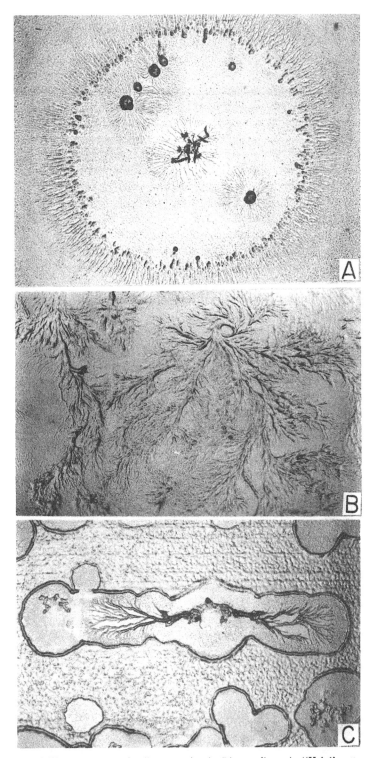

Fig. 7-9. Different patterns of cell aggregation in *Dictyostelium*. *A*. "Halo" pattern as commonly produced in relatively thin cultures of some smaller species of *Dictyostelium*, in this case *D. vinaceo-fuscum*. × 8. *B*. Pseudoplasmodium with spiral rather than fixed center, in this case *D. purpureum*, but often observed in other large species as well. × 8. *C*. A pseudoplasmodium of basic radial pattern but misshapen by location of contributing myxamoebae. The bounding ridges represent feeding fronts in a sparsely inoculated culture. ×
8

center is established (Fig. 7-9B). Whereas the pattern remains basically radial, the form and dimensions of an aggregation must depend upon the disposition of the myxamoebae contributing to its formation. Because of this, irregular patterns, often quite striking, are not uncommonly encountered (Fig. 7-9C).

Resulting, we presume, from physiological changes in the myxamoebae, or perhaps in some cases from alterations in the surface characteristics of the cells, the onset and progress of cell aggregation is clearly influenced by certain environmental factors as well. The examples cited apply particularly to *D. discoideum*, but insofar as studies have been made, comparable responses would also be expected of many other cellular slime molds.

DECREASED HUMIDITY. Cell aggregation is influenced strongly by the moisture content of the air contained within the culture vessels. In plate cultures where it is known that aggregation will normally begin in approximately 6 hours, if the covers of some dishes are opened slightly so that the humidity within is lessened, pseudoplasmodium formation will begin in such disturbed cultures 2-4 hours earlier than in closed control plates, other conditions remaining constant. In these cultures the pseudoplasmodia are smaller and more numerous than in the controls (Raper, 1940a).

INCREASED TEMPERATURE. Similarly, if cultures are grown at 20-24°C until approximately 6 hours before aggregation would normally occur and then are placed at a temperature of 27-28°C, aggregation will begin 2-3 hours earlier than in control cultures, other conditions remaining constant. As in the preceding experiment, the pseudoplasmodia in such cultures are smaller and more numerous than in control cultures continuously maintained at the lower temperature (Raper, 1940a). Konijn (1965) investigated the effect of temperature on chemotaxis in *D. discoideum* by depositing attracting and responding cells in small droplets at varying distances on specially prepared hydrophobic agar. As the temperature was decreased (1) the distance over which sensitive myxamoebae could be attracted increased, (2) the time interval from the onset and completion of aggregation increased, and (3) the time between cell deposition and the beginning of aggregation increased. Tests were run at 13°C, 18°C, 24°C, and 28.5°C.

LIGHT. Aggregations likewise develop 2-4 hours earlier in cultures grown in diffuse daylight than in similar cultures incubated in total darkness, other conditions remaining constant (1940a). Aggregations in lighted cultures are smaller and more numerous than in darkened ones, a result that corroborates the much earlier observations of Potts (1902) and Harper

(1932) that the fruiting structure of *D. mucoroides* and *Polysphondylium violaceum*, respectively, were larger when formed in darkness than in light. Neither Potts nor Harper noted whether pseudoplasmodia formed earlier in light than in darkness. Using pregrown and washed cells, Konijn and Raper (1965) found that in single strains of *D. mucoroides, D. purpureum, P. violaceum*, and *P. pallidum*, and one strain of *D. discoideum* (Acr-12), aggregations appeared first in cultures exposed to continuous light, whereas other strains of *D. discoideum*, including the type (NC-4), aggregated optimally after an initial dark period of 6-8 hours followed by 4 hours light. More recently and not surprising, Häder and Poff (1979) have reported that whereas the myxamoebae of *D. discoideum* are positively phototactic in relatively weak light, they are negatively phototactic and disperse from high intensity light. In experiments involving small droplets, Konijn and Raper (1966) found that the area of attraction of myxamoebe aggregating in darkness was greater than that of cells aggregating in light, hence the formation of larger aggregations. In a parallel study of *P. pallidum*, Kahn (1964) confirmed a pronounced light effect with the formation of more and smaller aggregations. In addition, he showed that the number of aggregations increased and their dimensions decreased also when preaggregative cells were exposed to activated charcoal or covered by a thin layer of heavy mineral oil.

Having noted that certain species of *Dictyostelium* share cAMP as an acrasin, whereas other dictyostelia and species of *Polysphondylium* do not (Traub and Hohl, 1976), we now consider the behavior of sharing and nonsharing species when their myxamoebae are intermixed, either by concurrent growth or manual admixture. Such a study was published decades ago by Raper and Thom (1941); the latest was reported recently by Gerisch and coworkers (1980). In the first of these, spores of *D. discoideum* and *D. mucoroides* were spot-inoculated about 5 cm apart in a freshly spread band of *Serratia marcescens* so that myxamoebae of the two expanding colonies would become intermixed along a common frontier. Since cells of *D. discoideum* retain the red pigment (prodigiosin), they become red; those of *D. mucoroides* destroy it and remain colorless. Similar cultures were prepared for *D. discoideum* and *P. violaceum*, except that *E. coli* was used as bacterial nutrient, since cells of both species retain the pigment of *Serratia*. The results were regarded as significant: myxamoebae of *D. discoideum* and *D. mucoroides* entered common aggregates but separated without admixture prior to sorocarp formation. Myxamoebae of *D. discoideum* and *P. violaceum* aggregated independently from the outset, and at times streams of the two slime molds were seen to overlap as their myxamoebae converged on separate centers (Fig. 7-10). When migrating pseudoplasmodia of *D. discoideum* and developing sorogens of *D. pur-*

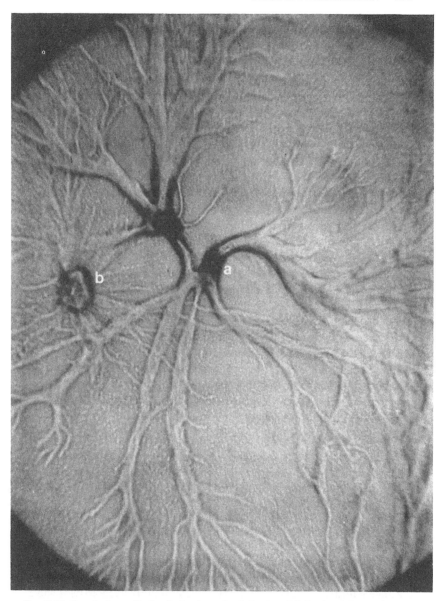

FIG. 7-10. Pseudoplasmodia developing in a mixed planting of *Polysphondylium violaceum* (a) and *Dictyostelium discoideum* (b). Note that inflowing streams overlap as cells of the two species aggregate to completely separate centers. × 30. (From Raper, 1940b)

pureum were crushed together and their myxamoebae blended, mixed cell aggregations occurred and some sorocarps containing spores of the two species were formed. When comparable structures of *D. discoideum* and *P. violaceum* were crushed and their cells intermixed, no common aggregation occurred. Even then was inferred what would be proved a quarter century later, that cells of the two genera "march to different drums."

In the latter study, Gerisch, Krelle, Bozzaro, Eitle, and Guggenheim (1980) investigated the behavior of myxamoebae of *D. discoideum* and *P. pallidum* when intermixed in various ways. On an agar surface cells of the two slime molds formed separate aggregations as previously reported for *D. discoideum* and *P. violaceum*. In gently shaken suspensions cell behavior varied with the developmental stage of the populations. Actively growing cells of the two species formed common aggregates that eventually sorted out into separate areas. In contrast, aggregation-competent cells when intermixed sorted out within an hour so that virtually monospecific aggregates of the two slime molds remained. Specificity of cell recognition and adhesion, mediated by contact sites A on older myxamoebae of *D. discoideum* and by their counterparts on the cells of *P. pallidum*, accounted in the main for the observed aggregative behavior. Additionally, the lectins discoidin I and II and palladins I and II, first reported by Rosen et al. (1973 and 1974, respectively), probably play a less clearly defined role (see Newell, 1977a, for full discussion).

The phenomenon of cell aggregation in the cellular slime molds is in itself a unique and intriguing phenomenon, but for many investigators, including ourselves, it is believed to have far broader potential significance, as the following quotations bear witness.

It is to be hoped that cell aggregation in Dd can be taken as a model for investigating the development of pattern in higher organisms. Possibly spatiotemporal patterns are organized by similar mechanisms in various developmental systems, and it is only because of its advantages that Dd has become the first known exponent of this group . . . (Gerisch, 1979).

The aggregation phase of the cellular slime moulds presents a tractable system with which to study cellular communication. While it is certainly true that such aggregation is a very unusual process and could involve a unique form of cellular interaction, it seems intuitively more reasonable that the interaction involved bears some resemblance to cellular communication in other situations.

At the risk of generalizing from slime moulds to man it is perhaps worth contemplating the possibility that a pulse relay system using nucleotides or small peptides may be a cellular communication system that is not just restricted to the cellular slime moulds. Had it not been for the stepwise movement of the *Dictyostelium* amoebae and the visualization of dark and light bands expanding from the aggregation centres, the pulsatile nature of the interaction would not easily have been observed, even in this system, and in the absence of such dramatic gestures it may easily go unnoticed in higher embryonic tissues (Newell, 1978).

CHAPTER 8

Fructification

Singular as is the phenomenon of cell aggregation in the dictyostelids, it is but a prelude to the even more remarkable process of fructification, wherein the increasingly interdependent myxamoebae construct multicellular sorocarps and produce the spores necessary for survival. While it is convenient to consider aggregation and fructification as separate developmental stages, as is done here, the reader should remember that they actually represent successive steps in a continuing progression. In fact, sorocarp formation in many species, if not in most, may be initiated well before cell aggregation is complete.

We also speak of an aggregation as typically producing *a* sorocarp, even though they commonly give rise to more than one. In fact, it is primarily in the larger species that the one-from-one relationship applies, and even there the exceptions may at times exceed the rule. Of basic importance in this matter are the genetic characteristics of the species, while superimposed on these may be a variety of constraints that arise from nutritional or environmental conditions. At all stages except for the terminal one we are dealing with cells unprotected by firm walls, and the remarkable thing is not that aberrancies occur but that so often they do not (Raper, 1940a).

As cell aggregation draws to a close, the inflowing myxamoebae characteristically crowd together to form an erect, or semi-erect, peg-like column of a size reflecting the number of collaborating myxamoebae. In most cellular slime molds stalk formation begins at this time (if not already in progress) near the base of the upright column and at the site where the myxamoebae converge. Consistent with the usage proposed by Harper (1926) the term *sorogen* is applied to such in-process-of-fruiting structures.

Postaggregative behavior varies substantially in different genera and species; hence we cannot present a single, all-inclusive picture of dictyostelid fructification. Instead, we shall select a particular species, *Dictyostelium discoideum*, and use its further morphogenesis as a basis for subsequent comparisons with other cellular slime molds. Although it differs from the pattern just described in some important particulars, such differences in fact commend it for the purpose intended. It has been studied most intensively in this laboratory and elsewhere, and its unique features render it especially favorable for observation and experimentation. Such advantages include freely migrating pseudoplasmodia (without stalk formation), delayed formation of sorogens, fruiting at locations other than aggregation sites, and the near constant proportions of its sorocarps. Since much of the basic information relative to the growth and development of

D. discoideum stems from our initial studies, the account that follows will draw heavily upon some of these early reports (Raper, 1940a,b; 1941a). At the same time we hope not to minimize or overlook more recent discoveries or current advances that significantly enhance our appreciation of these simple eukaryotes, or further demonstrate their usefulness for research in many areas of biology. We shall focus the discussion upon the central thread of morphogenesis from cell aggregation through sorocarp construction as this occurs in wild type strains growing in laboratory cultures in association with selected bacteria as nutrient. More often than not this will involve personal observations made upon species types.

There are vast amounts of important work on the physiology, biochemistry, genetics, and molecular biology of the dictyostelids for which we have neither the space in this volume nor the experience necessary to provide adequate coverage. Fortunately there are scores of publications from the laboratories of John T. Bonner, Gunther Gerisch, Brian Shaffer, M. and R. Sussman, Theo. M. Konijn, James H. Gregg, Barbara Wright, Hans R. Hohl, Michael F. Filosa, William F. Loomis, David Francis, James C. Cavender, Harvey F. Lodish, E. R. Katz, Peter Newell, A. J. Durston, Keith Williams, P. Brachet, David Cotter, Ikuo Takeuchi, Julian Gross, Danton H. O'Day, and many others who have given special attention to one or more of these areas. Three of the above have written books, Bonner (1959a and 1967), Wright (1973), and Loomis (1975); and others have written comprehensive reviews in which *D. discoideum* occupies center stage, for example, MacWilliams and Bonner (1979). A volume edited by P. Cappuccinelli and J. M. Ashworth (1977) includes reports from many of the aforementioned investigators together with papers from several other laboratories as well. A sixth and more recent book edited by Professor Loomis (1982) contains several chapters on the biochemistry, genetics, and molecular biology of *D. discoideum* with some information on other species and on *Polysphondylium* as well. It thus complements this monograph, which is focused primarily upon the natural history of the dictyostelids.

The Migrating Pseudoplasmodium

As cell aggregation in *Dictyostelium discoideum* is completed under optimal conditions, the upright column bends over and comes to lie flat on the agar surface. Then through the coordinated movement of its constituent cells it migrates as a unit for some distance, commonly 0.5 to 5.0 or 6.0 cm, prior to constructing a sorocarp. Such moving bodies are variously referred to as migrating pseudoplasmodia, slugs, or grexes, respectively, upon the basis of their cellular composition (Raper, 1934 et seq.), their form (Sussman and Sussman, 1961), or their origin (Shaffer, 1962). Structurally they are very simple and consist of compacted myxamoebae surrounded by a noncellular slime sheath or envelope that the cells themselves secrete. Since size is determined by the number of collaborating myxamoebae, this can vary greatly under different environmental and nutritional conditions. Even in the same culture migrating pseudoplasmodia often range from 0.5 to 2.0 mm in length × 0.1 to 0.25 mm in width, with larger structures generally rare and smaller ones not uncommon. We estimate there are approximately 125,000 cooperating myxamoebae in a large pseudoplasmodium 1.5 × 0.2 mm, whereas the number is approximately 10,000 in one 0.5 × 0.1 mm. While pseudoplasmodia vary somewhat in form due to differences in the ratio of length to breadth and may be either straight or somewhat sinuous (Fig. 8-1), the basic pattern remains the same irrespective of size. It may be described roughly as cartridge-shaped with the anterior end blunt and somewhat

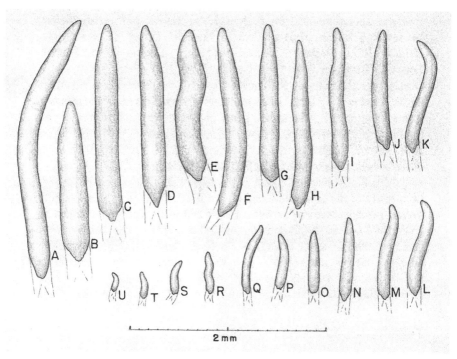

2 mm

FIG. 8-1. Camera lucida drawings of migrating pseudoplasmodia of *Dictyostelium discoideum* showing their generally constant pattern and extreme variation in size. Estimated number of collaborating myxamoebae in each pseudoplasmodium follows:

A: 189,000	F: 131,000	K: 24,900	P: 8,700
B: 174,000	G: 77,600	L: 21,200	Q: 7,900
C: 165,000	H: 62,500	M: 18,700	R: 2,200
D: 149,000	I: 48,500	N: 15,500	S: 1,850
E: 136,000	J: 44,500	O: 10,300	T: 1,450
			U: 770

(From Raper, 1941a, which see for method of estimating cell numbers) tapered, the main body fairly uniform in thickness, and the posterior end generally truncate and horizontally flattened. Left in the wake of the advancing body is the collapsed slime sheath, which remains on the agar as a ribbon that can be used to trace the prior path of migration. The rate of migration may be of the order of 1-2 mm/hour, but it is quite variable and clearly dependent upon the environment. Bonner et al. (1953) have shown that larger pseudoplasmodia move faster than smaller ones, and in his and our experience the thinner structures move somewhat faster than shorter ones of equal volume.

When first examined (Raper 1935, 1940a) the migrating pseudoplasmodia were thought to be essentially homogeneous throughout and to consist of elongate myxamoebae, all oriented with their long axes parallel to the direction of migration. This was found to be in error a few years later when Bonner (1944), using stained sections, showed that myxamoebae in the apical region were oriented in a more transverse direction and that positionwise they corresponded, in the main, to the region Raper (1940a) had found to produce the stalks in sorocarps from grafted pseudoplasmodia. Despite such presumptive demarcation, when a pseudoplasmodium is dissociated in water, or upon the moist surface of an agar plate, the component myxamoebae behave in a similar manner. They round up immediately, then draw apart and subsequently resume movement as separate and unrelated amoeboids. In the presence of bacteria they return to the vegetative stage, whereas in the absence of such nutritive material they reorganize after some delay into new and smaller fruiting units; but in either case the behavior is basically the same whether an entire pseudoplasmodium is involved or only a fractional part.

RESPONSES TO ENVIRONMENT

However designated—migrating pseudoplasmodia, slugs, or grexes—they are strongly influenced in their migration by atmospheric moisture (Raper, 1940a), free surface water (Bonner et al., 1982), and substrate composition (Bonner and Slifkin, 1949). Additionally, they are extremely sensitive to light and heat, moving in the direction of a light source and from cool to warmer temperatures on an agar plate.

When cultivated upon pregrown bacteria (e.g., *E. coli*) on nonnutrient agar plates in a saturated atmoshere at cool temperatures (17-20°C), pseudoplasmodia may continue to migrate for days and over distances of 20 cm or more. In richer cultures migration is progressively reduced, as it is also in agar plates that contain progressively greater concentrations of salts or nonelectrolytes. This was beautifully demonstrated by Slifkin and Bonner (1952), who grew *D. discoideum* on standard medium (Bonner, 1947) and at the onset of aggregation cut out disks bearing myxamoebae and placed these on plates of nonnutrient agar, each of which contained one of three mineral salts (NaCl, CaCl$_2$, or Na$_2$HPO$_4$) or one of three organic compounds (glucose, sucrose, or dl-alanine). Graded series of concentrations were tested, and in each case migration was reduced as the concentration of the added solute was increased. Not surprising, the salts were substantially more inhibitory than the nonelectrolytes at comparable molarities. Of the salts Na$_2$HPO$_4$ was the most inhibitory. The profound effect that buffers have on pseudoplasmodial migration was later demonstrated in dramatic fashion by Newell et al. (1969).

Light as a factor influencing the size and orientation of sorocarps in *Dictyostelium mucoroides* was noted by Potts (1902) and by Harper (1926), and in his early studies on *D. discoideum* Raper (1940a) emphasized its importance in determining the direction and extent of pseudoplasmodial migration. However, no quantitative measurements were attempted. Many years later, and using a monochrometer to emit selected wavelengths, Francis (1964) reported a maximum turning response at ca 425 μm and a lesser response at ca 550 μm for both the migrating pseudoplasmodia of *D. discoideum* and the prostrate stalk-forming sorogens of *D. purpureum* (Fig. 8-2). Positive responses were obtained at light intensities as low as 0.03 to 0.04 ergs/mm^2/sec. More detailed studies for *D. discoideum* by Poff and Loomis (1973) and Poff and Butler (1974) have confirmed the

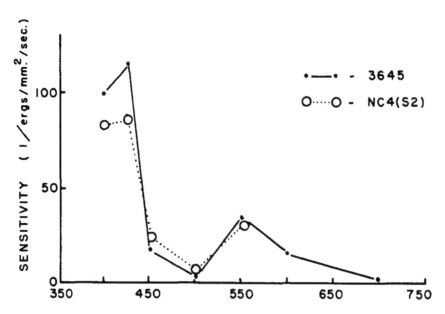

FIG. 8-2. Action spectra for phototaxis in *Dictyostelium discoideum*, strain NC-4(S2), and *D. purpureum*, strain 3645. Sensitivity is the reciprocal of the intensity required to give a specific directional response in ½ hour and is based upon the responses of 20-50 migrating pseudoplasmodia at each wave length. Abscissa denotes wavelength in μm. (From Francis, 1964)

results reported by Francis, while Poff et al. (1974) have isolated and identified the photoreceptor as a heme protein designated phototaxin. Also in agreement with Francis, Poff and associates concluded that light was focused by the cylindrical surface of the pseudoplasmodium so that its intensity was greatest on the distal side, causing the tip to turn in the direction of the light source, while overhead light of sufficient intensity could eliminate migration altogether and cause the pseudoplasmodia to form sorocarps in situ (Newell et al., 1969).

The response to temperature is equally interesting, for as the temperature is raised the distance traversed by the pseudoplasmodia declines progressively until at 27-28°C little or no migration occurs (Raper, 1940a). Bonner and Slifkin (1949) showed that pseudoplasmodial migration could be arrested by moving cultures from cool to warmer temperatures and, additionally, that smaller percentages of the constituent myxamoebae formed stalk cells at the higher temperatures, thus leaving a greater number to produce spores (see also Farnsworth, 1975; Forman and Garrod, 1977a,b). Especially noteworthy was the extreme sensitivity to temperature of the migrating structures. Employing thermocouples embedded in agar, Bonner et al. (1950) reported pseudoplasmodia to respond to a gradient of 0.05°C/cm. For a pseudoplasmodium 0.1 mm wide this represents a difference of 0.0005°C on the two sides of the migrating body—an almost unbelievable sensitivity. In fact, for a time Bonner suspected they might be measuring heat responses in his light experiments until a student, W. J. Gamble, found that the pseudoplasmodia would respond to the light from a culture of luminescent bacteria! Lest the reader still misbelieve, using more sophisticated techniques Poff and Skokut have since reported (1974) a sensitivity of 0.0004°C across a pseudoplasmodium of the same 0.1 mm diameter. Left unexplained is Poff and Skokut's observation that pseudoplasmodia formed from cells grown at 20°C were more thermotactically sensitive than cells grown at 23.5°C.

SLIME SHEATH

The slime envelope that surrounds the migrating body in *D. discoideum* is comparable to the slime sheath that E. W. Olive (1902) described for *Dictyostelium mucoroides* and *D. purpureum*. As in those species, it has its genesis when the center of the nearly completed aggregation rises peg-like into the air, and it represents an accumulation at the surface of the body of slime materials extruded by the myxamoebae within. According to Loomis (1975) the sheath contains mostly carbohydrate, including various sugars, a limited amount of proteinaceous materials, and substantial cellulose, the presence of which had been demonstrated earlier by staining (Raper and Fennell, 1952) and EM studies (Hohl and Jehli, 1973) of the collapsed and empty sheaths.

In a saturated atmosphere the slime sheath surrounding an actively moving pseudoplasmodium is slightly viscous but offers no apparent resistance to or sustains any evident damage from a fine glass needle or a microscalpel being passed through it. In fact a migrating pseudoplasmodium can under similar conditions be impaled on the agar by three or four needles of this type without noticeably slowing its forward movement. Thus the sheath that envelops an actively migrating pseudoplasmodium must be essentially fluid in character. If exposed for a few minutes to the dry air of the laboratory, however, the envelope becomes sufficiently tough so that a needle brought in contact with the side of a pseudoplasmodium noticeably depresses the body before the envelope breaks. Upon longer exposure the envelope becomes increasingly tough and will completely destroy the pattern of the pseudoplasmodium before it gives way. Although the envelope is noncellular and nonliving, it obviously plays a profound role in governing migration and allied phenomena. In fact the cessation of migration and the initiation of sorocarp formation can be brought about by reducing the fluidity of this envelope to a critical point (Raper, 1940a). Since the presence of cellulose was demonstrated in collapsed slime sheaths, or "slime tracks," it may have polymerized at a relatively late stage. In contrast to nascent sheath material, the desiccated slime tracks possess considerable strength (Francis, 1962) and in our experience cannot be rehydrated.

Whether all cells in a migrating body contribute to the surface sheath is not known, but the sheath is reported to accumulate at a graded level from front to rear of the pseudoplasmodium (Farnsworth and Loomis, 1975). We know that it is continuous over the whole body and stationary with regard to the substrate except for a narrow apical zone. This was first demonstrated many years ago by placing particles of carbon black on the upper surface of a migrating body and marking the positions of these on the agar beneath (Raper, 1941a). After the pseudoplasmodium had migrated for some distance and the slime sheath had collapsed behind it, the particles, except for those placed at the very tip, occupied the same positions relative to each other as they had when first deposited. Such studies were later confirmed by Bonner and other investigators. Francis (1962) in particular studied the displacement of markers (*Bovista* spores) mounted on the tips of migrating structures as a means of relating sheath origin and accumulation to pseudoplasmodial movement; Shaffer (1965) used similar methods (*Lycopodium* spores) for tracing the movement of underlying cells during pseudoplasmodial migration and in progressive stages of sorocarp construction. To what degree the slime sheath regulates pseudopod formation by the myxamoebae within, and hence pseudoplasmodial migration as suggested by Garrod (1969), is, we believe, still open to question, as is the definitive role in sorocarp formation attributed to it by Farnsworth and Loomis (1975).

Organization and Behavior

The migrating pseudoplasmodium is clearly a unit structure, and in responding to external stimuli it behaves in many ways as a true metaphyte. Yet its unity is very tenuous, for it may readily split into two smaller but functionally complete units; conversely, two migrating pseudoplasmodia of quite different origins may meet and coalesce to form a single larger structure (Fig. 8-3). In either case normal fructifications are subsequently produced that seemingly differ only in their dimensions. In pseudoplasmodial splitting the bifurcation is thought to be caused by a slight stiffening of the secreted slime in an apical position so that myxamoebae immediately behind this push out laterally on either side. In doing so they set up competing "directive centers" that rend the parent structure into two parts, which may but need not be of equal dimensions. Coalescence of previously separate migrating pseudoplasmodia is facilitated by initial contact and merger of their apices, which is then followed by longitudinal fusion of the entire bodies (Fig. 8-3E).

Migrating pseudoplasmodia may vary greatly in dimensions, not only with respect to total mass but in comparative proportions as well. While usually described as cartridge-shaped, they may be short and relatively thick to long and thin, the latter often being sinuous in outline (Fig. 8-1). Irrespective of size or form, they are strongly polarized with more or less pointed anterior ends (tips) that receive stimuli ("receptive centers") and from which the movements of the whole pseudoplasmodia are controlled ("directive centers"). The nature of the centers remains an enigma, and it is not known whether the two functions are performed by the same or closely coordinated corps of cells. What is known is that they reside somewhere within a small apical fraction ($\frac{1}{10}$ or less) of the total mass (Raper, 1940a; Rubin and Robertson, 1975). Accumulating evidence would seem to indicate that cells of the apical fraction acquire their dominant position by some method of "sorting out," and that this could result from or be associated with demonstrable differences among the cells. Bonner (1959c) observed that the anterior cells are somewhat larger than the more posterior ones, and Takeuchi (1969) reported that the anterior cells are denser than the light posterior cells which become spores, while Leach, Ashworth, and Garrod (1973) found that behavioral differences could be correlated with the composition of the substrates in which the cells had grown and might possibly be attributable to levels of cAMP emission. However accomplished, the phenomenon of cell realignment has occupied the attention of many investigators, and its reality is now generally conceded; albeit working with the migrating stage and using acridine orange adsorbed on cellulose particles as an ingestible marker, Farnsworth and Wolpert (1971) found no evidence of sorting out in grafted pseudoplasmodia after 4-6 hours.

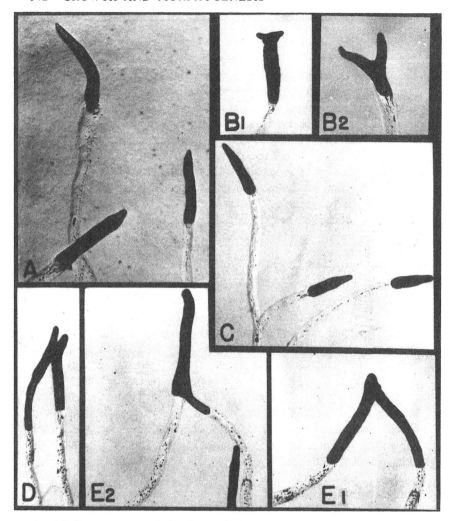

FIG. 8-3. Normal behavior of migrating pseudoplasmodia of *Dictyostelium discoideum*. *A*. Typical migrating bodies. *B1*. Pseudoplasmodium beginning to divide. *B2*. The same ½ hour later. *C*. Two daughter pseudoplasmodia migrating normally following dichotomous division. *D*. Contact without fusion. *E1*. Two pseudoplasmodia of different origin in contact at apices. *E2*. The same after fusion is nearly complete. × 12.5. (From Raper, 1940b)

Much has been written about cell sorting, tip dominance, and pseudo-plasmodial patterning in the years since 1940, and relevant papers include those of Bonner (1952 and 1957), Bonner and Adams (1958), Bonner et al. (1971), Francis and O'Day (1971), Müller and Hohl (1973), Yamamoto (1977), Forman and Garrod (1977a,b), Takeuchi et al. (1978), Inouye and Takeuchi (1979), Feinberg et al. (1979), Durston and Vork (1977, 1979), Matsukuma and Durston (1979), MacWilliams and Bonner (1979), Smith and Williams (1980), Sternfeld and David (1981a), Hayashi and Takeuchi (1981), Stenhouse and Williams (1981), Kopachik (1982a,b), and Gregg and Davis (1982). Of especial interest is a second paper by Sternfeld and David (1981b) wherein they reported using the vital dyes neutral red and nile blue sulfate as contrasting cell markers and concluded that "pattern formation in *Dictyostelium* consists of two processes: establishment of appropriate proportions of two cell types, and establishment of the pattern itself by the mechanism of sorting-out.'' In this they seemingly concur with Tasaka and Takeuchi (1981), who, using autoradiography and im-munochemistry, found that cell sorting was not a prerequisite for cell differentiation, although sorting out brought about an accumulation of prespore cells in a hemisphere to produce a prestalk-prespore pattern within the aggregate.

Migrating pseudoplasmodia are strongly phototactic, and their response to light is particularly striking. When cultures are grown in one-sided illumination, the pseudoplasmodia migrate toward the light by moving with their long axes roughly parallel with the beam and their apices pointed toward the source (Fig. 8-4A). If the position of the light is changed, the apical portion promptly turns toward it, and further migration is in that direction. If the light is shifted 90°, a pseudoplasmodium makes a right-angle turn (Fig. 8-4B) and subsequently pursues a course perpendicular to its original path; and two successive changes in the direction of illumination lead to responses as shown in Fig. 8-4C1-C3. Finally, if the light is shifted 180° so that it strikes a pseudoplasmodium from behind, it will continue migrating in the same direction until the apex veers to one side sufficiently to be influenced, whereupon the pseudoplasmodium makes a U-turn and so reverses its direction of migration (Fig. 8-4D). In each case the response is unmistakably initiated by the apical tip. Such experiments, however, do not reveal if the turns stem from an enhanced receptivity by the tip, or if perchance the tip is the only site where the slime envelope is sufficiently fluid for the cells within to shift their direction of movement.

Pseudoplasmodial Grafts

More positive evidence for the controlling influence exerted by the apical fraction may be obtained in other ways (Raper, 1940a). When the anterior portion of a pseudoplasmodium is removed, the myxamoebae comprising the decapitated mass (now void of a directive center) usually continue to move forward and crowd together into a rounded mass at the site of excision and, as a rule, build a sorocarp immediately. Less commonly a pseudo-plasmodium thus decapitated forms a secondary apical tip and resumes migration. In marked contrast to this behavior, the excised anterior portion when replaced on an agar plate regularly migrates for a period of time, depending largely upon temperature and other environmental factors.

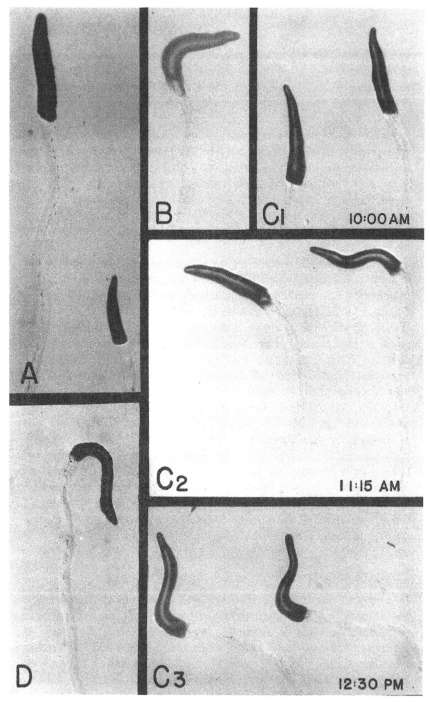

Fig. 8-4. Influence of light upon migration of pseudoplasmodia. *A*. Two pseudoplasmodia migrating toward fixed light source. *B*. Pseudoplasmodium responding to 90° shift in direction of light (1¼ hour interval). *C*. Two pseudoplasmodia showing uniform response to two shifts in direction of light; *C1*, light straight ahead; *C2*, 1¼ hour after light shifted 90° to left; *C3*, 1¼ hour after light returned to original position. *D*. Pseudoplasmodium 1½ hour after direction of light shifted 180°. × 15. (From Raper, 1940a)

The governing character of apical portions of pseudoplasmodia can be shown even more strikingly by experiments such as that illustrated in Fig. 8-5. To the sides of a typical migrating pseudoplasmodium the anterior portions of three additional pseudoplasmodia of approximately equal size were grafted, care being taken to establish intimate contact between the myxamoebae comprising the added fractions and the body to which they were applied. Within 15 minutes those portions of the original pseudoplasmodium immediately behind the added apices gave evidence of responding to their direction, and in less than an hour the identity of the original pseudoplasmodium was lost; in its stead were four daughter pseudoplasmodia, each characterized by its own organization. The original pseudoplasmodium was clearly a functional unit directed and controlled by a single directive center. The influence of this center, however, decreased progressively away from the apex, and the additional apices, being closer, exerted a stronger influence upon myxamoebae posterior to them than did the directive center to which they had formerly responded. Four pseudoplasmodia resulted that could in no way be distinguished from normally formed pseudoplasmodia of similar size.

The multiple graft experiment suggested also that the polarity of applied tips and recipient bodies might be especially significant. This was then confirmed in various ways. The directive center or tip of one pseudoplasmodium was found to be capable of dominating the much larger mass of a second pseudoplasmodium of similar polarity from which the original center had been previously removed (see also Smith and Williams, 1980). Would the original center when removed from its normal position but kept

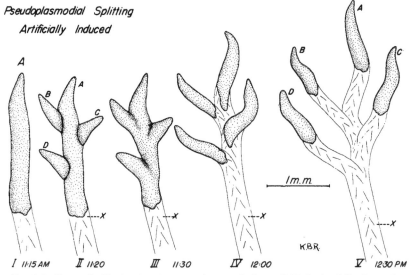

FIG. 8-5. Camera lucida sketches demonstrating pseudoplasmodial behavior following multiple tip grafts. A. Normal pseudoplasmodium. B,C,D. Apical fractions of foreign pseudoplasmodia grafted on A. (From Raper, 1940a)

in contact with the pseudoplasmodial body at some other site retain or lose its control? Or stated differently, which is of greater importance in the control of a pseudoplasmodium, the identity or the polarity and position of a directive center?

These questions were answered by experiments such as that illustrated in Fig. 8-6. Complementing fractions of different origin but like orientation coalesced to produce a single united migrating unit, whereas two fractions of like origin but of reverse orientation behaved quite independently. Thus it was seen that the prior identity of the directive center was of lesser consequence, whereas its polarity and position with regard to the recipient body were of prime importance (see also Bonner, 1950).

Other experiments showed that complementary portions of different pseudoplasmodia could be grafted experimentally and that the resulting pseudoplasmodia would subsequently behave in a normal manner. This was demonstrated by utilizing to particular advantage a striking character of *Dictyostelium discoideum*, namely the red coloration of its pseudoplasmodia when cultivated in association with *Serratia marcescens* upon selected substrata. In such cultures the myxamoebae appear red when massed in a pseudoplasmodium; and even separately they can be identified by the presence of persistent red vacuoles (Raper, 1937, 1940a). Apical fractions were removed from colorless pseudoplasmodia (Fig. 8-7A) produced in cultures with *Escherichia coli* upon lactose-peptone, dextrose-peptone, or

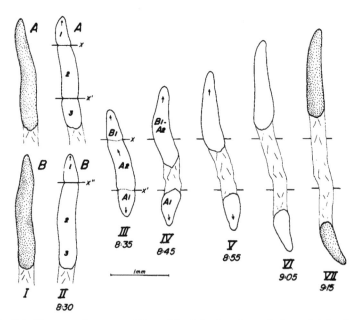

FIG. 8-6. Diagrammatic representation of the behavior of pseudoplasmodial fractions as influenced by inherent polarity vs. prior identity. (From Raper, 1940a)

carrot-peptone agar; and in their stead were immediately placed comparable fractions of red pseudoplasmodia (Fig. 8-7B) produced in cultures with *Serratia marcescens* upon buffered carrot-peptone agar (Chapter 4). In each case the orientation of the added red fraction coincided with that of the colorless body to which it was applied. Whenever a foreign anterior fraction was placed in close contact with a decapitated body without being crushed or otherwise mutilated, the two fractions promptly merged, such union being facilitated by a brief cessation of movement by the transplanted fraction and the continued forward advance of the decapitated body. The united pseudoplasmodium normally continued migration for a period and then formed a sorocarp characteristic in size and shape but unique in coloration (Fig. 8-7A), the stalk being pigmented as the apical part of the migrating structure and the sorus and basal disc as the posterior portion.

The myxamoebae of the two original fractions did not intermix to an appreciable degree, and a line of demarcation remained clearly visible

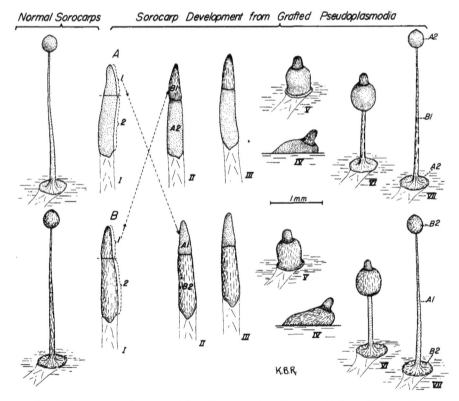

FIG. 8-7. Diagrammatic representation of sorocarp formation by grafted pseudoplasmodia. *A*. Colorless pseudoplasmodium (stippled) from culture with *Escherichia coli. B*. Red pseudoplasmodium (striated) from culture with *Serratia marcescens*. Apical fractions of the grafted pseudoplasmodia (*A1* and *B1*) form stalks, while nonapical fractions (*A2* and *B2*) form the sori and basal disks of bicolored sorocarps. (From Raper, 1940a)

during several hours of migration. The added red portion remained continuously in an apical position and obviously directed the movement and conduct of the whole mass, and in the end governed the building of the sorocarp and *contributed directly* to the formation of the stalk (Fig. 8-7A). In colorless pseudoplasmodia migration normally seems to merge almost imperceptibly into sorocarp formation, but in these grafted structures significant changes could be observed during this transitional period. The red anterior portion gradually ceased forward movement and became raised above the agar surface while the myxamoebae comprising the main body of the pseudoplasmodium continued to move forward, crowding around the colored anterior part. This came to occupy an axial position in the bulbous mass of colorless myxamoebae, projecting above it as a rounded, nipple-like tip and continuing downward into the center of the mass. The myxamoebae comprising the lower portion of this column then became vacuolate and compacted together forming the stalk initial. Subsequently the stalk was built upward by the progressive vacuolation of red colored myxamoebae in the axial region. Meanwhile the basal disc was formed by the similar vacuolation of colorless myxamoebae surrounding the base of the stalk and in contact with the substratum. As the stalk lengthened the main body of colorless myxamoebae ascended it en masse, the while becoming differentiated into spores progressively from the periphery of the mass toward its center. When all the myxamoebae had become transformed either into stalk cells or spores, the fruiting structure was complete. The red coloration of the stalk confirmed its origin from the anterior portion of the migrating pseudoplasmodium, while the absence of color in the sorus and in the basal disk indicated their origin from the posterior portion of the same (Fig. 8-7A).

In other experiments anterior fractions of colorless pseudoplasmodia were applied to previously decapitated bodies of red color (Fig. 8-7B). As expected, the results were the reverse of the preceding; but in general the bicolored character of the sorocarps was less striking (Raper, 1940a). In still other experiments red apical tips were placed against colorless truncated bodies but with unlike orientation. When light acted to draw the red tip more tightly against the decapitated pseudoplasmodium, the tip usually burrowd into the larger cell mass, soon assumed an apical position, and later took over direction of the combined structure, yielding a sorocarp with pigmented stalk and white sorus. When a foreign tip was placed directly behind a decapitated pseudoplasmodium of like orientation, but illuminated 90° from the axis of prior movement, the apical tip turned independently, moved toward the light, and, following a period of migration, constructed a small and uniformly red-pigmented sorocarp (Raper, 1940a).

The early experiments related above have been repeated and confirmed by several investigators, including Bonner and associates (1952 et seq.),

Gregg (1966 and 1967), Yamamoto (1977), Durston (1976), Durston and Vork (1979), Smith and Williams (1980), Sumino (1981), and others. In their work, however, vital stains such as nile blue sulfate, neutral red, or Bismarck brown have been used more often than *Serratia* to color pseudoplasmodia for grafting experiments, or to trace more directly the movement and fate of cell groups or clusters during pseudoplasmodial migration and sorocarp construction. Other methods have been used also to distinguish between prestalk and prespore cells, including selective staining for alkaline phosphatase and nonstarch polysaccharides (Bonner, Chiquoine, and Kolderie, 1955), immunological techniques (Gregg, 1965; Takeuchi, 1963; Takeuchi et al., 1977; Hayashi and Takeuchi, 1981), and electron microscopy of cell structures (Hohl and Hamamoto, 1969; Maeda and Takeuchi, 1969; Gregg and Badman, 1970; Müller and Hohl, 1973; and Yamamoto, Maeda, and Takeuchi, 1981). The last of these approaches has been especially rewarding because myxamoebae destined for spore formation contain a special kind of identifiable vacuole, designated prespore specific vacuole (PSV). Interestingly enough, these vacuoles do not appear in cells until sorocarp formation is imminent; and in small fruiting fragments, or isolated cells from pseudoplasmodia, they disappear as cells earlier committed to produce spores shift from the spore-forming pathway to become prestalk cells during terminal differentiation. Not surprisingly, cells first destined to form stalk cells but later needed for spore formation develop PSV vacuoles prior to differentiation (Sakai and Takeuchi, 1971; Takeuchi, 1972; and Sakai, 1973).

Fruiting by Parts of Pseudoplasmodia

This brings us to one of the more interesting and perplexing properties of *D. discoideum* and the other dictyostelids, namely the retention of a capacity by the myxamoebae to adjust and realign themselves to meet untoward contingencies that arise during morphogenesis. For example, so long as the cells remain amoeboid, if a fruiting organization is disassociated so that the myxamoebae are separated one from another they will in the presence of bacteria return to the vegetative stage, or in the absence of such nutrient they will reassemble, usually forming many small aggregations. If the disruption is less severe so that the overall organization is destroyed, but in limited fractions the still-amoeboid cells retain their previous contacts, they will generally transform themselves into small but nonetheless functional units and in due course produce sorocarps of normal proportions, the only exception being an apical fraction. These responses can be demonstrated in various ways, including that first published by Raper (1940a).

Young pseudoplasmodia approximately 2.0 mm long × 0.25 mm wide were each cut transversely into four approximately equal parts. The fractions were then removed separately with as little disturbance as possible and placed on fresh agar plates in previously marked positions, the fractions being designated 1 to 4 consecutively from anterior to posterior ends of the original pseudoplasmodia. The fractions were followed for a period of 24 hours. In no case did a nonapical fraction (2,3, or 4) resume migration. These rounded up promptly and soon began forming normal sorocarps with well-proportioned sori, evenly tapered sorophores, and flattened and expanded basal discs. No consistent difference was observed between sorocarps formed from fractions 2, 3, or 4 (Fig. 8-8) or between any of these and sorocarps of similar size formed from entire pseudoplasmodia. Most of the anterior fractions (1), however, resumed migration immediately, and many split dichotomously one or more times, giving rise to smaller fractions and eventually sorocarps. Sorocarps formed from apical fractions immediately after transfer were quite abnormal. When fractions migrated from 3 to 6 hours, the resulting structures showed relatively small sori, comparatively heavy stalks, and poorly formed, cone-shaped basal disks. Finally, when like fractions migrated for 18-24 hours before producing sorocarps, the resulting structures were essentially normal in pattern with well-formed sori, relatively thin stalks, and expanded basal disks (Fig. 8-8). As the apical fractions continued migration, a realignment and reorganization of myxamoebae was constantly and progressively taking place. From a body of myxamoebae formerly specialized for stalk production only, a seemingly normal ratio was eventually reestablished between stalk-forming and spore-forming myxamoebae; and a like relationship was also restored between the directive center and the body subject to it. The result was the emergence of a balanced organization, a new whole which except for size was not unlike the parent structure.

The nonapical fractions, on the other hand, presented a strikingly different picture, and it is noteworthy that their behavior was basically the same whether the bulk of a pseudoplasmodium was involved or only a small fraction of one. Nonapical fractions rarely established new centers or tips, able to initiate and direct migration. Yet they remained capable of instituting promptly the production of fruiting structures that appeared normal in pattern and generally of a size consistent with the mass of the fraction involved. More recently, J. Sampson (1976) has reported that the "stalky" character of fruits derived from front quarters persists even if the cells are dissassociated and hence have to reaggregate before fruiting. Additionally, he found that fruits derived from rear quarters had disproportionately larger sori, and that this effect became more pronounced with increasing age of the original pseudoplasmodia.

The migrating pseudoplasmodium is characterized by marked functional and regional specialization, which in turn depends upon a continuity of organization in the migrating body and the maintenance of particular spatial relationships among the myxamoebae comprising it. When the organization

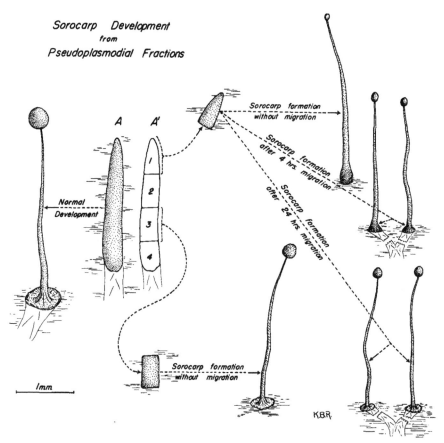

Sorocarp Development
from
Pseudoplasmodial Fractions

FIG. 8-8. Comparison of sorocarps formed from an entire pseudoplasmodium and from different fractions of one. All fractions except the apical one form normal sorocarps of reduced size. Only the apical fraction migrates (with or without splitting), and the degree of normality of the sorocarp(s) subsequently produced correlates with the period of migration. (From Raper, 1940a)

is disrupted the myxamoebae of the affected part set about immediately to establish new relationships, and the character and behavior of the new or modified organization will be determined largely by the portion of the pseudoplasmodium involved and the extent of disruption sustained.

Taken together, the observations cited would seem to indicate that cells in the excised apical fraction "sensed" the removal of the larger posterior part of the original pseudoplasmodium, and during the continued migration of that fraction the cells within made adjustments accordingly. This was accomplished by the gradual return of the more distal cells to a primitive state, whence they developed prespore specific vacuoles and other characteristics of spore-forming cells. This of course entailed a progressive reduction in the cells previously committed to stalk formation, since the finished sorocarp became more and more normal as the time of migration increased, up to about 18 hours. The situation with isolated nonapical fractions was quite different. Composed almost wholly of prespore cells at the outset, these fractions nonetheless rounded up, developed apical tips, and very soon initiated sorocarp formation. Since the fruiting process then proceeded at a normal rate and since the final products were of normal pattern and proportions, it would seem that the still-amoeboid prespore cells comprising the fractions had retained their potential for alternative differentiation and that a new balance had been established between prestalk and prespore cells in the absence of any migration.

Differing in purpose, but not unrelated, are a series of ingenious experiments performed by Inouye and Takeuchi (1980). Migrating pseudoplasmodia, and parts thereof, were transferred to agar capillaries attached to a manometer that enabled them to record motive forces. That of the anterior part was found to be far greater than that of the posterior part, albeit their weighted mean was approximately equal to that of an entire pseudoplasmodium. It is significant that after several hours, and concurrently with the conversion of cell types between prestalk and prespore cells, the motive force of either fraction regulated to reach the normal value for an intact pseudoplasmodium.

Culmination in *Dictyostelium discoideum*

The dictyostelid sorocarp is by any standard a special type of multicellular structure, for it arises not from the proliferation of a single cell (vegetative or zygotic) but by the progressive integration and subsequent differentiation of previously free-living cells. Using *Dictyostelium discoideum* as a basis for discussion, we find that, regardless of their dimensions, the sorocarps are basically similar in pattern when formed under optimal conditions. Each consists of an upright, tapering sorophore rising from an expanded basal disk and bearing at its apex a spherical to slightly citriform sorus (Fig. 8-9). The essential thing, then, is not the number of myxamoebae that make up the sorogen, but how the cells are organized to effect the divergent and still proportional differentiation required in sorocarp construction. Answers to this question have commanded the attention of investigators since van Tieghem's pioneering studies on *Acrasis* a full century ago (1880). Our comprehension is still woefully incomplete. Still, much has been learned, and it is now possible to relate in considerable detail what seems to occur even though we cannot understand why it should be so. Whereas outward manifestations vary among different species of *Dictyostelium* and between the different genera, the underlying steps in sorocarp formation or *culmination*, as Bonner (1944) aptly termed the process, are basically similar in all dictyostelids. Nevertheless, in considering culmination, we shall once again start with *D. discoideum*, for it has been studied in far greater detail than any other species.

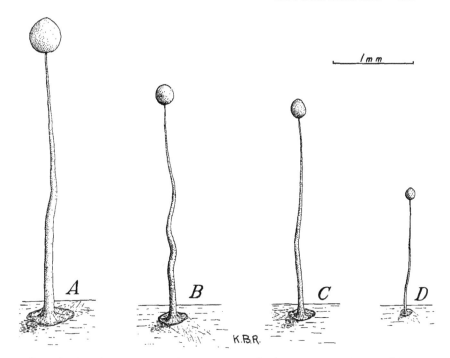

I m m

FIG. 8-9. Normal sorocarps of differing dimensions. In all cases under optimal conditions the parts of the structures—discoid bases, tapering sorophores, and sori—retain essentially the same proportions irrespective of size. (From Raper, 1937)

The first evidence of sorocarp formation consists of a rounding up of the pseudoplasmodial body and a shift of the apical tip from an anterior position to a vertical one, where it assumes the form of a definite papilla surmounting the body of cells (Fig. 8-10*A-C*). Shortly thereafter a delicate, hyaline membrane in the form of a short tube appears in the central position of the papilla. Following this the vertical axis of the cell mass decreases appreciably, thus reducing the distance the stalk initial must bridge. This downward shift, occasioned by a flattening of the whole mass, soon brings the lower end of the tube near to or in contact with the substrate, a stage that simulates in appearance a "Mexican hat" and is often so referred to in published papers (Fig. 8-10*D-F*). There is at this time no apparent differentiation of myxamoebae either within or outside the hyaline tube, of which an extricated example is shown in Fig. 8-11. Even so, this delicate structure may be appropriately called the "anlage" of the *sorophore sheath* or *tube*, since in situ it delimits cells that will later contribute to the formation of the basal portion of the stalk, or sorophore. The deposition of the sorophore sheath is of primary significance in sorocarp development, for it is in a very real sense the product of the whole community of coordinated cells acting in concert; and, under optimal conditions, its dimensions are correct for the total mass of interacting myxamoebae. Quite

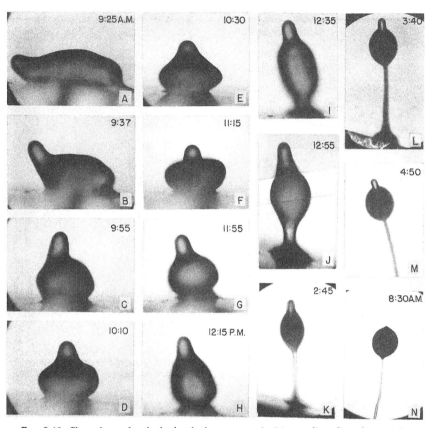

Fig. 8-10. Chronology of a single developing sorocarp in *Dictyostelium discoideum*. *A,B*. End of migration. *C,D*. Assumption of vertical orientation. *E,F*. Initiation of stalk formation. *G,H*. Beginning of sorophore elongation and basal disk formation. *I,J*. Initiation of sorogen ascent. *K*. Beginning of spore maturation. *L,M*. Continued spore maturation. *N*. Mature sorus—culmination having been completed the previous evening. *A-J* × 40; *K-N* × 26. (From Raper and Fennell, 1952)

remarkably it is laid down extracellularly and in just the right circular locus to achieve the proper proportional relationship between the sorophore and the body of cells to be supported during culmination.

The sorophore sheath is composed largely of cellulose. At this early stage it gives a typical blue color when stained with chloroiodide of zinc and shows strong birefringence when viewed with polarized light. The sheath and wall materials of mature sorocarps exhibit additional characteristics of true cellulose (Raper and Fennell, 1952; Gezelius and Ranby, 1957). Apparently cells adjacent to the critical site secrete enzymes which synthesize cellulose outside their own bodies, for the sheath forms no part of the wall of any cell. Whether such secretion occurs from within or outside the sheath at *this early state* has not been determined. Viewed under high magnifications with polarized light, or as seen in stained preparations with ordinary light, the young sheath appears to consist of a loose network of delicate, interlacing fibers oriented in the direction of the sorophore axis (Figs. 8-11, 8-12*D*). As the sheath lengthens, but prior to any obvious differentiation of cells within it, this network of cellulosic fibers contracts laterally, forming a continuous and compact cellophane-like tube. At the same time the uppermost, funnel-like portion, which represents the area of further sheath formation and extension, continues to exhibit the loose fibrous character that marked the earliest evidence of its development. After the sorophore contacts the substratum it begins to lengthen upward, and the mass of myxamoebae begins its slow ascent (Figs. 8-10*G-I*, 8-12*A,B*). The slime envelope, which continues to surround the whole body at this stage, now begins to invaginate near the base and in so doing isolates a somewhat conical mass of cells that becomes the basal disk (Fig. 8-13). As the much larger mass of undifferentiated cells moves upward, in part by their amoeboid movement and in part from the lifting effect of the lengthening stalk, a line of demarcation appears between cells that will form the sorus (prespore cells) and those that will extend the sorophore (prestalk cells). The former represent the main body of the mass, whereas the latter comprise a smaller body of cells that forms the apical papilla and extends down into the shoulder of the rounded mass

FIG. 8-11. Isolated stalk tube initial in *Dictyostelium discoideum*. Note the absence of any attached cells and the apparent fibrous nature of the cellulose. Photographed using polarized light. × 250

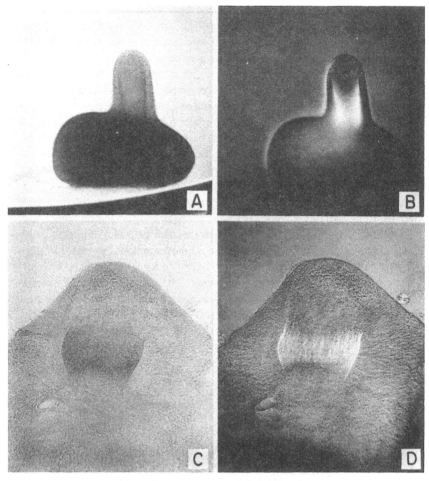

FIG. 8-12. Initiation of stalk formation in *Dictyostelium discoideum. A,B.* Intact fruiting masses fixed, stained, and photographed without distortion under normal and polarized light, respectively. × 70. *C,D.* Apical region at very early stage stained with iodine chloroiodide of zinc and somewhat compressed; photographed under normal and polarized light, respectively. × 150. (From Raper and Fennell, 1952)

(Fig. 8-14). Although cells in both areas are still amoeboid and plastic, differences in orientation are clearly evident. The prestalk cells show a generally horizontal orientation. The prespore cells in the lower portion of the mass adjacent to the sorophore show a diagonal to vertical orientation, whereas those in the uppermost region are more nearly isodiametric and show no consistent orientation.

As the mass of fruiting myxamoebae separates from the basal disk an important question arises: How are the prespore cells raised during culmination? About this there are differences of opinion. Brefeld (1869,

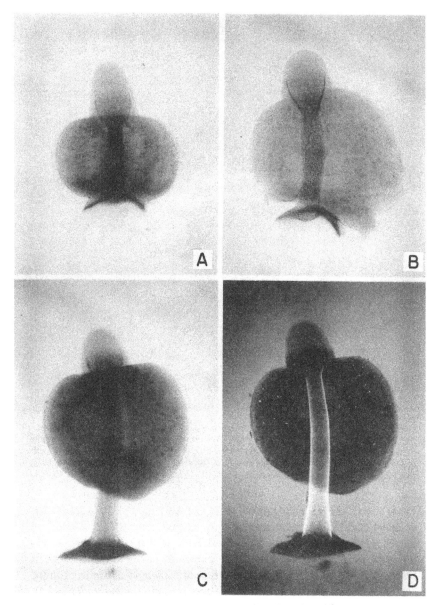

FIG. 8-13. Origin of the basal disk in *Dictyostelium discoideum*. Young sorocarps killed and stained with aceto-carmine. *A*. Slight invagination of slime sheath. *B*. Invagination has reached the sorophore, cutting off cells that differentiate as basal disk. *C,D*. Still later stage showing complete separation of sorogen as it begins its ascent. *A-C*, normal light; *D*, polarized light. × 75. (From Raper and Fennell, 1952)

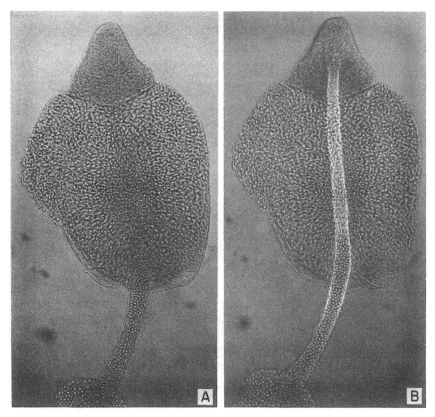

FIG. 8-14. Young sorocarp of *Dictyostelium discoideum* killed and stained with iodine-potassium iodide followed by zinc chloride. *A*, normal light; *B*, polarized light. Note the sharp separation of prestalk (above) and prespore cells (below) and the orientation of cells in each area. The slime envelope, still present at this stage, is distended and evident in basal area. × 260. (From Raper and Fennell, 1952)

1884), Olive (1901), and Harper (1926), all working with *D. mucoroides*, believed that the sorogen arose simply as the result of the cumulative amoeboid movements of the assembled cells. In that species such an explanation is probably adequate until a mid- to late-culmination state is reached. It is probable also that coordinated amoeboid movements in the central and basal portions of the young sorogens of *D. mucoroides* and *D. purpureum* constitute a substantial "lifting force." Likewise, it is probable that such movements contribute materially to sorogen uplift in *D. discoideum* during the period when a large proportion of the prespore cells are still amoeboid. Such a view is supported by the orientation of myxamoebae in the lower central area of sorogens such as that shown in Fig. 8-14. As culmination progresses, however, and as more and more of the prespore cells become transformed into firm-walled, immobile spores, it becomes increasingly doubtful if amoeboid movements of the remaining undifferentiated cells could constitute a sufficient motive force. This was the view expressed by Raper and Fennell (1952), who looked for and believed they identified other forces of equal or greater importance.

The theory of sorocarp construction then advanced was based upon a number of readily confirmable observations. (1) The enlargement of vacuolating stalk cells within a lengthening, firm-walled sorophore sheath (or tube), closed and resting on the agar below, created a substantial lifting force. (2) This force was exerted upon the still slime sheath-encased apical tip, and through cohesive forces between this and the body of prespore cells the latter was raised also. (3) Within the expansible funnel-shaped apical end of the sorophore tube the lower undifferentiated myxamoebae were arched upward as if under pressure from below (Fig. 8-15). (4) Prestalk-forming cells between the still intact portion of the slime sheath and the upper end of the sorophore tube were oriented in a diagonal to nearly horizontal direction, suggesting that cells in this area, albeit still amoeboid, were being pushed over and into the open end of the funnel, after which they differentiated as stalk cells. (5) The source of the latter pushing force, probably intermittent, was believed to stem from adjustments in the shape of the semi-fluid prespore mass as it changed form to attain a minimal surface area. This assumption has been challenged by Shaffer (1962) and likewise by L. S. Olive (1975), who cites as evidence

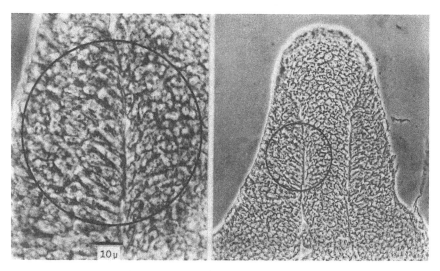

Fig. 8-15. Histological section through apical portion of a developing sorocarp of *Dictyostelium discoideum*. Right: overall view showing characteristic expanded terminus of stalk tube (cellulose) and the contrasting orientation of myxamoebae outside and at different levels within the tube. Left: the encircled area at right enlarged 3 × to show greater detail. (From Gregg, 1966)

an electron micrograph published by George, Hohl, and Raper (their Fig. 16, 1972).

When the full explanation of how the sorogen rises in *D. discoideum* is finally known, we suspect it will continue to involve a combination of remarkably coordinated myxamoebal movements together with substantial physical forces engendered by the dramatic enlargement of myxamoebae that differentiate as stalk cells within an open-ended but otherwise rigid sorophore tube. It is now generally recognized that the tube is formed by cells on the outside, namely the diagonally oriented myxamoebae that act, according to Bonner et al. (1955), "almost as a secretory epithelium."

The formation and functional role of the sorophore tube during culmination deserves more than passing comment. First, the deposition of cellulose outside the walls of cells is very rare in nature (see Raper and Fennell, 1952). Second, it is laid down *between cells* so that the tube is correctly placed and is of a diameter proportional to the size of the whole cell community, indicating a high degree of intercellular communication and response. Third, the synthesis and deposition of cellulose (or its precursors) as a nonrigid tube of one diameter and the subsequent condensation or polymerization of this material into a firm-walled tube of lesser *and* tapered diameter would seem to require continuing community control, albeit by a steadily decreasing body of cells. Fourth, and perhaps most remarkable of all, it is the sorophore tube, and not the entrapped myxamoebae, that determines the ultimate form assumed by the vacuolate stalk cells as they enlarge during differentiation.[1]

We have noted previously the nearly constant proportions of the sorocarps in *D. discoideum* irrespective of their size when these develop under optimal culture conditions. We have noted also that migrating pseudoplasmodia of different dimensions are strikingly similar in pattern and behavior. These two important characteristics are clearly illustrated in Figs. 8-1 and 8-16. Shown in the former figure are camera lucida drawings of 21 pseudoplasmodia taken from a single series of cultures and ranging from quite large to very small, while the latter figure contains camera lucida tracings of 5 typical sorocarps of different dimensions together with details of cellular composition at four marked levels in each structure. The similarity of sorocarp pattern independent of overall size is clearly apparent. Shown also in Fig. 8-16 is a second characteristic, not only of *D. discoideum* but of all dictyostelids, namely variations in cell shape as this relates to the diameter of the sorophore. This relationship stems, of course, from properties noted above: the dimensions of the sorophore sheath, or tube, are determined in some manner by the whole body of myxamoebae acting in concert, whereas individual cells within the tube assume whatever form

---

[1] Farnsworth (1973, 1974) has made the interesting observation that tipless masses of myxamoebae differentiate into stalk-like cells when these are placed in tubes of cellulose nitrate.

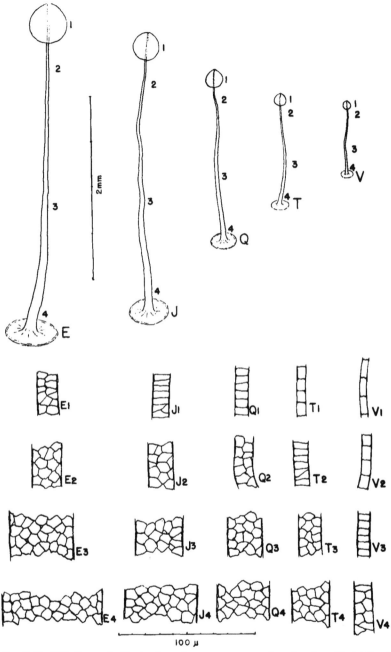

FIG. 8-16. *E-V*. Camera lucida sketches of mature sorocarps of varying size with details of their cellular composition. E_{1-4} through V_{1-4}. Detail of cellular patterns in sections corresponding to numbered positions on the sorocarps above. Note that the patterns of the sorocarps remain essentially constant irrespective of size and that individual myxamoebae in contributing to stalk formation assume whatever shape is necessary to attain this pattern. (From Raper, 1941a)

is required to fill the lumen, and by doing so give added strength to the lengthening sorophore. In *D. discoideum* some fairly definite relationships can be cited: if the sorophore exceeds 12-13 μm in diameter it consists in cross section of a number of essentially isodiametric cells, the number being proportional to the cross-sectional area of the stalk. In appearance they resemble the pith cells of many higher plants. If the tube is approximately 10 μm in diameter it usually contains a single tier of horizontally flattened cells. If it is 7-8 μm the cells are again essentially isodiametric, but in a single tier. In still thinner sorophores the cells appear as vertically elongate cylinders, one above the other, and their length is progressively greater as the diameter of the sorophore tube is reduced. Similar information concerning *Dictyostelium mucoroides* and *Polysphondylium violaceum* may be obtained from Harper's papers of 1926 and 1929, respectively. Although comparable analyses have not been published concerning other species insofar as we know, our observations indicate that similar relationships prevail throughout these genera.

In all the larger dictyostelids sorocarps vary greatly in size even in the same culture. Rich cultures normally contain substantially greater numbers of large sorocarps, but regardless of the number produced the dimensions of individual sorocarps do not exceed a certain maximum. For *D. discoideum* these would be about as follows: sorophores 4.2-4.5 mm in height, flattened basal disks 500-650 μ wide, and newly formed sori 400-500 μm in diameter. Sorocarps very rarely attain these dimensions, however. In contrast, there may be no minimal size. Working with a clone of strain NC-4, Konijn and Raper (1961) observed some very small sorocarps (as few as 15 cells) and illustrated one showing 14 or 15 stalk cells and about 70-75 spores (Fig. 7-6). Even earlier M. and R. Sussman had reported (1961) minute sorocarps of a "fruity" mutant (Fty-1) to contain as few as "12 cells or less," and had published photographs of one such structure containing 9 spores, 2 stalk cells, and a basal cell. The relative stability in the proportionality of parts exhibited by this species is truly amazing! For other species there are undoubtedly maximum dimensions as well, but these would be very difficult to catalogue because of the large variables introduced in most species by the concurrence of migration and stalk formation. In these also there would seem to be no minimal size (Bonner, 1959a; Bonner and Dodd, 1962a).

The role of cAMP in cell aggregation in *D. discoideum* has been profusely documented, and as early as 1949 Bonner was able to show that the anterior end of the migrating pseudoplasmodium of this species was a potent source of "acrasin" (now equated with cyclic AMP in *D. discoideum*). More recently various investigators have implicated cAMP as influencing cellular differentiation, if not, in fact, exercising a substantial measure of control over culmination. Of interest in this connection is a paper by Clark and Steck (1979) wherein they propose an orbital hypothesis for morphogenesis in *D. discoideum* in which intercellular AMP signaling emanates from the apical tip and operates throughout the developmental cycle. Although much additional work and confirmation is needed, their hypothesis of sustained circumferential cell movement may help to explain the observed behavior of the rising sorogen. Could it be that what appear as horizontally oriented cells in the papilla are in fact vertically compressed myxamoebae moving in near circular orbits?

If one is so inclined, a certain poetic justice can be detected in the life of *Dictyostelium*; for whereas the cells in the anterior tip govern migration of the pseudoplasmodium, and later seemingly control sorocarp formation, it is also the anterior cells that fashion the sorophore tube and are later entrapped by it. Once within the tube the fate of the cells is sealed as they gradually become more and more vacuolate and compacted together to form a parenchyma-like tissue that Brefeld, even in his account of discovery (1869), declared to be nonviable. This view has been contested only once (Sussman, 1951), and that objection was later shown to be in error (Whittingham and Raper, 1960). So in this organism one can say with assurance, "the last shall be first, and the first shall be last," for the posterior cells end up as propagative spores while the anterior ones become nonviable stalk cells. But enough of clichés. There is still the matter of when the stalk-forming myxamoebae lose their capacity to revegetate, and this is of some interest.

Working with *D. purpureum* for technical reasons, Whittingham and Raper (1960) showed that the cells within and near the apex of the sorophore tube retained their regenerative capability so long as they remained amoeboid. They lost this when they became conspicuously vacuolate but still not mutually compressed, or, stated differently, not until they had become partially differentiated as stalk cells. This point of no return was determined by (1) selecting prostrate stalk-forming sorogens with well-defined funnel-like sorophore tips (Fig. 12-14G); (2) gripping the protruding stalks with fine-tipped forceps and withdrawing them from the sorogens to expose the naked funnels; (3) using very thin glass rods to tease the still free cells from the expanded open ends (much as one would extrude the spores from an ascus of *Neurospora*); (4) separating the cells so they could be examined individually with the compound microscope at 440 ×; and (5) transferring single cells to small droplets of an *E. coli* suspension on nonnutrient agar with incubation at a temperature suitable for revegetation, if this capability still remained. In other tests stalks were broken by various means in an effort to free individual cells. In this we were quite unsuccessful, for the breaks generally occurred *through* the walls of cells rather than between them; and in no case did we obtain growth from any fragment of a mature stalk when this was freed of extraneous spores and myxamoebae.

Although we used *D. purpureum* because of the ease with which we could obtain expanded terminals of sorophores completely free of external spores or myxamoebae, we are convinced that similar cells extruded from the sorophores of *D. discoideum* would behave in a comparable manner. In fact, the general observations recounted above would, we think, be applicable to all species of *Dictyostelium* if it were possible to test them in an equally rigorous manner.

CULMINATION IN OTHER DICTYOSTELIA

Sorocarps of a different pattern are formed in *Dictyostelium mucoroides*, *D. giganteum*, and *D. purpureum* by modifications of the above processes. None of the species possesses a freely migrating pseudoplasmodium, and in each case sorocarp formation begins at the site of cell aggregation and is commonly initiated before aggregation of the myxamoebae is complete. As in *D. discoideum*, the first evidence of sorocarp construction is the deposition within a nascent sorogen of a centrally placed hyaline tube that heralds the beginning of stalk formation. No basal disk is formed. Sorophore elongation follows the course outlined above, namely the upward extension of the sorophore sheath that delimits a column of myxamoebae and the vacuolation and fitting together of these cellular units to form a rigid stalk which supports the rising sorogenic mass.

Marked similarities are apparent between responses of the sorogens of these slime molds and the migrating pseudoplasmodia of *D. discoideum*. Of these the most striking is the response to light, which is illustrated in Figs. 4-6 and 8-17. Whereas the pseudoplasmodium of *D. discoideum* moves freely toward the light for an indefinite period and then builds an erect sorocarp, the young sorogens of *D. mucoroides*, *D. giganteum*, and *D. purpureum* often build their sorophores toward the light for a like period before the initiation of spore formation. In side-illuminated cultures, a sorocarp starts to develop perpendicular to the substratum but soon shifts its orientation toward the light source (Figure 8-17*B*). The sorophore, which is relatively thinner than that of *D. discoideum*, becomes unable to support the mass of myxamoebae in an angular position and the body falls to the agar surface, where it continues to build in the direction of light and in contact with the substratum. Shortly before the beginning of spore maturation, however, its orientation shifts again, and further development is more nearly perpendicular to the culture medium and strikingly similar to that of *D. discoideum* (cf. Figs. 8-17*A* and *B*). In cultures incubated in darkness the general behavior of the young sorocarp is the same except that it is not built in any consistent direction unless governed by some other stimulus, such as temperature. In cultures incubated under full illumination sorophores are shorter and sorocarps are more nearly vertical in orientation.

A young sorogen of *D. purpureum*, *D. giganteum*, or *D. mucoroides*, as illustrated in Figure 8-17*B*, is structurally comparable to the migrating

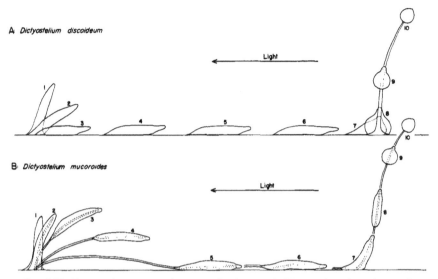

FIG. 8-17. Comparative patterns of sorocarp development in *Dictyostelium discoideum* and *D. mucoroides*. Note that the later stages (7 - 10) are essentially comparable in the two slime molds. (From Raper, 1941a)

pseudoplasmodium of *D. discoideum* except for the presence of the developing sorophore, for spore formation has not yet begun and the main body of myxamoebae remains undifferentiated (Raper and Fennell, 1952; Bonner, 1949). The behavior of their constituent myxamoebe is likewise comparable. If blocs of myxamoebae are separated from a young sorogen, each such mass will form a separate fruiting structure; if the myxamoebae are dissociated completely in the presence of bacteria, they will individually return to the vegetative stage; if they are similarly dissociated in the absence of bacteria, they will reaggregate into minute pseudoplasmodia and subsequently produce diminutive fruiting structures.

Whereas the sorogen of *D. mucoroides* consists of prespore and prestalk cells, the line of demarcation between the two remains fluid until sorocarp formation is nearly complete. This unsettled state stems from the fact that sorophore length in this species is indeterminate and allowance must be made for cells to move in whatever direction the cultural environment dictates. For a prostrate sorogen of *D. mucoroides* to extend its stalk beyond a limited distance requires that some cells which were previously committed to spore formation must lose their prespore characteristics (e.g., degrade prespore vacuoles, etc.) and take on the attributes of prestalk cells. That such a change does occur, and quickly (average 1.5 minutes), has been reported and strikingly illustrated by Gregg and Davis (1982; see also Gregg and Karp, 1978). Sorogens of *D. mucoroides* were labeled with L-(6-^{3}H) fucose and cell transformations followed by autoradiography, electron microscopy, and Richardson's stain to detect specific biochemical and ultrastructural changes that accompanied redifferentiation. Conversely, under altered circumstances cells with prestalk characteristics can revert to the prespore pattern and subsequently form spores.

Fructification in *D. dimigraformum* and *D. intermedium*, isolated and described by James Cavender from soils of Trinidad (1970) and Java (1976a) respectively, appears to be intermediate between that of *D. mucoroides* and *D. discoideum*. The sorocarps are basically similar to those of *D. mucoroides*, yet each species at times produces freely migrating pseudoplasmodia that are superficially indistinguishable from those of *D. discoideum*. In well-lighted cultures pseudoplasmodia of *D. dimigraformum* rarely migrate and sorocarps bearing yellow sori are generally erect or semi-erect; in one-sided light sorogens build toward the light and often move in contact with the substrate, forming continuous stalks as they advance; in complete darkness pseudoplasmodia are generally stalkless and seemingly move at random. As in *D. discoideum* culmination can be induced within an hour or two by exposures to overhead light. Pseudoplasmodial migration or extended sorophore construction is favored by temperatures of 20°C or lower.

In *D. intermedium* sorocarps tend to be long and flexuous, bear globose white sori, and so resemble those of *D. mucoroides*. Pseudoplasmodia fruit directly from sites of cell aggregation in bright overhead light but are induced to migrate in darkness or one-sided illumination. The colorless migrating bodies are superficially indistinguishable from those of *D. discoideum*, and during culmination a small conical mass of cells surrounding the base remains in contact with the substrate, suggesting, but in no sense duplicating, the basal disk of *D. discoideum*. As in that species pseudoplamodial migration is favored by unbuffered media of low nutrient content.

Stalkless migration of pseudoplasmodia may be seen occasionally in cultures of *D. purpureum* and even in some strains of *D. mucoroides*, but in neither species is it a character of special significance. Among cultures of *D. sphaerocephalum*, however, some isolates consistently produce migrating structures when incubated at lower than usual temperatures such as 15-18°C. Migration is always limited, and the pseudoplasmodia soon cease forward movement and culminate in sorocarps having short, heavy sorophores that rise in arcs from the substrate and bear relatively large sori. The species is closely related to *D. mucoroides*.

Sorocarps of *Dictyostelium rosarium* appear strikingly different from any of the above, yet they are formed in much the same way except for the manner in which prespore cells become distributed and subsequently differentiate (Raper and Cavender, 1968). As aggregation is completed the cells collect into irregularly spaced mounds, several being formed by the severance of inflowing streams. There is no stalkless migration. One or more papillae then appear on the surface of a mound and these quickly develop into low vertical columns, each of which develops its own centrally positioned sorophore initial. This is deposited and extended in the manner described for *D. discoideum*. As it lengthens, the elongate sausage-like sorogen moves upward by continued amoeboid movements and by the lifting forces generated by stalk cell enlargement. What happens next is without parallel in species such as *D. mucoroides* or *D. discoideum*. As the sorogen rises some distance above the substrate, posterior portions of this are periodically abstricted and remain as globular masses adherent to the stalk. Each such mass is at first surrounded by a slime envelope, but this soon disappears as the cells differentiate and the globose spores lie free, forming an unwalled sessile sorus. With time, repeated abstrictions occur so that one may examine on a single axis lateral sori in progressive stages of maturation (Fig. 12-38*H*). Such sori may or may not be evenly spaced or of uniform size, and may range in number from a very few to 50 or more on a single sorophore. True branching occurs frequently and is effected through splitting of the aerial sorogen in much the same manner as that of the migrating pseudoplasmodium in *D. discoideum*. Sorocarps of *D. rosarium* approximate those of *D. mucoroides* in general dimensions, and the myxamoebae aggregate in like manner in response to cAMP but with long intervals between pulses (Konijn, 1972).

Another striking variation is seen in *Dictyostelium polycephalum*, where there are true but very different migrating pseudoplasmodia and where synchronous culmination yields coremium-like clusters of sorocarps (Raper, 1956a). As cell aggregation nears completion a variable number of thin, elongate, and very tenuous migrating pseudoplasmodia develop at the primary center; and seemingly unaffected by light, these move outward across the agar surface. They are commonly 5-10 mm or more in length but seldom exceed 50-60 μm in diameter, hence are very fragile. Although they are surrounded by a very delicate slime envelope, they lack the continuing apical dominance that is so characteristic in *D. discoideum*. They may be straight or variously twisted, of uniform diameter throughout or somewhat nodular, and are easily severed. There is seemingly little coordinated movement of the body as a whole, and yet a pseudoplasmodium can move intact through a mycelial maze touching only three or four hyphae in its entire length of several millimeters (Fig. 12-27I). Upon cessation of migration the pseudoplasmodium changes shape to form a rounded mass. Soon this becomes papillate and beneath each of several papillae a diminutive sorocarp begins to form. The amazing thing is how these separate and seemingly divergent centers of sorophore construction effect a synchronous development, and how the several stalks are kept together as the process advances. Obviously, a considerable measure of interpopulational communication is required for the construction of the coremiform fructification. To a degree greater than in most species, optimal fructification in *D. polycephalum* is dependent upon a favorable physicochemical environment wherein the amount of atmospheric moisture is of special importance (Whittingham and Raper, 1957).

Other patterns of culmination are found in *Dictyostelium rhizopodium* Raper and Fennell (1967) and in *D. mucoroides* var. *stoloniferum* Cavender and Raper (1968). The first of these is representative of several species, primarily of Central American origin, that produce sorocarps with digitate bases believed to approximate the crampons described for *Coenonia* by van Tieghem a full century ago (1884). Unfortunately this cannot be fully confirmed, since his genus still awaits rediscovery; and our slime molds seem to fit his description in few other particulars. *D. rhizopodium* is the largest of the crampon-based species and it has been studied more thoroughly than other members of that group. Quite aside from any resemblance to *Coenonia*, the formation of such a sorophore base is in itself significant, and the manner in which it arises is most intriguing. Although still imperfectly understood, the developmental sequence, as worked out some years ago with the help of June Shaw, would seem to be as follows. The aggregating myxamoebae collect into a rounded conical mass much as in *D. discoideum*, but then, instead of first forming an open-ended cellulose tube, a hyaline structure believed to be of similar composition but in the form of a diminutive cap is laid down just beneath the subapical surface (Fig. 8-18*A*). Soon thereafter, pointed rhizoid-like projections in the form of lacunae extend downward through the cell mass until they reach the substrate (Fig. 8-18*B*). Now for the first time differentiating myxamoebae can be detected as enlarged, irregular, and strongly vacuolate cells that occupy distal positions within the cellulose-lined crevices (Fig. 8-18*D*). We do not know whether the rhizoid-like projections entrap cells at the time of their deposition or whether cells from the outside enter after these are formed. In any case, other myxamoebae adjacent to those already vacuolated now differentiate in a similar manner (Fig. 8-18*C,E*), and this upward transformation proceeds apace until all the cells within the "rhizoidal" structure have assumed a parenchyma-like pattern. Meanwhile, from the surface of the cap, which now appears to have opened, a tubular cellulose sheath begins to develop upward into the peg-like apex. From this point onward sorophore construction proceeds as in *D. mucoroides*. At first glance one might presume that the crampon of *D. rhizopodium* and the basal disk of *D. discoideum* are homologous structures differing only in pattern. But such is not the case. The latter develops not within the cell mass as just described, but from a residue of myxamoebae left at the base of the rising sorogen by the invagination of the slime envelope that completely covers the nascent sorocarp (see Fig. 8-13).

Dictyostelium mucoroides var. *stoloniferum* is no less interesting, but for entirely different reasons. When the sorus of *D. mucoroides* or *D. purpureum* collapses upon the agar surface and the spores remain together,

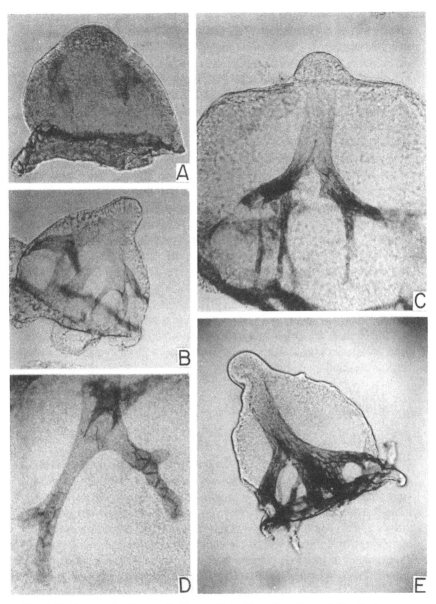

FIG. 8-18. Crampon development in *Dictyostelium rhizopodium*. Young fruiting bodies were killed, cleared in clove oil, and stained with chloroiodide of zinc. *A.* A very young stage showing the first evidence of stalk (crampon) formation. Cellulose is deposited in the form of a diminutive "umbrella." *B.* A later stage: a very short tube is now evident from which rhizoid-like projections in the form of lacunae extend downward to the substrate. *C.* An older stage, now showing a definite tube beneath the apical tip. *D.* Stage comparable to *C* somewhat enlarged; cell differentiation is beginning in the distal portions of the lacunae. *E.* A fully formed crampon, or holdfast. *A,B,C,E* × 225; *D* × 480

forming a dense circular film, they do not germinate. This is due to the presence of an autoinhibitor(s), for Russell and Bonner (1960) have shown that the percent of germination among spores of *D. mucoroides* varies inversely as the density/mm². In contrast, when a sorus of *D. mucoroides* var. *stoloniferum* touches the agar and the spores remain contiguous, these germinate en masse within a matter of 2-3 hours, and at 3½-4 hours the formation of small aggregates can be detected (Cavender and Raper, 1968). Each forms a separate sorocarp, and in groups up to ten or more these are spaced at approximately equal distances one from another. Three points of special interest may be noted: (1) the ability of the spores to germinate in the absence of any real dispersal; (2) the rapidity and uniformity with which the process of germination occurs, for few spores remain unchanged; and (3) the ability of the newly emerged protoplasts to effect aggregation without an intervening period of vegetation or substantial cell reorientation, which in *D. mucoroides* requires 3-4 hours and in *D. discoideum* about twice that time. The species name reflects the striking manner in which the secondary sorocarps arise from sites that are stoloniferous in their inception.

Other Dictyostelia possess distinctive patterns of culmination as well, and it is upon such bases that species differences are recognized. However, such differences are largely matters of scale, pigmentation, or spacing and need no special consideration here, since they will be considered in the species descriptions.

CULMINATION IN OTHER GENERA

Polysphondylium

A more complex pattern of culmination is presented by the genus *Polysphondylium*, of which two species are widely distributed in nature: *P. violaceum* Brefeld (1884) and *P. pallidum* E. W. Olive (1901, 1902). Of these *P. violaceum*, with pigmented sori, is the more robust and generally thrives under conditions that favor the larger species of *Dictyostelium*. *P. pallidum*, with colorless sori, is typically more delicate and attains optimal development in less rich cultures and under somewhat more acidic conditions. The latter slime mold will be given greater emphasis here because (1) it produces, on the whole, more uniformly symmetrical sorocarps than *P. violaceum*, (2) it has been studied more intensively (Kahn, 1964; Francis, 1965, 1975a, 1977, and 1979; Hohl et al., 1977), and (3) it can be cultivated readily on comparatively simple substrates free of bacteria if the investigator so desires (Hohl and Raper, 1963a,b; Sussman, 1963). Unlike *D. discoideum*, where presumptive prespore and prestalk cells are delineated in the migrating pseudoplasmodium, or *D. mucoroides*, where such a division becomes evident but less striking in the sorogen, there is virtually no demarcation in *Polysphondylium*. Instead stalks cells are continuously recruited by the transformation of prespore cells that occupy almost the whole sorogen (Bonner et al., 1955; Hohl et al., 1977). Considering the pattern of the resultant sorocarp it could hardly be otherwise.

Cell aggregation is triggered in *P. pallidum* by a founder cell (Francis, 1965) in the manner described and illustrated earlier for *P. violaceum* by Shaffer (1961b). Except that inflowing streams arise more abruptly, as a rule, and may tend to develop fewer interruptions as they converge, aggregation outwardly appears to follow the basic pattern we have described for *Dictyostelium*, although a dipeptide rather than cAMP is the effective acrasin.

Sorocarps of *Polysphondylium* may be regarded as typically erect or semi-erect in orientation. The sorogens are phototrophic, however, and in one-sided light some strains construct long sorophores toward the light and in contact with the substrate. Such behavior is especially true of *P. violaceum*. Branching does not normally occur unless the sorogen is "airborne," although this attitude is frequently discontinuous.

What outwardly distinguishes the genus are the whorls of side branches that lend the completed sorocarp a pattern simulating "a miniature pine tree," to use Harper's descriptive terminology (1929). The main axis of the sorocarp is constructed in the same manner as the stalk of *D. mucoroides* and involves the deposition of a sorophore sheath or tube of cellulose, followed by the vacuolation within it of myxamoebae to provide more rigid support. As the sorogen moves upward on the lengthening sorophore, however, small globular masses of myxamoebae are periodically abstricted from the posterior end of the main body of cells. These remain stationary on the sorophore and soon become vertically segmented. Each segment of a doughnut-like ring then develops into a lateral branch and together they form a characteristic whorl (Fig. 8-19). On a diminutive scale, each branch is formed in the same manner as the main axis and represents, in a sense, a secondary fructification anchored to the sorophore by congealed slime just as the sorophore is anchored to the substrate. Whereas numbers are in no sense fixed, it is not uncommon in *P. pallidum* for sorocarps to bear 10-12 or more whorls, of which the first one or two as a rule and always those nearest the top contain fewer branches and bear smaller sori than those in the central region. The sorocarp terminates in a single sorus that is typically larger than those on the branches below. Occasional isolates of *P. pallidum* may show branches bearing subbranches, a capacity that seems to diminish with continued laboratory cultivation. In contrast, in *P. filamentosum* Traub, Hohl, and Cavender (1981a), an uncommon species first isolated in Switzerland, branching at multiple levels and a virtual absence of terminal sori are the principal diagnostic characters. The base of the sorophore is slightly bulbous, just as in *D. mucoroides*, and may contain few to several cells in cross section, while the main shaft is lightly

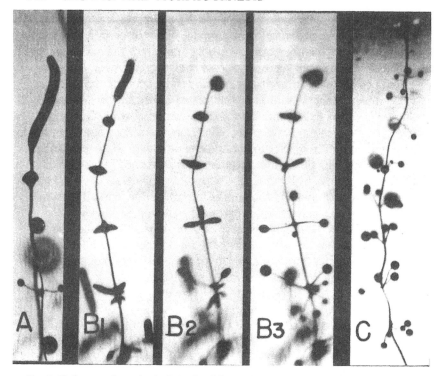

Fɪɢ. 8-19. Sorocarp formation in *Polysphondylium violaceum*. *A*. Developing sorocarp show-ing large terminal sorogen and smaller masses of abstricted cells attached to the sorophore below. *B1-B3*. Progressive stages in sorocarp development. Note that masses of abstricted cells subdivide and that whorls of side branches then develop from the resulting segments. *C*. Mature sorocarp with evenly spaced whorls of side branches. × 10. (See also Figs. 1-2 and 1-3)

tapered and commonly consists of a single tier of cells throughout most of its length, the width of this gradually diminishing upward. Under altered conditions, such as one-sided illumination or complete darkness, stalks may be substantially longer, in which case whorls tend to be less regular both in symmetry and in spacing.

One cannot but be impressed by the beauty and symmetry of a well-formed sorocarp of *Polysphondylium*; and in viewing such a structure the regularly spaced whorls are at once the most obvious and puzzling attribute of the fructification. How is the pattern achieved, and what underlies this orderly arrangement? Can it in some way be related to the wave-like movements that characterize the aggregative process, representing as it were a continuing but interrupted response? We know for example that distal portions of inflowing streams commonly disintegrate as soon as one or more sorogens arise from an aggregation center, presumably due to a loss of apical dominance. Within the rising sorogen, does the periodic abscission of posterior fractions result from a progressive reduction in apical guidance, which periodically reaches a level where posterior fractions of the cell mass no longer respond, hence lag behind? This is indicated by a recent cinematographic analysis of sorocarp development in *P. pallidum* (O'Day and Durston, 1980).

Acytostelium

Of the remaining genera of the Dictyostelidae, *Acytostelium* is the only one of which we have either cultures or illustrations. As the name implies, the genus is unique among the dictyostelids, for it builds its sorocarps without sacrificing any myxamoebae in stalk formation. Concerning the process we now have substantial information, even though we question how some of it should be interpreted.

Four species of *Acytostelium* have been reported, and since *A. leptosomum* (Raper and Quinlan, 1958) is the generic type and the best known, it will serve as the focal point for the account that follows. Pseudoplasmodium formation resembles rather closely that of *Dictyostelium* and *Polysphondylium* except that several to many sorogens regularly arise from a single wheel-like aggregation (Fig. 14-1C), and some of these often reach maturity before the influx of cells is complete. Sorophores are formed in a manner markedly different from that in *Dictyostelium*, and it is the character of these structures (Fig. 14-2A-F), in particular, that accords the genus and family taxonomic status separate from the Dictyosteliaceae.

Since the stalk is acellular, it cannot be constructed in the manner postulated for *Dictyostelium*: there are no stalk cells to vacuolate and enlarge, hence no comparable lifting force to raise the ripening sorogen. But rise it does; so one must look elsewhere for a likely explanation. Perhaps this is to be found in the form, orientation, and arrangement of the still undifferentiated but highly integrated myxamoebae during this developmental phase—an arrangement with the form of an elongate, sheathencased structure, closed at the top, with comparatively heavy cellular side "walls" and an exceedingly narrow central canal within which the sorophore is continuously being formed (Fig. 8-20). The cells near to but not part of the apical cone are believed to be most active in secreting materials for stalk construction (Fig. 8-20A). Such materials are discharged into a confined space, and their accumulation and polymerization may provide a part of the necessary lifting force. Since the lower part of the central canal is filled by the already formed sorophore and since this in turn rests upon the substrate, any pressure developed within the canal could be applied against the still plastic apical area, thus raising the whole mass because of its cohesion. This hypothesis advanced by Raper and Quinlan (1958) is admittedly speculative, but the cell mass does form a cell-free stalk and it does ascend the stalk as this is formed. Shaffer (1962) tends to discount the above as a realistic explanation, believing instead that the myxamoebae

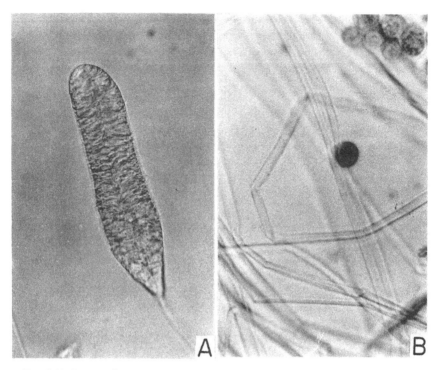

FIG. 8-20. Sorocarp formation in *Acytostelium leptosomum*. *A*. A blunt sorogen building its stalk into the air. Note how the constituent cells are oriented horizontal to the stalk axis. *B*. Detail of the acellular stalks, which are less than half the diameter of most spores (center and upper right). A × 600; B × 1150. (See also Figs. 14-1 and 14-2)

actively propel the mass upward despite their flattened, horizontal posture. There should be no conflict in either case concerning the secretion of stalk-forming materials and their concentration within the central canal, whether the constituent cells are pulled upward by a lengthening sorophore or advance solely via coordinated amoeboid movements. If it could be shown that cells in the sorogen of *Acytostelium* move in an orbital pattern, as Clark and Steck (1979) have proposed for *D. discoideum*, perhaps this might provide a partial explanation.

However formed, the sorocarps of *Acytostelium leptosomum* are remarkable structures. Consider that the erect or semi-erect sorophores are commonly 750-1500 μm in length but only 1.0-2.0 μm wide and bear globose sori 30-50 μm in diameter. Consider also that the stalk is less than half the diameter of a *single* globose spore (Fig. 8-20*B*), yet supports scores of these suspended in a droplet of slime. As in other dictyostelids, a principle strengthening material in the sorophore and in the spores is cellulose (Raper and Quinlan, 1958).

The sorophores were originally thought to represent very narrow tubes, for when bent and viewed microscopically they presented pictures simulating soda straws bent in similar fashion. But whether hollow or solid is a moot point—Gezelius (1959) concluded that they had no lumen and Shaffer (1962) concurred in this view, while Hohl, Hamamoto, and Hemmes in a later study (1968) reported stalks to have "a central lumen devoid of cellulose fibrils." Perhaps both are correct, for if stalk-forming materials are secreted from the ends (or sides) of cells lining a canal, whether the canal becomes filled or not might well depend on the amount of secretion and the diameter of the channel. It is not, we think, a matter of special significance. What is significant is that *Acytostelium* builds an upright sorocarp, expends no myxamoebae in constructing the stalk, and conserves the entire cell population for producing spores.

GAS-INDUCED ORIENTATION IN CULMINATION

Sorocarps of *Dictyostelium discoideum* are normally constructed approximately perpendicular to the surface from which they develop, and generally this holds true for other species as well in the absence of directive stimuli such as light. But this is not always the case, and several years ago two of Bonner's students, John Rorke and George Rosenthal, made a most interesting observation with possible implications of basic significance regarding the cause of such deviation. They examined the process of sorocarp formation in *D. discoideum* in a film, and in one sequence a migrating pseudoplasmodium was cut into three parts and the fractions freed from each other so that they would culminate separately (Raper, 1940a; Bonner, 1944 et seq.; Sampson, 1976). It was observed that the central fraction fruited in a nearly vertical direction, whereas the anterior and posterior parts constructed sorocarps at substantial angles from the vertical and in opposite directions. Little was thought of this, as it seemed reasonable for the apical portion to develop in a forward direction and the posterior portion rearward. But the behavior was not to be explained so casually when Bonner and his students took a second look. They then found that when a migrating pseudoplasmodium was so partitioned and the fractions placed less than 0.8 mm apart, the two sorocarps at either side of the central one diverged at substantial angles irrespective of the order in which front, center, and rear fractions were replaced on the agar surface (Fig. 8-21). They postulated that a gas of unknown composition was given off by the rising sorogens, that a gradient was established, and that given a "choice," culmination was in the direction of lower concentration. Many other experiments were performed that gave support to this possibility. For example, a sorocarp developing adjacent to a block of agar would lean away from the agar face; a sorocarp developing beneath a coverglass positioned at an angle of 40-45° would build toward the glass but not touch it, then adjust its direction of development to bisect the space

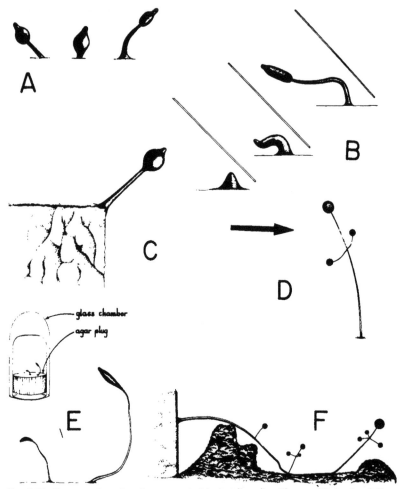

FIG. 8-21. Gaseous orientation of sorocarps in *Dictyostelium discoideum* and *Polysphondylium pallidum*. *A*. Sorocarps developing from segments of a single pseudoplasmodium bend away from one another. *B*. A sorocarp of *Polysphondylium pallidum* avoids touching a coverglass. *C*. A sorocarp of *D. discoideum* culminates away from an agar support. *D*. A sorocarp of *P. pallidum* bends toward an air current in a wind tunnel. *E*. *P. pallidum* fruiting in a small confined space without identifiable orientation. *F*. Sorocarp of *P. pallidum* seemingly attracted to charcoal. (From Bonner and Dodd, 1962b)

between the agar from which it arose and the glass surface which it adroitly avoided; sorocarps developing in a minute wind tunnel leaned into the wind. Significantly, a rising sorogen did not build away from a surface covered with activated charcoal, where, owing to absorption, no primary gradient or echo effect could be anticipated (Bonner and Dodd, 1962b).

Attempts to identify the gas that might be responsible for the directed orientation of developing sorocarps was unsuccessful. They did show, however, that the effect was not species specific, for culminating sorogens of different cellular slime molds showed a similar repulsion when placed in close proximity. Extending their observations to single sorocarps of *Polysphondylium*, Bonner and Dodd noted that in any single whorl the branches were usually spaced equidistant from one another, and where only two branches were present they occupied positions on opposite sides of the stalk. The same explanation can probably be offered for the divergence of sorophores in *D. polycephalum* just prior to spore maturation, for the sori typically occupy positions of maximal separation.

We are indebted to Bonner and Dodd (1962) for another and serendipitous observation as well, namely that sorocarps of some species are attracted to activated charcoal, presumably because it absorbs self-produced inimical gases. Soon thereafter Bonner and Hoffman (1963) and Kahn (1964) showed that the addition of activated charcoal greatly increased the numbers of aggregations formed in *D. mucoroides* and *P. pallidum*, respectively. This in turn has led to an increasing use of activated charcoal as a means of enhancing fructification in different cellular slime molds, particularly the smaller species, and nowhere has its beneficial effects been demonstrated better than in *Dictyostelium polycephalum* and *Acytostelium leptosomum*.

We have long known that the apical portion of a migrating pseudoplasmodium produces substantially greater amounts of acrasin than more posterior regions, and that a relatively high level of productivity characterizes the apical tip of the rising sorogen as well (Bonner, 1949). More recently investigators have turned their attention to the influence of cAMP upon culmination in general terms, and upon cellular differentiation in particular. As early as 1970, Bonner showed that isolated, unaggregated myxamoebae of *D. discoideum* (NC-4) when deposited in small drops on agar containing cAMP (10^{-3} M) would differentiate into vacuolate cells indistinguishable from stalk cells within 24-48 hours. Fluorescent brighteners, cellulose stains, and birefringence served to confirm cellulose as a primary component of their heavy, angular cell walls. More recently studies of factors promoting stalk cell formation have been greatly extended by C. D. Town and other investigators at the Mill Hill Laboratories in London (Town, Gross, and Kay, 1976; Town and Stanford, 1977; Town and Gross, 1978). They have shown that not just cAMP is involved but a diffusible inducer of low molecular weight as well, and that the latter acts to reinforce the former. The inducer has been identified as an oligosaccharide moiety containing sialic acid, L-fucose, and N-acetylgalactosamine (Town and Stanford, 1979). This work, together with subsequent studies of the differentiation-inducing factor (DIF), has been summarized recently by Gross et al. (1981) in relation to the broader phenomenon of cell patterning in *Dictyostelium*.

CHAPTER 9

Macrocysts

In addition to the asexual fructifications, or sorocarps, which consist of upright sorophores bearing one or more sori, some dictyostelids show an alternative, sexual mode of reproduction—the macrocysts. Although probably seen and illustrated by Brefeld in 1869, these structures were not so designated until the middle of this century (Raper, 1951), and their potential importance in the life cycle of cultures that produce them was not recognized until much later (Clark, Francis, and Eisenberg, 1973; Erdos, Raper, and Vogen, 1973). The history of how this recognition came about is worth recounting.

In his description of *Dictyostelium mucoroides*, Brefeld (1869) described and illustrated structures identified as "dwarf sporangia," of which we are reasonably sure some would today be regarded as macrocysts (Fig. 1-1D). He spoke of these as being formed *under* the normally completed fruiting structures, hence in contact with the substrate, and illustrated some as containing extremely rudimentary stalks. In others there was no stalk at all. These were surrounded by much thicker "membrans" that stained as cellulose[1] (violet with chloro-zinc iodine, blue with iodine and sulphuric acid). He did not elaborate further except to say that the cells within the sporangium arose by division of the spore plasm, which was thought by him at that time to be a transient plasmodium.

Dwarf sporangia were not discussed in his subsequent paper (1884), in which he reviewed his earlier work with *D. mucoroides* and described the new genus and species *Polysphondylium violaceum*, nor were comparable structures mentioned in the extended reports of Potts (1902), E. W. Olive (1902), Skupienski (1918), Harper (1926), Arndt (1937), or in the papers of any other early investigator. In fact, they were not again encountered, apparently, until 1937, when Raper detected and photographed what appeared to be globose, multicellular cysts in a culture of *D. mucoroides*, strain NC-12 (Fig. 9-1). Incidentally, this slime mold was isolated from a sample of soil collected in the Craggy Mountains of western North Carolina concurrently with the one which yielded strain NC-4, the type culture of *D. discoideum*.

Strain NC-12 was then and for some years thereafter regarded as unique, for no other culture in our possession produced comparable structures under

---

[1] We have never seen a macrocyst with a stalk enclosed within the primary wall; we have seen cases where a single macrocyst and a small sorocarp developed from a single small aggregation.

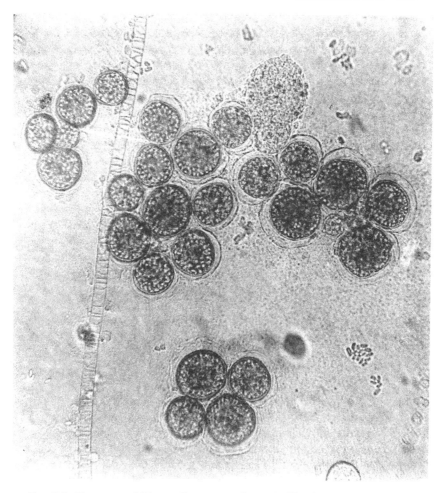

FIG. 9-1. Macrocysts of *Dictyostelium mucoroides*, strain NC-12, as first observed and photographed in 1937. × 200

the culture conditions then being employed. In fact, the idea that the structures represented an alternative pattern of morphogenesis in *D. mucoroides* was not immediately obvious, and it was not until we had made repeated cultivations from carefully collected spores that the large, seemingly multicellular cysts were accepted as an integral morphogenetic phase of that particular strain. Their significance in the life cycle was completely unknown, but it did seem even then that structures so singular should have some identifiable function. Clearly they were not sporangia unless the slime mold produced spores of two quite different types—a condition not unknown in the fungi.

Although questions about the origin and significance of the cysts persisted, these could not be explored until some years after World War II. Meanwhile strain NC-12 was retained by periodic transfer and occasionally accorded cursory examination to confirm its continued cyst production. The next cyst-forming culture encountered was a strain of *D. minutum* isolated from soil collected in a mixed deciduous forest near Purdue University in the spring of 1950. It was in writing about *D. minutum* the following year that the term *macrocyst* was introduced (Raper, 1951) to distinguish between such structures of multicellular origin and the small, unicellular microcysts formed by the encystment of individual myxamoebae (Chapter 6). In this year also the second of many subsequent macrocyst-forming strains of *D. mucoroides*, strain S-28b, was isolated from soil collected in Holly Hill, South Carolina; and this, together with strain WS-47, obtained later from arboretum soil (University of Wisconsin), became the objects of intensive study a few years later (Raper, 1956b; Blaskovics and Raper, 1957).

Much was learned about the macrocysts of *D. mucoroides* in the latter study, albeit their unique origin and potential significance still eluded us. The macrocysts were seen to be somewhat flattened, irregularly circular to ellipsoidal, multicellular structures generally ranging from 25-50 μm in diameter. Contributing myxamoebae appeared normal in every respect and formed aggregations in a typical manner (Fig. 9-2A), except that these were usually of limited dimensions. However, instead of producing upright sorocarps the myxamoebae soon collected into compact heaps, which, if limited in size, then became encircled by thin fibrillar walls or membranes; if the aggregations were larger, some subdivision normally occurred and each fractional part laid down its own encircling wall. One macrocyst typically developed within each cluster of encircled cells, although two or rarely more might be formed within what we regarded as the *primary wall* (Fig. 9-3C-F). The strains differed appreciably. In some the cysts tended to develop singly or in small bunches, while in others they often formed raft-like clusters (Fig. 9-2B), depending in large measure upon the size of the parent aggregation. What then followed was not understood, but at the same time it appeared that the cells comprising the incipient cyst developed defractive and seemingly semi-rigid walls so that each cell stood out as a discrete entity, to which we applied the term *endocyte*. It was observed that the transformation of cells from amoeboid to "firm walled" progressed from the center of the cyst outward, and that when all the peripheral cells had become transformed, a comparatively heavy wall rich in cellulose was laid down just beyond the outermost endocytes (Fig. 9-2C and 9-3E). This was designated the *secondary wall* (Fig. 9-3H). There was yet a third, *tertiary wall*, or membrane, and this could be easily demonstrated by subjecting endocyte-filled macrocysts to plasmolysis by immersion in a concentrated sucrose solution (Fig. 9-3A). We assumed upon the basis of visual observations that the endocytes were in some way physically con-

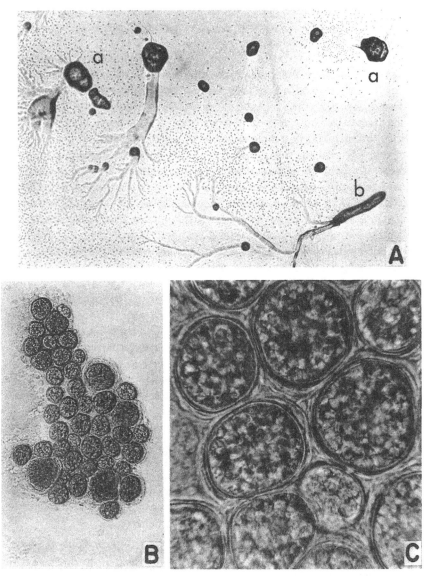

FIG. 9-2. Macrocysts of *Dictyostelium mucoroides*, strain WS-47. *A*. Cell aggregations leading to macrocyst formation (a) and to sorocarp formation (b) within the same microscopic field. × 22. *B*. A cluster of macrocysts developed from a single aggregation. × 115. *C*. Detail of some of the macrocysts shown in *B*. × 550. (From Raper, 1962)

tinuous with the contributing myxamoebae, but did note their smaller dimensions, the difference being 6-8 μm for rounded myxamoebae versus 3.6-4.8 μm for endocytes.

What we considered as the endocyte stage usually lasted for 2-3 weeks, with considerable variation between strains and even among cysts in the same culture. After this a puzzling thing occurred: the endocytes seemingly disappeared so that the contents of the macrocysts became finely granular and essentially homogeneous in appearance. If a macrocyst was ruptured prior to this stage, its content emerged as small, rounded bodies, or as a mixture of these and myxamoebae depending upon the stage of development. If ruptured at the so-called homogeneous stage, the content flowed out freely and appeared finely granular with no trace of a residual cellular structure. If, however, a similar "homogeneous cyst" was placed in a cellulose solvent such as Schweitzer's reagent for several minutes, the content was released intact as a single membrane bound body (Fig. 9-3B), while the empty and crumpled secondary wall, now devoid of cellulose, was no longer birefringent under polarized light.

Now back to the living cysts. About 2-4 weeks after attaining a homogeneous state, the content of the cyst shrank away from the heavy secondary (cellulose) wall, became slightly brownish in color, and came to lie loosely within the cyst cavity (Fig. 9-3G). Despite their withered appearance, such cysts could be rehydrated and germinated in limited numbers by covering them with distilled water and incubating at lower than usual temperatures (10-15°C). Significantly, the myxamoebae that emerged from these cysts seemed normal in size and appearance and were immediately able to renew the vegetative cycle.

What should we have concluded from the foregoing observations? Looking backward from the vantage point of 1983, some of the uncertainties and anomalies of the late 1950s seem startlingly clear. Nonetheless, I think it appropriate to insert a few excerpts from the Blaskovics and Raper report (1957) to illustrate, if for no other reason, how researchers grope for answers and may at times come close but still miss the mark. (What follows should be read in the context of the completed chapter.)

Intriguing questions are posed by the macrocysts of *Dictyostelium* with regard to their morphogenesis and their probable primary function in the life-cycle of these slime molds. Do they represent a normal but generally unrevealed stage in the developmental history of the Acrasieae, i.e., could they be demonstrated in all members of the group if we but knew the conditions required to evoke them? Do they provide a resting stage whereby these micro-organisms survive otherwise impossible environmental conditions? Do they perhaps constitute some unanticipated manifestation of a sexual stage? . . . Does the identity of the contributory myxamoebae remain unchanged during the formation of the endocytes,

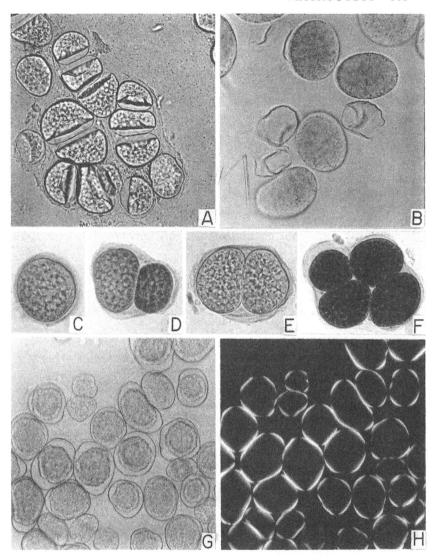

FIG. 9-3. Macrocysts of *Dictyostelium mucoroides*, strain S-28b. *A*. Young endocyte-filled macrocyst exposed to a saturated sucrose solution. Note how the tertiary wall pulls away from the rigid secondary wall due to plasmolysis. × 250. *B*. Two-month old macrocysts following treatment with Schweitzer's reagent. Note that the cyst contents (still bounded by a tertiary wall) have become freed of the now wrinkled and partially dissolved cellulose-rich secondary wall. × 250. *C,D,E,F*. Preparations stained with chloroiodide of zinc to reveal 1, 2, 2, and 4 macrocysts within single primary walls. × 250. *G*. Two-month old macrocysts photographed with normal light. Note the shrunken character of the cyst contents and the absence of endocytes. × 220. *H*. The same photographed with polarized light to reveal the strongly birefringent secondary walls. × 220. (Blaskovics and Raper, 1957)

and do the latter in some altered form persist to once again emerge during germination as myxamoebae capable of perpetuating the species?

Much evidence supports the belief that macrocysts arise through an orderly and natural morphogenetic process, and hence in no wise represent aberrant fruiting structures. For those strains which produce them, they would appear to be no less normal than the sorocarps which regularly develop under similar or, in some instances, altered conditions. A measure of homology is suggested by the basically similar aggregative process which precedes the formation of both types of structure.

We have obtained convincing evidence that macrocysts germinate under certain circumstances, emitting amoeboid cells which then reinitiate vegetative growth. But we cannot say with confidence that the macrocysts represent a vital resting stage, as their appearance might suggest. Heat tolerance tests indicate that they can withstand appreciably higher temperatures than vegetative myxamoebae, but they are in turn less resistant than true spores. While the matter has not been explored under conditions that exist in nature, it is possible that they might be produced under circumstances which would preclude the formation of sorocarps and spores.

Possibly they are endowed with other unique properties, a suggestion presently based less upon fact than fancy. We find it difficult to dismiss lightly a structure of multicellular origin which appears to be so highly organized as the older macrocyst. We cannot say with absolute certainty that its content represents a single homogeneous multinucleate protoplast, but such tests as we have applied would seem to support this notion. If the endocytes do actually lose their identity, as appears to be the case, the acellular content of the aged macrocyst would represent a coenocyte, or to use a term more commonly associated with slime molds, perhaps a plasmodium, albeit one that is enclosed within a heavy cellulose wall. Such a plasmodium would of course be quite unlike that which Brefeld (1869) once thought to be present, or that which Skupienski (1920) envisioned as an accompaniment to reported sexuality in *Dictyostelium*. Needless to say, it would represent quite a different structure from the large vegetative body that occupies so conspicuous a place in the life-cycle of the Myxogastrales. Clearly, the two could not be regarded as homologous.

Here the matter must rest for the present, and definitive information regarding the true nature and ultimate significance of the macrocysts must await further research.

The next substantial advances were made by a graduate student in this laboratory, Robert A. Hirschy (1963), which are here formally published for the first time, albeit an abstract of a paper presented at a meeting of the American Society for Bacteriology was included in *Bacteriological*

Proceedings, 1964, p. 27. Starting where Blaskovics and Raper left off, Hirschy included five additional macrocyst-formers in his study: the Purdue strain of *D. minutum* and four strains of *D. purpureum* that had been freshly isolated by J. C. Cavender from soils of Costa Rica. Of the four isolates of *D. purpureum*, special attention was given to strain Za-3b; and in studies centered upon this slime mold, Hirschy broadened appreciably our knowledge of conditions that favor or impede macrocyst formation and maturation. More specifically, he made two very important contributions and barely missed a third that could have accelerated macrocyst research by several years, had the significance of a recorded observation been recognized either by him or by his professor (KBR).

In the course of cultivating strains of *D. purpureum*, a chance observation led Hirschy to investigate the effect of light upon the relative balance between sorocarps and macrocysts in conventional cultures, and then to establish the definitive role of light (daylight and white fluorescent) in promoting the formation of sorocarps and of darkness in promoting the formation of macrocysts. Especially interesting was the discovery that light exerted its major influence not during aggregation or culmination but during the period of myxamoebal growth;[2] that is, the highest yields of macrocysts were obtained from cells grown in darkness, whereas cells grown and aggregated in light produced only sorocarps (Table 9-1). This led him to

TABLE 9-1. Effect of light and dark periods during growth on the morphogenesis of *Dictyostelium purpureum* Za-3b in small populations of pregrown cells (Hirschy, 1963)

Growth in					
Dark (hrs):	20	24	29	31	34
Light (hrs):	13	9	5	3	0
Total populations	151	135	140	142	283
Aggregated	150	131	127	108	216
Unaggregated	1	4	13	34	67
Macrocysts	0	12	100	93	216
Sorocarps	150	119	114	74	0
% aggregation	99	97	91	76	76
% macrocysts	0	9	79	86	100

NOTE: The cells were grown in association with *Escherichia coli* in culture tubes on a rotary shaker (250 rpm) at 25°C, washed by centrifugation to remove excess bacteria, and deposited as small drops on 1% Difco Purified Agar in Bonner's salt solution (1947). Populations usually ranged between 300 and 900 cells/drop and at densities never less than 700 myxamoebae/mm². The light period followed the dark period; incubation in the dark was at 20°C.

[2] A decade later, and working with populations of mixed mating types of the heterothallic species *D. discoideum*, Erdos et al. (1976) showed that light exerted its maximum influence upon the ultimate direction of morphogenesis (whether toward sorocarps or macrocysts) during the vegetative stage at 24-28 hours, long before any sign of aggregation, and that the maximum effect of temperature was delayed only a few hours.

speculate about the role of macrocysts in nature and to question with Bonner (1959a and 1967) if they might not represent a resistant stage formed in darkness below the soil surface or underwater, sites known to be inappropriate for sorocarp construction. It was shown that a very small percentage (1%) of cells grown in the dark could occasionally induce myxamoebae grown predominantly (99%) in light to form macrocysts when cells of comparable age were intermixed.

Hirschy's "missed contribution" is in retrospect the most interesting of all, for in his study he gave serious attention to the cytology of developing macrocysts. We did not have access to or experience with an electron microscope, and so the work was done using paraffin sections and a variety of stains then in common use. Comparisons were made between the cysts of *D. mucoroides* and *D. purpureum*, noting that they were basically similar but that the outer wall of the *D. purpureum* cyst was looser and thicker than that of *D. mucoroides*, and that its endocytes tended to lose their identity more slowly. For a more important observation we should let him speak:

> The histology of the macrocysts has been one of the more interesting aspects of the study. The first preparations examined showed an inner cavity amongst the mass of endocytes of a young macrocyst of *Dictyostelium purpureum*. The cells within the cavity stained more darkly with hematoxylin. Figures 9-4A through 9-4E are serial sections of a macrocyst, beginning at one edge and moving toward the center. The orientation is the same in all the photomicrographs. Since the sections are 8 μm, the distance shown is about 40 μ. The macrocysts from which these photomicrographs were obtained were developed under submerged conditions, so a logical question was whether these cavities were a product of this preparatory method. Macrocysts of the same age were taken from 0.1% lactose-peptone agar plates and the cavities were still present. Furthermore, the cavities were found to be present in young macrocysts of *D. mucoroides* as well (Hirschy, 1963, pp. 26-27).

Thus is shown here the first record, verbal and photographic, of an early stage in the development of a dictyostelid macrocyst—the so-called phagocytic cell of Filosa and Dengler (1972) containing ingested and partially digested myxamoebae, then and now referred to as endocytes.

The pity of it is that neither Hirschy nor his mentor comprehended the significance of what he saw. Still, the image was not thoughtlessly dismissed, for in his concluding page Hirschy (1963, p. 55) returns to wonder about this and other questions:

> Various aspects of the entire study raise a multitude of additional questions. We still do not know the ecological significance of the macrocyst, nor the conditions which encourage germination to levels where

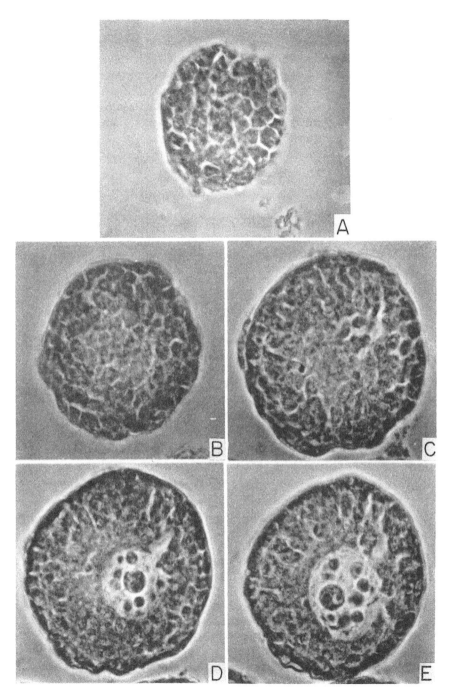

Fig. 9-4. Serial sections through a developing macrocyst of *Dictyostelium purpureum*, strain Za-3b. *A-E*. Sections are 8 μm thick and represent the first through the fifth cuts, respectively. Note the conspicuous cavities in *D* and *E* representing but not then identified as a giant (or phagocytic) cell. × 650. (Photographs by Robert A. Hirschy)

it would be significantly useful to the organism. Morphologically, what is the significance of the cavity amongst the endocytes? And what is the significance of the disappearance of the endocytes during their continued maturation? Is there truly fusion within the macrocyst? The light response raises a number of further questions. What wave length and intensity will be most effective in blocking macrocyst production? What is the responsible mechanism? If the cells have a "memory," how does it operate? The answers to these and many other questions must await further study.

For some of Hirschy's questions we now have answers; others still stand as challenges.

Unveiling the Macrocyst

The first evidence to suggest the true nature of the macrocyst came in a brief paper, "Ultrastructural aspects of macrocyst production in *Dictyostelium mucoroides*," presented by R. E. Dengler, M. F. Filosa, and Y. Y. Shao at the annual meeting of the Botanical Society of America, Bloomington, Indiana, in August 1970. Their abstract follows:

> Some strains of the cellular slime mold *Dictyostelium* form aggregates of amoebae which, instead of producing pseudoplasmodia and eventually fruiting bodies, produce a macrocyst surrounded by a cellulose wall. . . . Shortly after aggregation each aggregate of amoebae subdivides into several spherical masses. In the center of each mass a large cell appears. As this giant cell enlarges it ingests amoeba cells peripheral to it, enclosing each within a vacuole. These phagocytized cells are equivalent to the endocytes previously described in light microscope studies. The nucleus of the giant cell becomes greatly enlarged. Eventually all the peripheral amoebae are ingested, and a cellulose wall is laid down adjacent to the plasma membrane of the giant cell. Subsequently, numerous smaller nuclei appear in place of the large nucleus of the giant cell thus forming a multinucleate cyst. The endocytes eventually break down into granules. Our studies have not gone beyond this stage (145 hr of development) (p. 737).

Two years later a comprehensive and beautifully illustrated paper on macrocyst ultrastructure appeared under the authorship of Filosa and Dengler (1972). Myxamoebae were grown on agar plates, harvested at about 36 hours, washed by centrifugation, redispensed at 1.75×10^7/ml in saline, and incubated on a reciprocal shaker at 19-20°C. Cells began clumping within a few hours and macrocyst initiation followed some hours later, with giant or phagocytic cells, as they were then called, typically evident at 20 hours. Using both conventional light and electron microscopy, the further development of the macrocysts was clearly described and illustrated. Several questions raised in the work of Blaskovics and Raper (1957) and in that of Hirschy (1963) were clarified. The endocytes were shown to be not persistent cells but ingested myxamoebae undergoing slow disintegration. The heavy cellulose (secondary) wall was shown to be produced adjacent to and snythesized by the plasma membrane of the single phagocytic cell, while the "inner membrane," demonstrable by cyst plasmolysis, was the plasma membrane of the same cell. The maturing cyst was found to be a coenocyte of sorts, since it became multinucleate, but this state arose from the division of the single giant cell nucleus, not the continuity of many nuclei from initially separate myxamoebae.

The paper by Filosa and Dengler (1972) went far toward explaining the initiation and subsequent development of the macrocysts of *Dictyostelium mucoroides*, and by analogy those of other species as well. Clearly shown were progressive stages in cyst development and maturation from young phagocytic cells containing few endocytes, through mid-stages with increasing numbers of endocytes, to mature cysts with heavy walls and seemingly homogeneous and finely granular content (Fig. 9-5). Especially informative were the electron micrographs showing a developing cyst with a single enlarged nucleus (Fig. 9-6) and a mature cyst with multiple nuclei derived from it. The report did not, however, clarify the origin of the cytophagic cell; it did not cover the germination of the macrocysts; and it did not address cultural conditions that especially favor or depress macrocyst production. Neither did it stress the significant role that the macrocyst might play in the life cycle of the slime mold.

Answers to some of these questions were soon forthcoming. In the following year Nickerson and Raper published investigative studies that defined as precisely as possible the environmental conditions that (1) enhance or accelerate macrocyst formation (1973a), and (2) those that induce or stimulate macrocyst germination (1973b) among the dictyostelids. Results previously reported by Weinkauff and Filosa (1965) for *Dictyostelium mucoroides* and by Hirschy and Raper (1964) for *D. mucoroides* and *D. purpureum* were generally confirmed, and in certain particulars substantially extended. The importance of darkness and of wet agar surfaces for maximum cyst production (Hirschy, 1963) were reexamined and amplified, and the probability of a temperature effect were carefully investigated. Relatively few strains had been examined in earlier studies, so the number was trebled to include several additional isolates of *D. mucoroides*, four new cultures of *D. minutum*, and the sole macrocyst-producing strains of *D. discoideum* and *Polysphondylium violaceum* then in our possession. The sources and cyst characteristics of the cultures investigated are listed in Table 9-2. In testing the relative effects of light (cool white fluorescent) and of temperature, replicate plates were incubated in light and in darknesss at five different temperatures. The data are presented in Table 9-3. For

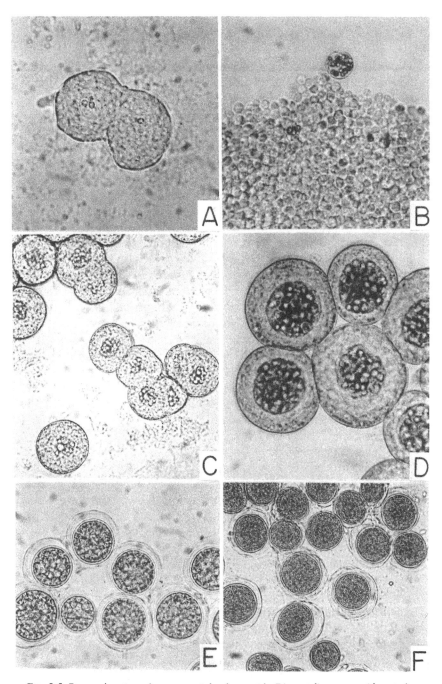

FIG. 9-5. Progressive stages in macrocyst development in *Dictyostelium mucoroides*, strain DM-7. *A*. Very early stage—centrally positioned giant, or phagocytic, cells evident as two or three endocytes (engulfed myxamoebae). × 325. *B*. Young cyst at slightly later stage compressed to release giant cell (above). × 350. *C*. Clusters of cysts around which primary walls are forming. × 250. *D*. Cysts containing giant cells filled with endocytes and surrounded by peripheral cells (myxamoebae) that await engulfment. Primary walls now complete. × 400. *E*. Giant cells have engulfed remaining peripheral cells and deposited secondary (cellulose) walls; endocytes still appear intact. × 250. *F*. Aged cysts in which endocytes have been digested; cyst contents now appear finely granular and brownish in color. × 250

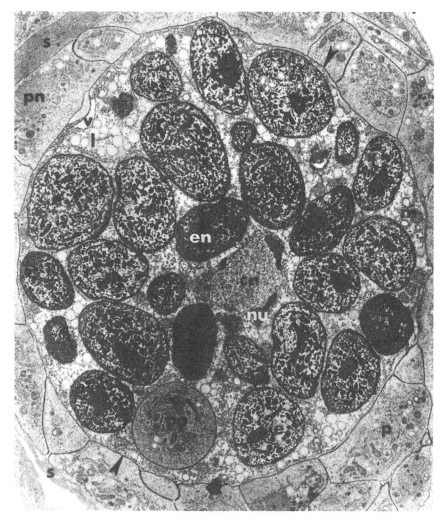

FIG. 9-6. Developing macrocyst of *Dictyostelium mucoroides* showing fibrillar sheath (s), peripheral myxamoebae (p), and large central cytophagic or giant cell (delimited by arrows) containing engulfed myxamoebae, endocytes (e). Note how a recently engulfed peripheral cell (pp) differs in appearance. Nuclei of peripheral cells (pn) are less electron dense than those of typical endocytes (en). The nucleus of the cytophagic cell (cn) with nucleoli (nu) is greatly enlarged. Mitochondria (m) of the cytophagic cell appear larger than those of the peripheral cells. Lipid inclusions (l); vacuole (v). × 3600. (Electron micrograph courtesy of M. F. Filosa, from Filosa and Dengler, 1972)

comparative tests the slime molds were cultivated in association with *Escherichia coli* B/r on 0.1L-P agar. Inoculations were made by spreading mixed suspensions of bacteria and slime mold spores over the entire surface of freshly poured plates.

Of the *D. mucoroides* strains, WS-47 stood apart by forming some macrocysts at all temperatures in the light, even at 10°C. Two other strains,

TABLE 9-2. Identity, source, and cyst characteristics of macrocyst-producing dictyostelids investigated by Nickerson and Raper (1973a)

Species	Strain	Source and date of isolation		Days to maturity	Dimensions of macrocysts
D. mucoroides	S-28b	K. B. Raper, S.C.	1951	18-21	15-35 μm
	WS-47	K. B. Raper, Wis.	1953	18-21	20-50 μm
	WS-116D	K. B. Raper, Wis.	1954	12-14	20-55 μm
	DM-7	M. F. Filosa, N.J.(?)	1954(?)	5-7	25-50 μm
	IW-41	I. A. Worley, N. Zea.	1965	>21	20-50 μm
	WS-520	J. M. McCleary, Ill.	1966	10-12	15-40 μm
	WS-524	T. Kurzynski, Wis.	1967	7-10	15-40 μm
	WS-525	E. Tash, Wis.	1967	10-12	15-35 μm
	CH-2	J. C. Cavender, Mex.	1967	18-21	15-35 μm
	WS-558	D. Castener, Mo.	1970	12-14	15-35 μm
	DC-2	D. Castener, Mo.	1970	10-12	25-55 μm
D. purpureum	Za-3b	J. C. Cavender, C. Rica	1962	42-50	30-85 μm
	Za-2a	J. C. Cavender, C. Rica	1962	35-42	30-100 μm
	TR-II-1	J. C. Cavender, C. Rica	1962	18-21	30-70 μm
	ES-9	J. C. Cavender, C. Rica	1962	42-50	30-65 μm
D. minutum	WS-572	J. Sutherland, Wis.	1970	—	15-35 μm
	DC-1	D. Castener, Mo.	1970	—	20-50 μm
	DC-4	D. Castener, Mo.	1970	—	15-35 μm
	DC-43	D. Castener, Mo.	1970	—	20-50 μm
D. discoideum	AC-4	J. C. Cavender, Mex.	1964	35-42	35-90 μm
P. violaceum	ID-10	J. C. Cavender, Phil.	1970	21-28	25-70 μm

NOTES: Macrocysts produced in the dark at 22.5°C; D. minutum grown on ½ strength 0.1L-P agar, other species on 0.1L-P agar, with Escherichia coli B/r as the food source.

WS-116D and S-28b, differed from most in producing macrocysts at 22.5 and 25°C in the light. Only at 22.5°C, a temperature not included by Hirschy (1963), did strains of D. purpureum form any macrocysts in the light. As the data suggests, strains of D. minutum were erratic in their fruiting behavior and showed no substantial difference between incubation in the light and the dark, while irregularities in development precluded citation of specific periods necessary for cyst maturation (Table 9-2). The single culture of D. discoideum followed a pattern generally similar to that of D. purpureum, while Polysphondylium violaceum, although forging its

TABLE 9-3. The effect of light and temperature on macrocyst production in dictyostelids investigated by Nickerson and Raper (1973a)

Species and strain	Incubation in light					Incubation in dark				
	10°C	15°C	20°C	22.5°C	25°C	10°C	15°C	20°C	22.5°C	25°C
D. mucoroides										
WS-47	1	1	1	1	1	4	4	4	4	4
WS-116D	0	0	0	4	4	3	4	4	4	4
S-28b	0	0	0	4	4	1	3	3	4	4
DC-2	0	0	0	1	1	4	4	4	4	4
WS-520	0	0	0	1	1	4	4	4	4	4
WS-524	0	0	0	1	1	3	4	4	4	4
WS-558	0	0	0	1	1	2	3-4	4	4	4
DM-7	0	0	0	1	1	2	3	3-4	4	4
CH-2	0	0	0	0	0	4	4	4	4	4
WS-525	0	0	0	0	0	3	4	4	4	4
IW-41	0	0	0	0	0	3	4	4	4	1
D. purpureum										
Za-3b	0	0	0	0-1	0	0	1-2	3	4	4
Za-2a	0	0	0	0-1	0	0	0-1	3-4	4	4
ES-9	0	0	0	0-1	0	0	0	0	4	4
TR-II-1	0	0	0	0-1	0	0	0	0	4	4
D. minutum										
WS-572	0	0-1	2	1-2	1	0	1	2-3	2-3	3
DC-1	0	1	2-3	2-3	2-3	0	2	2-3	3	3
DC-4	0	1	1-2	1-2	1-2	2	2	2-3	2-3	2
DC-43	0	2-3	3-4	3-4	3-4	2-3	3-4	4	4	4
D. discoideum										
AC-4	0	0	0	0-1	0-1	0	4	4	4	4
P. violaceum										
ID-10	0	0	0-1	2	3	0	0-1	3-4	4	4

KEY: 0 = No macrocysts formed, sorocarps formed exclusively
1 = Limited macrocysts formed, sorocarps predominant
2 = Approximately equal numbers of macrocysts and sorocarps
3 = Macrocysts predominant, some sorocarps formed
4 = Nearly all macrocysts, very few sorocarps formed

NOTE: Cultures grown with *Escherichia coli* B/r on 0.1L-P agar.

own line of morphogenesis, resembled *D. minutum* more than any other species surveyed (Fig 9-7).

In tests to determine patterns of development underwater, plates were set up in the usual way and flooded with distilled water to a depth of 1 mm. In *D. mucoroides* macrocysts formed on the underlying agar whether or not the cultures were illuminated. In *D. purpureum*, however, only macrocysts formed in darkness at temperatures of 22.5°C and 25°C, while loose cell aggregates but few macrocysts developed at temperatures of 15°C and 20°C.

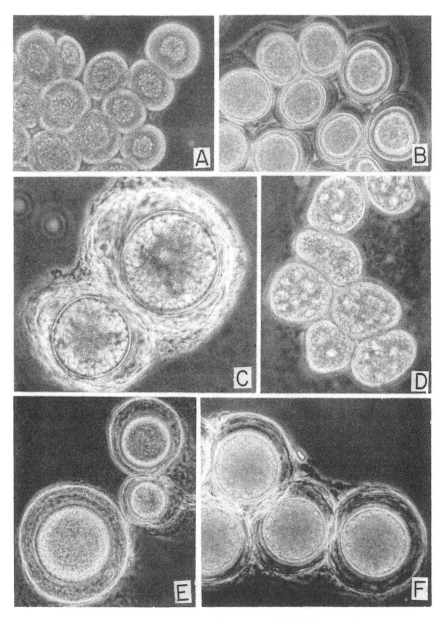

FIG. 9-7. Macrocysts of five species of the Dictyosteliaceae. *A*. Early stage of development in *Dictyostelium mucoroides*, strain DM-7. Note the centrally positioned, endocyte-filled giant cells and the absence of a secondary wall. × 250. *B*. Mature macrocysts of the same strain. Wall structures completed and cyst contents contracted away from secondary walls in some cases. × 250. *C*. Endocyte-filled macrocysts of *D. purpureum*, strain Za-3b. Note the multilayered character of the primary wall. × 400. *D*. Immature macrocysts of *D. minutum*, strain DC-43. × 335. *E*. Mature macrocysts of *D. discoideum*, strain AC-4. Again note the multilayered composition of the primary wall. × 200. *F*. Mature macrocysts of *Polysphondylium violaceum*, strain ID-10. × 300. All strains illustrated are homothallic or self-compatible. (From Nickerson and Raper, 1973a)

Fewer macrocysts developed when agar plates stood for a day or more prior to inoculation, due perhaps to lowered surface moisture. To investigate the effects of drying, plates inoculated as above were covered with porous clay (Coors) lids and incubated as before in light and in darkness. No strains of *D. mucoroides* produced macrocysts in the light, while in dark-incubated cultures cyst production was completely suppressed in some strains, greatly reduced in others, and seemingly unaffected in one, strain CH-2. No macrocysts were ever produced in clay-covered plates in light, while in darkness production was severely curtailed in some strains and moderately so in others.

The repressive influence of phosphate ions was in many strains almost as dramatic as the response to clay covers. In cultivating dictyostelids it is quite commonplace to buffer the substrate with phosphate salts and adjust the pH to 6.0 or 6.5. However, when proven macrocyst-forming strains were implanted on such substrates, the results were very disappointing. Where one would have expected to secure virtually all macrocysts, the numbers of these were often severely reduced with a concomitant increase in sorocarps. Plates containing 0.05 M phosphate salts and adjusted to pH levels between 5.8 and 7.8 all showed reduced levels of macrocysts and increased numbers of sorocarps. Similar concentrations of KCl or NaCl when added to 0.1L-P agar were without obvious effect. Upon the evidence obtained, we concluded that the suppression of macrocysts on phosphate buffered media was due to the specific presence of the phosphate ions rather than to any general pH or osmotic effect.

In considering the effects of light, temperature, free water, porous plate covers, or phosphates, the reader must recognize that the extent to which any one of these factors influences macrocyst production depends upon the species and strain under observation, for isolates are not all equally responsive to these factors. Additionally, the potential effect of one factor may be either strengthened or masked by another, and the end product of morphogenesis then reflects a cumulative or competitive rather than a separate response. If asked to identify the factor most important in favoring cyst production, one would have to say the presence of free water (Bonner, 1959a). This principle was successfully employed by Filosa and Chan (1972) in screening primary isolates of *D. mucoroides* for macrocyst production, for of 70 isolated from 83 soil samples 35, or 50%, were found to produce macrocysts. Three years later, and still working with *D. mucoroides*, Filosa et al. (1975) reported that in a majority of cases (78%) cells in the process of aggregating would form macrocysts when transferred to saline in small plastic dishes, while of 116 pseudoplasmodia (sorogens) so transferred 24% produced macrocysts in 48 hours. (Similar transformations have been observed in the sorogens of a cyst-forming strain of *Polysphondylium violaceum*.) Of more special interest was the observation that carefully collected and concentrated peripheral cells (i.e., myxamoebae surrounding phagocytic cells in developing macrocysts) would form macrocysts promptly when transferred to an air-agar interface. However, if such cells were allowed to return to the vegetative stage, they later reaggregated and formed sorocarps, as one might expect.

Since the formation of either macrocysts or sorocarps can be controlled in most strains by altering environmental conditions and since both types of structures arise from cell aggregations and represent the end products of contrasting morphogenetic processes, it would seem that the control of morphogenesis and terminal differentiation in the cellular slime molds could be studied to advantage with macrocyst-forming strains. For example, what emergent differences in cell surfaces, cell structure, metabolism, etc. direct a population of myxamoebae to form sorocarps in one case and macrocysts in another? And what differences in enzyme patterns accompany the contrasting modes of morphogenesis once the commitment to sorocarp or macrocyst formation has been made? The use of selected cultures grown under appropriate conditions can provide invaluable material for the resolution of such questions in these simple eukaryotes, and if answered here might have relevance to higher organisms. Recent papers by Saga and Yanagisawa (1982 and 1983) and Saga, Okada, and Yanagisawa (1983) represent significant steps in this direction.

Germination

In the earlier work of Blaskovics and Raper (1957) limited success in germinating macrocysts had been achieved. However, it was not possible to follow the sequence of events that preceded myxamoeba emergence. Because of this Nickerson and Raper (1973b) set for themselves a threefold goal: (1) to determine the morphological events that occur during the transition from a dormant macrocyst to the liberation of myxamoebae at germination; (2) to identify a strain in which the process of germination could be studied advantageously; and (3) to use this strain to examine carefully the factors influencing macrocyst germination. Less extensive study would be accorded macrocyst germination in other strains and species (Fig. 9-7).

Since tests of macrocyst germination can hinge on small differences and may at times be very frustrating, we shall relate in some detail the procedures devised and employed by Nickerson and Raper (1973b).

The organism most used in the study was *Dictyostelium mucoroides*, strain DM-7. Limited studies were also made of *D. mucoroides*, strains S-28b, CH-2, IW-41, WS-47, WS-520, and WS-525; *D. purpureum*, strains Za-3b, Za-2a, and TR-II-1; *D. minutum*, strains DC-1 and DC-43; *D. discoideum*, strain AC-4; and *Polysphondylium violaceum*, strain ID-10 (Table 9-2).

Macrocysts were produced on 0.1L-P agar inoculated with 0.1 ml of a mixed suspension of slime mold spores and *Escherichia coli*, strain B/r. The inoculum was spread over the entire surface of freshly poured plates, and the plates stacked in metal cans to exclude light and incubated at 22.5°C. Plates were undisturbed until the macrocysts reached the age desired for testing germination. Drying of the agar was retarded by storage in metal cans, thus permitting macrocysts to be harvested during many months.

Macrocysts were harvested at the desired age by flooding the plates with sterile distilled water and gently rubbing the agar surface with a sterile bent glass rod to dislodge the macrocysts. The resulting suspension was washed five times via centrifugation with sterile distilled water and the macrocysts replated on nonnutrient agar (1.5% Difco agar). The plates were placed in lighted incubators at 10, 15, 20, 22.5, and 25°C. Plates containing macrocysts to be tested in the dark were wrapped in aluminum foil. The plates were removed at predetermined times, scored for germination, and discarded. For experiments in which the progress of germination was followed for several days, a replicate plate was scored each day. The interval between replating and examination was generally 7 days. Observations of macrocyst germination were made with a Zeiss phase microscope at 500×. From 200 to 500 macrocysts were scored for germination on each plate.

The following account of germination is condensed from Nickerson and Raper (1973b).

> The first identifiable change leading to germination of the dormant macrocyst was a swelling of the contracted protoplasmic content until it once again filled the space within the heavy secondary wall (Fig. 9-8A). Under some experimental conditions the swelling was reversible without apparent loss of viability, but at no later stage could the germination process be reversed.
>
> Shortly after swelling occurred, slow tumbling movements within the granular mass became visible. These were quite slow, but nonetheless distinct. Individual cells could not be distinguished, although the possibility that these existed cannot be excluded.
>
> As the content of the microcysts showed increasingly turbulent movements, the dark coloration gradually faded, leaving an essentially colorless mass of granular protoplasm. A number of changes then occurred almost simultaneously. Particularly striking was the reappearance of individual cells within the heavy wall. Vaguely seen at first (Fig. 9-8A,B), they gradually became more distinct as small contractile vacuoles slowly appeared.
>
> The final stages of germination began with the partial dissolution and breaking of the heavy secondary wall and ended with the liberation of the myxamoebae . . . (Fig. 9-8C,D). As a general rule, the myxamoebae

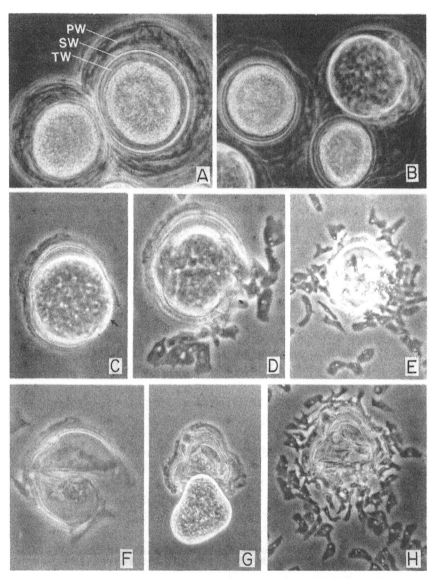

FIG. 9-8. Macrocyst germination in *Dictyostelium mucoroides*, strain DM-7. *A*. Two macrocysts, one dormant revealing primary (PW), secondary (SW), and tertiary (TW) walls and with cyst content still contracted; the other with cyst content swollen and filling the walled chamber. × 300. *B*. Dormant cyst with homogeneous content (upper left) and cyst with newly formed myxamoebae (upper right). × 300. *C-F*. A sequence of photographs taken over a 6-hour period showing the liberation of myxamoebae from a single cyst. *C*. Primary and secondary walls have broken preparatory to germination. × 500. *D*. Myxamoebae beginning to emerge through broken tertiary wall. × 500. *E*. Late stage in germination after nearly all myxamoebae have emerged. × 300. *F*. Empty cyst case. × 500. *G*. Liberation of myxamoebae en masse by rupture of the secondary wall—tertiary wall still intact. *H*. The same 4 hours later after the tertiary *wall* has ruptured and nearly all myxamoebae have emerged. × 300. (From Nickerson and Raper, 1973b)

emerged through the broken walls of the cysts individually over a period of several hours (Fig. 9-8*E,F*). On occasion, however, they remained within the tertiary wall and the whole mass, apparently under some pressure, was extruded intact (Fig. 9-8*G,H*). A variable period of time then elapsed before the myxamoebae were liberated.

Once released, the myxamoebae reinitiated the life cycle at whatever point environmental conditions dictated. If bacteria were present, they became vegetative and resumed normal growth and division. In the absence of bacteria, they immediatly reaggregated and soon formed either new macrocysts or fully differentiated sorocarps. Myxamoebae liberated from a single cyst often gave rise to a diminutive sorocarp if the germinating macrocysts were widely separated.

Whereas the foregoing account is based upon *Dictyostelium mucoroides* strain DM-7 in substantial part, our experiences then and subsequently indicate that it reflects with reasonable accuracy the process of germination in many other strains and species as well. It was chosen for special study because its macrocysts, unlike those of many other strains, did not require prolonged aging in order to germinate at satisfactory levels. Paradoxically, whereas darkness enhances macrocyst production in all strains, and is a requirement in some, light is needed to induce rapid germination in young cysts (Fig. 9-9). It is not, however, an absolute requirement. Germination can occur even in young cysts if given sufficient time, but the period required may be a month in darkness versus a week in light. Young cysts also germinate best at relatively low temperatures. For example, only 5-10% of 22 day old cysts germinated in 7 days at 22.5°C in the light, while 80-90% germinated within the same period when incubated at 15°C in light. As may be seen from Fig. 9-10, light becomes less important for germination as cysts age; concomitantly, the capacity to germinate more rapidly and at higher percentages increases over a period of 2½-3 months. Differences in germination rates of 23 and 75 day old cysts when incubated in light are graphically shown in Fig. 9-10; the same figure shows just as clearly the erasure of such light effect in aged cysts. Such curves raise interesting questions: Does light trigger some reaction that accelerates the aging processs, hence hastens the time a cyst can germinate? Or does the natural process of aging per se act to overrun and in time offset any stimulatory effect by light?

Germinating macrocysts appear to follow the same sequence of morphological changes in all species examined. However, the rates and percentages of germination vary widely. Some other strains of *D. mucoroides* approached DM-7 in their ability to germinate whereas others, such as strains CH-2 and IW-41, never achieved a level greater than 5%. In *D. purpureum* cysts of strains Za-2a and Za-3b (Table 9-2) required a longer period of aging, and little germination occurred among cysts of less than 50 days; and even these germinated best at 15°C. Cysts of *Polysphondylium*

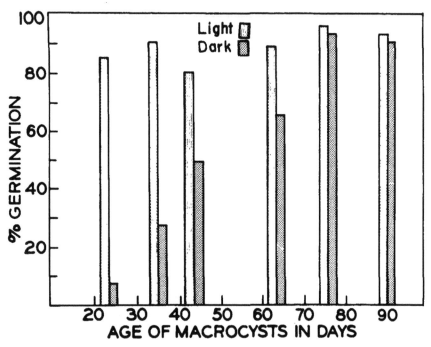

Fig. 9-9. Germination of macrocysts of various ages when tested in the light and in the dark at 15°C (*D. mucoroides*, DM-7). Cysts were produced in association with *E. coli* on 0.1L-P agar at 22.5°C in the dark and held there until removed for testing. Percentages recorded 7 days after cysts were harvested, washed, and replated on nonnutrient agar and reincubated at 15°C in light and in darkness. Results at 10°C and 20°C were comparable. (From Nickerson and Raper, 1973b)

violaceum likewise required prolonged aging, but once they had reached 50 days, germination at 60-70% could be obtained in 7 days at 15° or 20°C. In *D. minutum* cysts could germinate almost immediately after attaining maturity but showed great variability in reaching this stage. Of all the species examined, the greatest difficulty was encountered in securing germination of *D. discoideum* macrocysts, even at very low percentages. The only culture then known to produce macrocysts was the homothallic or self-compatible strain, AC-4. Cysts later derived from paired mating types of heterothallic strains do not perform much better (Wallace and Raper, 1979), which is most unfortunate since this species has been studied so intensively. A breakthrough would be most welcome.

As suggested by Filosa and Dengler (1972) and as shown by Erdos, Nickerson, and Raper (1972, 1973), dormant cysts undergo a protracted series of cytological changes before any myxamoebae are liberated during germination. But the timing of such changes is in no case defined or presently predictable. As already mentioned, light may stimulate the ger-

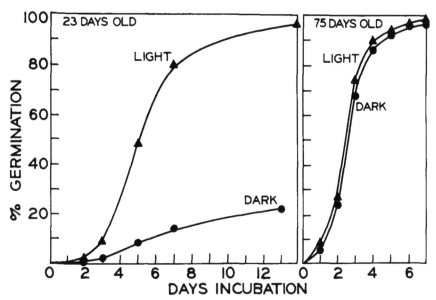

FIG. 9-10. Rates of germination among cysts 23 and 75 days in the light and in the dark at 20°C. Cysts were produced and processed as noted for Fig. 9-9. (From Nickerson and Raper, 1973b)

mination of younger cysts by hastening some step(s) in the maturation process, or older cysts may germinate more readily because of the loss or decline of some germination inhibitor. We know little about the interrelationship between the cyst content and its restraining wall, nor do we know what causes the final rupture of the latter: whether gradual dissolution with age or the increase of pressure from within coincident with rehydration. Several years ago Clegg and Filosa (1961) made the interesting observation that the walls of macrocysts contain trehalose in the amount similar to that of spores, and prophetically suggested that restriction of the sugar to these structures implied its utilization during germination, a role that was subsequently verified for spores by Cotter and other investigators. There is much still to be learned.

SEXUALITY

The possibility that macrocysts might represent a sexual stage in the life cycle of a dictyostelid was first mentioned by Blaskovics and Raper (1957) with reference to these structures in two strains of *Dictyostelium mucoroides*. The same possibility was reiterated, if somewhat obliquely, by Hirschy (1963) in his study of macrocysts in *D. purpureum*. Also, Filosa and Dengler (1972) in questioning the origin of the cytophagic cell considered the possibility of cell fusion and noted, "if there is fusion of cells, then there is a possibility for genetic recombination." It remained, however, for Erdos, Nickerson, and Raper (1972) to publish the first tangible evidence that the single, enlarged nucleus in the developing macrocyst might be that of a zygote, and that the initial steps toward achieving a multinucleate state, as described and illustrated by Filosa and Dengler and ourselves, might be meiotic (Figs. 9-11,12,13). The pictorial evidence was not overwhelming, but in a series of sections taken from a developing macrocyst of *Polysphondylium violaceum*, strain ID-10, structures were present that Erdos and other cytologists identified as axial elements of a synaptonemal complex. These structures are illustrated in Figs. 9-12 and 9-13, concerning which Erdos, Nickerson, and Raper (1972) wrote as follows:

The presence of single axial elements of synaptonemal complexes have only been reported for meiotic nuclei (Moses as cited [12]). These are generally thought to indicate late leptotene [15]. Single as well as apparently paired axial elements are found in the early dormant macrocyst (Fig. 9-13). Possible explanations for the absence of the central element in those that appear to be paired, include: first, these may have not yet formed the central element; second, the rigorous methods needed to fix this stage may have destroyed the central element. The second may be more reasonable since those lateral elements that seem paired are properly spaced suggesting that they are in pachytene. There are also single axial elements that are not near any others indicating that pairing may have just begun and that the central element would have not yet formed. We believe there is good cause to assume that this evidence places meiosis in the macrocyst stage in the developmental cycle of the cellular slime molds, a possibility suggested earlier by Blaskovics and Raper [1]. (Numerals in brackets refer to citations in Erdos et al., 1972.)

Taken together, the report by Filosa and Dengler on *Dictyostelium mucoroides* (1972) and that of Erdos, Nickerson, and Raper (1972) on *Polysphondylium violaceum* give a reasonably clear picture of the emergent giant, or cytophagic, cell and its further growth at the expense of the surrounding myxamoebae. Furthermore, both teams followed cyst development to a multinucleate stage, during which time the endocytes (ingested myxamoebae) were reduced to fragments in a finely granular cytoplasm (Fig. 9-14). Whereas neither report covered cyst germination or the changes that precede the reappearance and emergence of myxamoebae, the story was completed in less than a year with the publication of a further study by Erdos, Nickerson, and Raper (1973) on cyst maturation and germination. For this investigation we used *D. mucoroides*, strain DM-7, since Nickerson's prior studies enabled us to predict with some assurance the

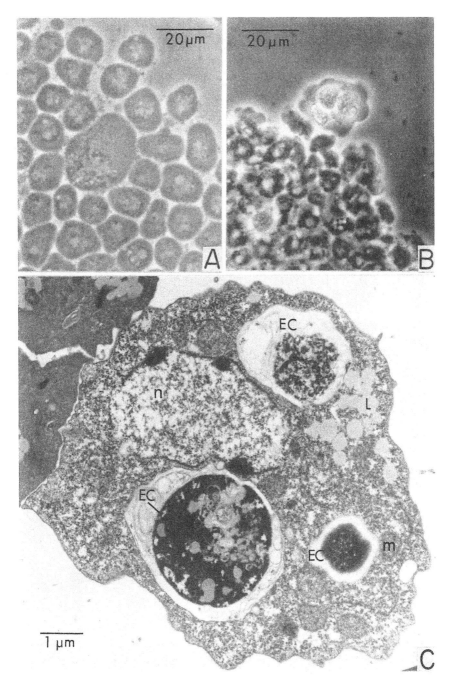

FIG. 9-11. Giant cells of *Polysphondylium violaceum*, strain ID-10. *A*. Giant cell among normal myxamoebae from a very young, broken macrocyst. *B*. Young giant cell containing four endocytes. *C*. Electron micrograph of giant cell containing three endocytes in progressive stages of digestion: (EC), endocyte; (L), lipid; (m), mitochondrion; (n), nucleus. (From Erdos, Nickerson, and Raper, 1972)

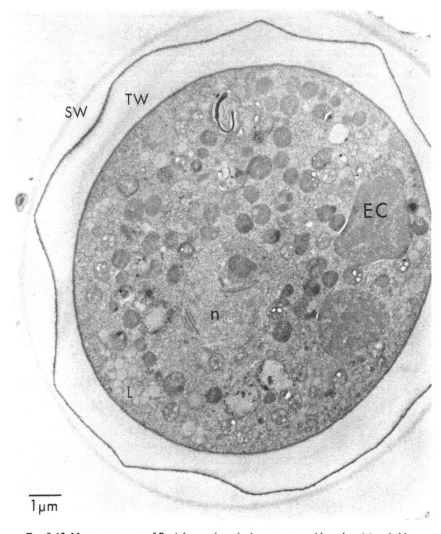

Fɪɢ. 9-12. Mature macrocyst of *P. violaceum* in early dormant stage with nucleus (n) probably in meiotic prophase. Few endocytes (EC) still present, (L) lipid, (SW) secondary wall, (TW) tertiary wall. (From Erdos, Nickerson, and Raper, 1972)

periods of cytoplasmic cleavage within the macrocyst, the activation of newly formed myxamoebae, and the approximate time of their escape from the confining walls. The process of cyst germination and the critical events preceding this were described in this manner:

The cytoplasm of the dormant macrocyst is filled with endocyte fragments which are generally 0.2 μ to 1.0 μ in size. They are usually very

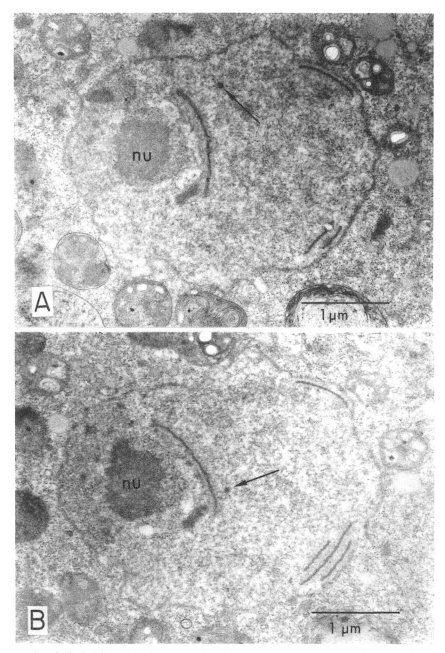

FIG. 9-13. Serial sections of nucleus of *P. violaceum* in meiotic prophase. *A,B*. Apparently paired and unpaired axial elements seen in longitudinal section: single axial elements in cross section (arrows); nucleolus (nu). (From Erdos, Nickerson, and Raper, 1972)

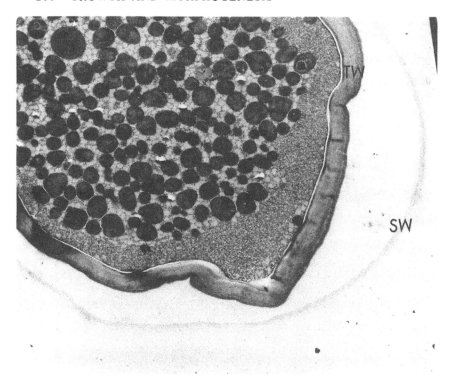

FIG. 9-14. Dormant macrocyst of *Dictyostelium mucoroides*, strain DM-7. Note finely granular cytoplasm, endocyte fragments, and shrinkage of cyst content away from secondary wall (SW), (TW), tertiary wall. × 7500. (From Erdos, Nickerson, and Raper, 1973)

dense and though some substructure can be seen there are no identifiable organelles. These fragments are still bound by the vacuolar membranes.

The first ultrastructural indication of germination is the splitting of the tertiary wall into two separated parts. This splitting occurs in the inner dense layer, and is thought to occur at the time of the initial swelling of the cyst content. A thin portion of this inner layer remains appressed to the cyst plasma membrane. The cyst then cleaves into large uninucleate segments which are larger than myxamoebae (Fig. 9-15). These pro-amoebae still contain many endocyte fragments and the cytoplasm is electron dense. The pro-amoebae then proceed to digest the greater part of the endocyte fragments before continuing to divide. The number of organelles increases during this time, and the cytoplasm becomes less dense. This stage probably corresponds to loss of color and the increase in protoplasmic activity noted by Nickerson and Raper (1973b). The secondary wall seems less compact at this stage, and it together with the separated part of the tertiary wall break away leaving only the inner part of the tertiary wall still intact. The pro-amoebae

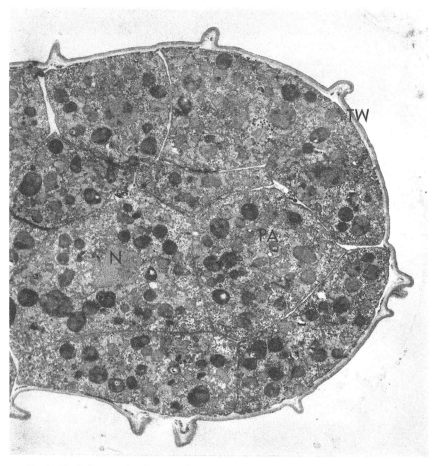

FIG. 9-15. Early stage in cleavage of macrocyst protoplast in *D. mucoroides:* (PA), pro-amoebae; (N), nucleus; (TW), tertiary wall. × 7000. (From Erdos, Nickerson, and Raper, 1973)

divide several additional times, presumably accompanied by concurrent nuclear division, to form the emergent myxamoebae (Fig. 9-16). These cells still contain some evidence of endocyte fragments but are otherwise ultrastructurally indistinguishable from trophic myxamoebae. Finally, the myxamoebae, which no longer contain endocyte fragments, escape by breaking through the previously intact portion of the tertiary wall (Fig. 9-17). They then behave as normal trophic myxamoebae and proceed to multiply (if bacteria are present), subsequently forming sorocarps or macrocysts depending on the conditions (Erdos, Nickerson, and Raper, 1973).

As germination approaches, the secondary wall becomes less compact

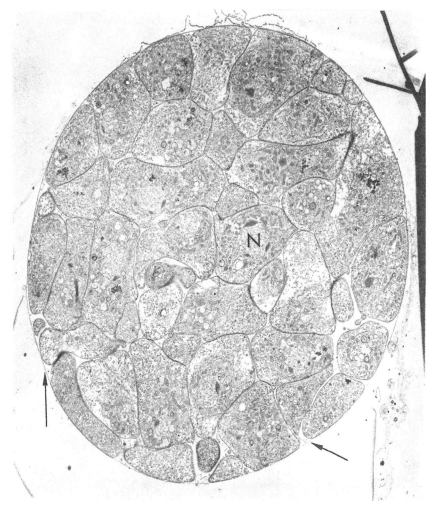

FIG. 9-16. Macrocyst of *D. mucoroides* following cleavage into more nearly myxamoeba-sized protoplasts. Inner part of tertiary wall (arrows) still intact. (N), nucleus. × 4500. (From Erdos, Nickerson, and Raper, 1973)

and is broken, apparently at some random location. The tertiary wall then splits with the outer and middle layers breaking before the inner part. Since the process can be stopped at this point by treatment with cycloheximide, it suggests that breakage of this innermost layer requires the synthesis of an enzyme to effect rupture. Cleavage of the cell content is of interest as well, since this seems to follow the pattern seen, for example, in sporangia of the myxomycete *Didymium nigrum* (Schuster, 1964).

What was not presented by Filosa and Dengler (1972), or by Erdos et

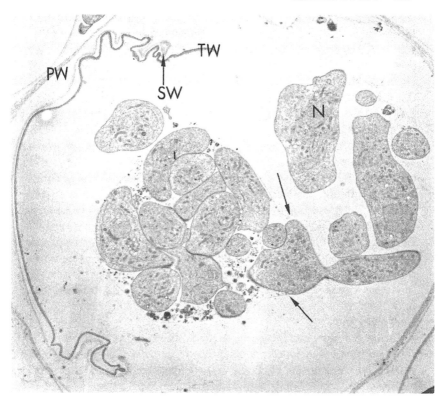

FIG. 9-17. Myxamoebae in process of emerging from broken cyst walls (PW,SW,TW) during germination. Fragments of the innermost layer of tertiary wall still faintly evident (arrows). (N), nucleus. × 4500. (From Erdos, Nickerson, and Raper, 1973)

al. (1972, 1973), or by anyone since that time, was a complete picture of the nuclear history of the macrocyst, particularly the acts of cell and nuclear fusion and the details of the meiotic process. Skill and patience will be required to accomplish these ends, but they are goals worth seeking.

The cytological evidence then at hand clearly pointed to the macrocysts as the sexual stage of the dictyostelids. Still something was missing: genetic proof. The first step toward attaining this objective came also in the spring of 1973, which together with the autumn of 1972 might be referred to in dictyostelid lore as "the year of the macrocyst." We have already noted several landmark papers that appeared in this brief period, but perhaps even more important were two additional papers that for the first time reported the existence of different mating types in the dictyostelids—mating types that when paired appropriately led to the formation of macrocysts. For more than a century the Dictyostelidae had been considered to be asexual, except for an unverified report by Skupienski (1918). Very much later evidence was found indicating that the macrocysts might be the sites of true sexuality, albeit all known cyst-producing strains were self-compatible. Then in 1973 mating types were discovered in two genera and several species, including *Dictyostelium discoideum* and *Polysphondylium violaceum*, for each of which only single homothallic strains were previously known. The papers submitted within a week of each other were published two months later.

The first of these papers, by Clark, Francis, and Eisenberg (1973), reported the presence of two mating types in *D. purpureum*, of which strain DEP2 was designated mating type I while three strains, DEP1 and two others, were designated mating type II. Four other isolates made no macrocysts in any combination. Of 11 isolates of *D. discoideum* tested, strain De-5, representing mating type I, produced macrocysts when paired with De-1 (and 6 others) representing mating type II. Three other strains did not respond to either mating type. The picture in *Polysphondylium violaceum* was more complicated. Of 10 strains examined, mating types were found in each of two groups, or syngens, but showed no mating between syngens. Two strains made no macrocysts in any combination. In *D. mucoroides*, 5 of 25 strains proved to be homothallic but no mating types were found.

The second paper, by Erdos, Raper, and Vogen (1973), dealt only with *D. discoideum* but included a total of 32 strains isolated over a period of years from many localities in the United States and Central America. Sixteen strains, including NC-4, represented one mating type, which was designated as mating type A_1; 8 strains, including V-12, represented a second mating type, A_2, which produced macrocysts when paired with A_1; 2 strains formed macrocysts when paired with strains of either mating type but not alone or with each other; several strains did not form macrocysts under any circumstance. One strain, AC-4, previously studied by Nickerson and Raper (1973a,b), was confirmed as self-compatible. Paired strains that formed macrocysts did so in a graded pattern as indicated in Table 9-4 and as illustrated in Fig. 9-18. The finding that strains NC-4 and V-12 were of different mating types and strongly compatible was viewed with special interest, since the former is the type for the species and has been most studied throughout the world, while the latter was sent to Professor Gerisch many years ago and has been studied quite extensively by him and his associates. Mating types A_1 and A_2 of Erdos et al. (1973) are equivalent to I and II of Clark et al. (1973), respectively, in *D. discoideum*.

Other investigations soon followed. The first of these, by Macinnes and Francis (1974), was based upon *Dictyostelium mucoroides* strain DM-7, the homothallic strain previously investigated by Nickerson and Raper (1973a,b) for macrocyst formation and germination and by Erdos et al. (1973) for EM studies of macrocyst maturation and germination. Three contrasting mutants and the wild type were used in making crosses for

TABLE 9-4. Macrocyst formation among paired isolates of *Dictyostelium discoideum* as reported by Erdos, Raper, and Vogen (1973)

	NC-4	WS-582	WS-583	WS-584	WS-5-1	WS-7	WS-10	WS-11-1	WS-28-1	WS-57-6	WS-112b
NC-4		+++	—	—	—	+++	—	+	++	—	++
	WS-582	+++	—	+++	—	+++	—	—	—	—	++
		WS-583	—	—	+++	—	+	++	—	+	
			WS-584	—	—	—	—	—	—	—	
				WS-5-1	++	—	+	+	—	+	
					WS-7	+	—	—	—	+	
						WS-10	+	+	—	+	
							WS-11-1	—	—	—	
								WS-28-1	—	—	
									WS-57-6	—	
										WS-112b	

KEY: —, no macrocysts; +, fewer than 4 macrocysts per aggregate; ++, more than 4 macrocysts per aggregate and generally on the surface; +++, aggregate almost entirely converted to macrocysts

NOTE: Results obtained with strains paired in all possible combinations.

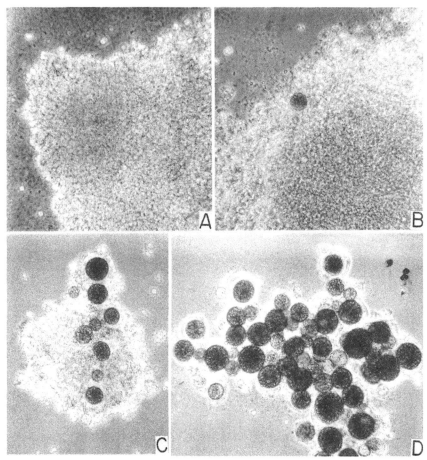

FIG. 9-18. Levels of macrocyst formation among paired isolates in survey of mating types in *Dictyostelium discoideum*. *A*. No mating, no macrocysts. *B*. Little mating, few macrocysts. *C*. Moderate mating, substantial numbers of macrocysts. *D*. Strong mating, virtually all myxamoebae have contributed to macrocyst formation. *A,B,C,D* correspond to −, +, + +, and + + +, respectively, in Table 9-4. (Photographs by G. W. Erdos)

progeny and analyses. In considering possible explanations of observed behavior, Macinnes and Francis concluded "that in *D. mucoroides* (strain DM-7) the assortment of mutant traits among progeny of single macrocysts is consistent only with meiosis." While the possibility of multiple meioses was not excluded, evidence was presented that argued against parasexual haploidization as a part of the macrocyst cycle.

Further evidence that the macrocysts are the sites of true sexuality in the dictyostelids was reported the following year by Erdos, Raper, and Vogen (1975). This study was based upon *Dictyostelium giganteum*, a species clearly related to *D. mucoroides* but with larger sorocarps. The species is heterothallic, but unlike *D. discoideum* it was found to contain four mating types (A_1, A_2, A_3, A_4) rather than two or three. Four wild type strains, all isolated from forest soil and each representing a different mating type, together with temperature-sensitive and drug-resistant mutants obtained from them, were investigated. Crosses were made between the wild types, between the wild types and mutants, and between selected double mutants of different mating types. Some 1400 individual cysts from such crosses were germinated and analyzed. It was of special interest that from any single cyst only one type of progeny was recovered, while among the total progeny from many cysts all expected types were present. This was taken to indicate that three of the four possible meiotic nuclei were lost in a random fashion. Similar situations have been found in the fungus *Rhizopus nigricans* (Gauger, 1961) and in species of the myxomycete *Physarum* (Aldrich, 1967), and it was suggested that the missing products of meiosis may have degenerated in a fashion similar to that reported for *Physarum*. In summary, the results of many crosses seemed to indicate that mating and macrocyst formation in *D. giganteum* is controlled by a single locus, multiple allele incompatibility system.[3]

Additional investigations confirming the central role of macrocysts in cellular slime mold genetics have been published more recently. Beginning about 1974, M. A. Wallace undertook a broad survey of the isolates of *Dictyostelium discoideum* then in our collection with the hope of finding compatible strains which could produce macrocysts that would, after appropriate aging, germinate at high percentages. This goal was never reached, but two strains were identified, WS-10 of mating type *matA* (formerly A_1) and WS-582 of mating type *mata* (formerly A_2), which when paired gave good production of cysts (Fig. 9-19) that at least germinated better than those involving the classic strains, NC-4 (*matA*) and V-12 (*mata*). Mutants resistant to methanol and to cyloheximide were produced, which with their mating type gave three markers each for WS-10 and WS-582. The procedures employed were basically similar to those employed by Erdos et al. (1975) except that cysts were produced in agar plates overlaid with Bonner's salt solution (1947), and plates for cyst germination were incubated in the light at 21-22°C for 2 weeks or more.

Over 6000 progeny from 263 individual germinated macrocysts from

[3] Since some readers may be interested to know how the above tests were conducted, the procedure follows herewith. Cysts 5-6 weeks old produced in the dark on 0.1L-P agar in association with *E. coli* B/r were harvested, washed repeatedly in a streptomycin solution (250 µg/ml), deposited thinly on nonnutrient agar containing streptomycin (to preclude bacterial growth), and incubated in light at 25°C until they germinated (3 days plus). Emergent myxamoebae were allowed to aggregate and produce sorocarps. Sori from single sorocarps were harvested by needle and spores suspended in 0.6 ml of a heavy bacterial suspension. Spore-bacterial suspensions were subdivided into five parts and each part was added to 2.5 ml semi-liquid molten growth medium; aliquots were agitated and poured over plates of solid growth medium. Clones were isolated after 3 days and subcultured to test for mating type and mutant characteristics.

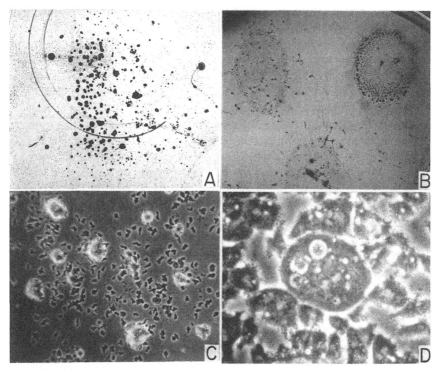

Fig. 9-19. Mating types and macrocyst formation in *Dictyostelium discoideum*. *A*. Macrocysts formed at the frontier between colonies of WS-10, mating type *matA* (left), and WS-582, mating type *mata* (right). × 4. *B*. Test for mating type: spores of unknown mating type were intermixed with those of tester strain, *matA*, and a small drop of the mixture implanted with *E. coli* on 0.1L-P agar. Formation of sorocarps only indicates incompatible reaction (upper left and lower center); presence of abundant macrocysts indicates compatible reaction and shows mating type of unknown to be different from that of tester strain (upper right). × 1.2. *C*. Nascent macrocysts forming around giant cells that have arisen from a mixed planting of *matA* and *mata* spores. × 135. *D*. Newly formed giant cell in a population of mixed mating types. Note that the giant cell contains three endocytes and appears to be binucleate. × 450. (Photographs by M. A. Wallace)

four single-factor crosses, five two-factor crosses, and one three-factor cross were examined and characterized (Wallace and Raper, 1979). In most cases the progeny from a single macrocyst were of one genotype, although in the population of macrocysts from any two-factor cross all possible parental and recombinant genotypes were recovered, as Erdos et al. (1975) had found in *D. giganteum*. There was no evidence of linkage between any of the markers examined. No selection against progeny carrying the methanol or the cycloheximide resistance markers was found in the two-factor crosses, but selection against progeny carrying both resistance markers was found in the three-factor cross. Germination of macrocysts in all crosses was poor, only once exceeding 2.5% of the total macrocyst population. A variety of crosses and back-crosses with different parental strains indicated that germination might be influenced by both extrinsic (environmental) and multiple genetic factors. About 10% of the macrocysts yielded progeny that were ambivalent in their mating reactions, but after extensive recloning these populations could be resolved to the normal *matA* and *mata* mating types and might therefore have represented aneuploids. The results obtained with *D. discoideum* macrocysts differ from those obtained with other cellular slime molds and are reminiscent of the results reported for germinated zygospores of *Phycomyces blakesleeanus* (see Eslava et al., 1975).

Many of the cultures that had been investigated by Erdos et al. (1973), including strong maters, ambivalent strains, and nonmaters, together with a second weakly homothallic strain (Za-3a), were subsequently sent to Keith L. Williams in Canberra, Australia, for further study. Papers by Robson and Williams followed. The first (1979) was a comprehensive report on vegetative incompatibility, wherein they reported the formation of parasexual diploids at a frequency of 10^{-6} to 10^{-5} between compatible haploid strains, while incompatible strains yielded only about 10^{-8} diploids. A single vegetative incompatibility site was thought to be located at, or closely linked to, the mating-type locus. A second study (1980) was centered upon the mating system per se. Their summary of the latter follows:

> The mating reaction (macrocyst formation) and vegetative compatibility (which is believed to be associated with only the mating type locus in *D. discoideum*) of asexual, bisexual and homothallic strains of *Dictyostelium discoideum* were examined. Three asexual strains were vegetatively compatible with *matA* tester strain and vegetatively incompatible with a *mata* tester strain, so we propose that these asexual strains are in fact strains of *matA* mating type with defective mating capacity. Two bisexual and two homothallic strains were vegetatively incompatible with both *matA* and *mata* tester strains, indicating that they either express both mating alleles or that they have a third mating type allele. On the basis of observations on the relative mating capacity with *matA* and *mata* tester strains, we propose that bisexual strains are closely related to homothallic strains and are not strains carrying a third mating type allele. Hence we suggest that *D. discoideum* has a one locus, two allele mating system. The homothallic strains AC-4 and ZA-3A may express both established mating type alleles (*matA* and *mata*). They are, however, normal haploid strains with seven chromosomes.

In their initial paper Clark, Francis, and Eisenberg (1973) confirmed cyst formation in *Dictyostelium mucoroides, D. purpureum, D. discoideum*, and *Polysphondylium violaceum* (see also Nickerson and Raper, 1973a; Erdos, Raper, and Vogen, 1973). More significantly, they showed that except for *D. mucoroides* these species were basically heterothallic and required the pairing of compatible mating types to effect macrocyst formation. Furthermore, in *P. violaceum* the active isolates, eight of ten, seemingly constituted two syngens, or breeding groups; that is, a strain belonging to one syngen could not mate with any member of the other syngen. A further and more comprehensive study by Clark (1974) confirmed this breeding pattern. By careful analyses of 49 isolates, 20 from Delaware and Maryland and 29 from Texas, she was able to identify two syngens, designated 1 and 2, each with two mating types, I & II and III & IV, respectively. Ten stocks were nonreactive and might have represented a third syngen if a suitable cohort had been found, although, as Clark recognized, other explanations were possible. Disparate growth rates seemed an unlikely explanation, since in other studies Clark and Speight (1973) had found macrocysts at times in cell mixtures where one mating type represented 0.1% of the total cell population, a figure even lower than that reported by Hirschy (1963) for macrocyst induction by dark-grown cells of *D. purpureum*.

Clark et al. found no evidence of macrocyst formation in *Polysphondylium pallidum* in 1973, but cysts were found soon thereafter and became the subjects of investigations by Francis (1975b) and by Eisenberg and Francis (1977). Once he had discovered two mating types in *P. pallidum*, designated I & II, Francis studied the species in much the same manner as Macinnes and Francis had done for *D. mucoroides* (1974). Morphogenetic mutants were again employed, coupled in this case with different mating types. Of special interest were mutants which alone did not develop beyond the aggregation stage, but which when intermixed were capable of producing macrocysts that could be germinated after 6 weeks aging. The frequency of germination was 5-10%, and the progeny when cloned yielded several classes of nonparental phenotypes. While recognizing other possibilities, the recovery of wild type progeny from the dihybrid cross between morphogenetic mutants was taken as further support for the idea that the macrocysts are zygotes and meiosis is the means of genetic exchange in the dictyostelids. As isolated from nature, strain Pp 28s, which produced substantially larger spores and was homothallic as well, was thought to represent a diploid heterozygous for mating type. This was subsequently examined by Robson and Williams (1980) and found to have 13 chromosomes, the haploid number in *P. pallidum*.

In a further study of macrocyst formation and its significance in *Polysphondylium pallidum*, Eisenberg and Francis (1977) analyzed for mating potential a large number of isolates from widely separated collection sites. Of 231 strains, mostly from northern Delaware, 78% represented one or the other of two mating types, 18% failed to mate in any pairing, 3% were homothallic, and two isolates from a Florida soil proved to belong to a second syngen. The latter strains were judged initially to be nonmaters but later found to form macrocysts when paired, but only with themselves.

We have learned recently via correspondence with, and have received cultures from, Hiromitsu Hagiwara, who has isolated compatible mating types of *Polysphondylium pallidum* from soils of Japan. Of two isolates used as tester strains, one or the other is reported to have produced macrocysts when paired with isolates from Formosa, New Guinea, The Seychelles, and Canada. Mating types capable of producing macrocysts have also been isolated for a relatively new and especially striking nonpigmented species, *P. tenuissimum* Hagiwara (1979).

Macrocysts are known to occur in two additional cellular slime molds, *Dictyostelium rosarium* and *D. mexicanum*, and may in fact be produced in all dictyostelids when cultural conditions are propitious, and when compatible mating types are present, if such are required. When first described by Raper and Cavender in 1968, *D. rosarium* was thought to be rare, but in screening for cellular slime molds in soils of southern California a decade later, Benson and Mahoney (1977) found *D. rosarium* to be the most prevalent species. Many of their isolates were sent to us for comparison, and some 20 of these, together with the 6 strains then at hand, were investigated for their capacity to produce macrocysts by Ming Tu Chang, then a doctoral candidate in our laboratory. In summary, she found *D. rosarium* to be heterothallic with no homothallic strains among those examined. Three mating types were identified and designated A_1, A_2, and A_3. All strains were self-sterile but cross-fertile with strains belonging to each of the other mating types. The results of this study were reported by Chang and Raper (1981). Conditions favoring macrocyst formation were generally similar to those found to be optimal for most other species: *Escherichia coli* or *E. coli* B/r as a source of food, equal numbers of spores of paired cultures spread on freshly poured unbuffered 0.1L-P agar plates, dark incubation, free surface liquid in plates, and incubation at 22-23°C. The macrocysts were generally smaller and matured more slowly than those of *D. mucoroides, D. purpureum*, and *D. discoideum* studied previously. The cysts failed to germinate even at 2 months of age.

Limited information is available relative to macrocyst production in *Dictyostelium mexicanum*, a species isolated from Central American soils and only recently described by Cavender, Worley, and Raper (1981). In tests conducted by Dr. Franz Traub (1977) some small, yellowish macrocysts were formed when pairings were made between one strain and each of two others, thus indicating the presence of two mating types. Three isolates were inactive, apparently, and no homothallic strains were reported.

Although no suggestion of close kinship is intended, it is of interest that in several genera of fungi, such as *Mucor* and *Absidia*, most species are heterothallic while a minority are homothallic. A comparable situation appears to prevail among the dictyostelids, except that the divergent behavior occurs within single species, as is true also of a few plasmodium-forming myxomycetes (Gray and Alexopoulos, 1968). Currently we think of *Dictyostelium mucoroides* and *D. minutum* as being homothallic whereas *D. giganteum* and *D. rosarium* are heterothallic, insofar as is known. Other species such as *D. discoideum, Polyphondylium violaceum, P. pallidum,* and probably *D. purpureum* are basically heterothallic but contain occasional self-compatible strains. For species such as *D. rhizopodium, D. lacteum, D. aureo-stipes,* and *Acytostelium leptosomum* macrocysts have not been reported. This is not to say, however, that such do not occur in nature, or that they might not arise in laboratory cultures if we but knew how to induce their formation and favor their development. In fact, when we view the subject broadly and objectively, it is painfully clear that we still know comparatively little about the macrocysts—how they arise, how they develop, and how they germinate, while of their possible significance in nature we know virtually nothing.

For example, how does the giant cell arise in a homothallic clone? Can we conclude that it is by cell and nuclear fusion as in the case of paired heterothallic strains, or between mutants such as those studied by Macinnes and Francis (1974)? And when does nuclear fusion occur—must it precede phagocytosis by the giant cell? What determines the size of the precyst? Is this conditioned by or independent of the giant cell within? If the former, how is the message transmitted; and if the latter, how are peripheral cells stimulated to lay down an encircling fibrillar (primary) wall? What happens to the supernumerary giant cells that do not develop into macrocysts? And what determines where the precyst boundaries will be within a cell mass that yields multiple macrocysts? These are but a few of the questions for which we have only limited information, or in some cases none at all.

Several studies have been published reporting the presence of diffusible or volatile substances of unknown nature that play a key role in initiating macrocyst formation. In the first of these, based upon *Dictyostelium mucoroides,* Weinkauff and Filosa (1965) described experiments wherein macrocyst formation was dependent upon the absorption by charcoal of some gas or volatile substance. This study has since been extended by Filosa, and in a later paper (1979) he concludes that the direction of morphogenesis (whether to sorocarps or macrocysts) is influenced by the interplay of at least three factors: light, possibly CO_2, and CAS (charcoal-absorbable-substance). Other investigators have turned their attention to *Dictyostelium discoideum* with two papers appearing in 1975, one by O'Day and Lewis, the other by Machac and Bonner. In each case the historic strains NC-4 (mating *matA*) and V-12 (mating *mata*) were studied, and in each case a substance was reported to pass through a dialyzing membrane from cultures of NC-4 to V-12 and to induce macrocyst formation in the latter strain—there was no reverse response. Thus the myxamoebae of NC-4 were portrayed as the inducers and those of V-12 as the responders. Two years later, O'Day and Lewis (1977) reported a volatile sex hormone, or pheromone, to be produced by cultures of NC-4 that induced macrocyst formation in V-12. For this experiment small petri dishes containing the two mating types were placed within a larger petri dish, sealed with black plastic tape, and incubated in darkness. Results were as before; macrocysts developed in cultures of V-12 but not in NC-4. Additionally, more macrocysts were formed when the larger dish contained two cultures of NC-4 and one of V-12 rather than one of each, suggesting a dosage effect of the volatile pheromone. The results appear unequivocal as reported and in essence not unlike behavior found in certain fungi (see van den Ende, 1976). Yet there is lingering doubt, for students in our laboratry and in Bonner's (personal communication) have failed to confirm macrocyst production when cross contamination of mating types is rigorously excluded. But the picture is still not clear. Working with the same mating strains, Donna Bozzone of Princeton has reported recently (1982) that cells of NC-4 must be present for those of V-12 to form, successively, giant cells and macrocysts. At the same time the number of giant cells and macrocysts bears a direct relationship to the number of V-12 myxamoebae when the two strains are intermixed. Especially interesting is her report that at high V-12:NC-4 ratios some macrocysts are formed without the inclusion of any NC-4 cell. For such behavior, a possible explanation was offered by Bozzone and Bonner (1982): "Somehow, live NC-4 cells must induce V-12 amoebae to proceed on the macrocyst development pathway. Perhaps they must come into contact with the V-12 cells and in this way affect their cell surface. Alternatively, cell contact per se might not be required but cells must be in close proximity so that short-range interactions necessary for development can take place (Gross et al., 1981)."

Once the giant cells have formed in *Dictyostelium discoideum* there is substantial evidence that they chemotactically attract myxamoebae in the immediate environs, although there may be limited outward movement as well. This was first observed by Wallace (1977) in time-lapse sequences and subsequently confirmed in like manner by O'Day and Durston (1979) and by O'Day (1979) with tracings of individual cell movements. In the latter paper O'Day implicated cAMP and noted that macrocysts up to the mid-endocyte stage acted as strong centers of attraction for aggregation-competent myxamoebae, whereas macrocysts more advanced in their development became less attractive. When myxamoebae of different mating types are permitted to grow as a mixed population in the dark, one may expect giant cells to appear after a day or more, depending upon the substrate and the culture conditions employed. Chagla, Lewis, and O'Day (1980) have made the interesting observation that the number of giant cells increases dramatically with the addition of CA^{2+} up to a level of 1.0 mM as $CaCl_2$, the proportion of these in the overall population reaching more than 14% at this concentration and representing a 57-fold enhancement above the control. At similar concentration other salts such as MgCl, NaCl, and KCl had little effect.

Limited studies have been performed with giant, or cytophagic, cells per se, and of these the more significant include those of Yoshio Fukui (1976) and of Saga and Yanagisawa (1982). Fukui dissociated nascent macrocysts of *D. mucoroides* (DM-7) enzymatically and separated the giant cells by density gradient centrifugation. Giant cells recovered in this manner remained active, and when combined with the smaller peripheral cells in liquid shake cultures reconstituted macrocysts within a few hours. A giant cell would attach to an agglutinate and then move to its center. When two agglutinates came in contact a dumbbell-shaped mass was formed, after which one of the giant cells moved through the peripheral cells and phagocytized the other before macrocyst formation proceeded. Saga and Yanagisawa (1982) based their study upon two strains of *D. discoideum*, NC-4 nd HM_1, a derivative of V-12. Myxamoebae of the two strains were suspended at a ratio of 1:1 in Bonner's salt solution containing 1×10^{10} bacteria and incubated with agitation at 22°C in the dark. After 20-22 hours giant cells were recovered by straining the mixture through a fine nylon mesh (10 μm). By recombining the giant cells with washed nonaggregative cells at predetermined levels, or by omitting them, synchronous production of macrocysts or of sorocarps could be achieved. This in turn permitted them to perform various analyses and to show that the specific activity of several developmentally regulated enzymes differed significantly between the two contrasting morphogenetic modes. Gerisch and Huesgen (1976) reported the formation of macrocysts by aggregation-deficient mutants and also recorded the presence of giant cells that remained actively motile in a free state. The mutants were derived from NC-4 and V-12, hence were of different mating type. That such "free" giant cells could subsequently initiate macrocysts was inferred but not so stated. Emphasized was the fact that the mating reaction could occur in the absence of full aggregative competence. Very recently Szabo, O'Day, and Chagla (1982) have reported in detail concerning cell fusion, nuclear fusion, and zygote differentiation during sexual development in *D. discoideum*—noting, for example, that cell fusions begin about 11 hours after mating type cells are intermixed, and that some giant cells may later contain multiple nuclei, only one of which is zygotic.

Since light was known to enhance the formation of sorocarps and to suppress the formation of macrocysts, it became of special interest to determine what wavelength(s) were most effective in influencing postaggregative development. To resolve this question Dr. Chang of our laboratory enlisted the cooperation of Professor Kenneth Poff of Michigan State University, and together they investigated the effects of twelve different monochromatic light sources at four different intensities ranging from 0.007 to 0.778 μW/cm². Blue light at 425 μm was found to be most effective in inhibiting macrocyst formation, whereas infrared was totally ineffective (Chang, Raper, and Poff, 1983). It was thought that the primary photoreceptor was possibly a type b or type c cytochrome, and that suppression of macrocyst formation could be mediated by the oxidized form of the photoreceptor. It is considered significant that wavelength 425 μm is also especially effective in directing the photoaxis of the migrating pseudoplasmodia of *D. discoideum* and the sorogens of *D. purpureum*, as previously reported by Francis (1964).

Now that the macrocysts are known to represent the sexual stage of the dictyostelids, and now that we have at hand the information necessary to produce them at will and in quantity for several species, it seems unreal that they have figured prominently in so little research to date. Unreal, that is, until one recalls the long periods still required for cyst maturation and the difficulties encountered in securing even the present low levels of cyst germination in all but one strain of *D. mucoroides*, DM-7. Care and patience will be required; but there should be ways of accelerating maturation and enhancing germination, if we could but devise them.

PART II
SYSTEMATICS

CHAPTER 10

Acrasiomycetes

The Acrasiomycetes all have a trophic phase consisting of small, free-living myxamoebae, all show some type of cell aggregation, and all undergo some degree of cellular differentiation incident to fructification. Such differentiation is limited in some cases but substantial in others, and upon this and other bases two subclasses can be recognized. Of these, the Acrasidae (acrasids) represent the less differentiated forms and are presumed to be more primitive, while the Dictyostelidae (dictyostelids) represent more strongly differentiated forms and are considered to be more advanced. Historically, the two have been studied together, sometimes with subtle implications of progressive phylogeny (E. W. Olive, 1901 and 1902). This view is now considered untenable (Raper, 1973; L. S. Olive, 1975), and it is doubtful if the two groups constitute a natural class. Nonetheless, and despite the fact that this monograph relates specifically to the Dictyostelidae, we think it proper that the Acrasidae be included in a summary way to illustrate additional patterns of development that have emerged through the aggregation, coordination, and differentiation of previously independent myxamoebae.

Investigators disagree concerning the proper classification of the Acrasiomycetes. Most mycologists have regarded them as nonhyphal fungi and have included them in a special division currently designated the Myxomycota. Other mycologists and the protozoologists have considered them to be colonial protists and have assigned them to the Rhizopoda or Sarcodina among the protozoa. Arguments can be advanced in support of each position, and we shall not attempt to resolve the question here. In the interest of nomenclatorial consistency, however, we shall continue to classify them as primitive fungi and will, insofar as possible, adhere to the International Code of Botanical Nomenclature when applying names for families and higher taxa.

The broad classification to be followed is that proposed by Ainsworth (1973), while classification within the class will follow Raper (1973) with such additions as are necessary to include the different pseudoplasmodial slime molds presently known. With reference to the Dictyostelidae it may be noted that relatively recent electron microscopic studies of nuclear divisions in *Polysphondylium violaceum* (Roos, 1975a) and *Dictyostelium discoideum* (Moens, 1976) have revealed, according to Kubai (1978), "characters that one would hope to find in the prototype of fungal divisions." Olive (1975) holds a different view and in recent correspondence has emphasized especially the ingestion of food organisms as an animal characteristic.

The first organism to be described as an aggregative slime mold was reported by van Tieghem in 1880 as an isolate from a paste of brewer's yeast, and for this he selected the genus name *Acrasis* to denote a complete lack of fusion among the convergent myxamoebae. He studied species of *Dictyostelium* also and, noting the absence of cell fusion in that genus as well, created the family Acrasiées to include both genera, along with *Guttulina* previously described by Cienkowski (1873). Van Tieghem emphasized particularly that these organisms were characterized by "Plasmode agrégé" and contrasted such associations with the "Plasmode fusionné" formed by myxomycetes, to which Brefeld (1869) had assigned *Dictyostelium* initially. Few taxonomic studies have been published, but these have consistently recognized some adaptation of the name *Acrasis* as a proper designation for the class, subclass, or order that included the dictyostelids and other pseudoplasmodial slime molds. That precedent is continued here.

In addition to those noted above, additional genera have been described by Brefeld (*Polysphondylium*, 1884), van Tieghem (*Coenonia*, 1884), Zopf (*Copromyxa*, 1885), E. W. Olive (*Guttulinopsis*, 1901), Raper (1946) and Raper and Quinlan (*Acytostelium*, 1958), Traub (*Heterosphondylium*, 1977), Raper, Worley, and Kurzynski (*Copromyxella*, 1978), and Worley, Raper, and Hohl (*Fonticula*, 1979). These are shown in one of several possible arrangements in the class key that follows and are discussed briefly and illustrated in the accompanying text. Genera of the Acrasidae are included primarily as supplemental material; for this reason, descriptions of these cellular slime molds are brief and comments limited, in comparison with the attention to be given to genera and species of the Dictyostelidae that follow.

ACRASIOMYCETES AND OTHER SLIME MOLDS
Class Key Adapted from Raper (1973)

Kingdom Plantae (plants)
Kingdom Mycetae (fungi)
I. Division Myxomycota (mycetozoa or slime fungi). Vegetative stage amoeboid; plasmodia or pseudoplasmodia present
  A. Class Acrasiomycetes (cellular slime molds): Acrasiées van Tieghem. Vegetative (trophic) stage small amoeboid cells; multicellular fructifications formed from pseudoplasmodia; plasmodia lacking; flagellate cells reported for undescribed genus (L. S. Olive)
    1. Subclass Acrasidae (acrasids). Myxamoebae with lobose pseudopodia and nuclei with compact, centrally positioned nucleoli; cell aggregation without

stream formation; cellulose rarely present

Order Acrasiales—same as the subclass

a. Family Acrasiaceae Raper (1973). Fructifications consist of differentiated cellular stalks bearing spores in chains or sori; some walls reported to contain cellulose

   (1) Genus *Acrasis* van Tieghem (1880). Stalk cells 1.5-2.0 times longer than broad; basal cell(s) forming a palm-like holdfast. *Acrasis* as redescribed by Olive and Stoianovitch (1960) may differ

   (2) Genus *Pocheina* Loeblich and Tappan (1961). See *Guttulina* Cienkowski (1873)

b. Family Copromyxaceae Olive and Stoianovitch (in L. S. Olive, 1975). Fructifications lacking demarcation into stalks and spores, composed of sorocysts throughout; myxamoebae of limax type

   (1) Genus *Copromyxa* Zopf (1885). Fructifications robust columns, simple to very irregularly branched; myxamoebae commony 30 μm in diameter; habitat dungs

   (2) Genus *Copromyxella* Raper, Worley, and Kurzynski (1978). Fructifications delicate columns, simple or branched; myxamoebae rarely exceed 20 μm in diameter; habitat soil, dung, and decaying mushrooms

c. Family Guttulinaceae E. W. Olive (1901). Fructifications with definite stalks and sori; sori globose to subglobose, unwalled

   (1) Genus *Guttulina* Cienkowski (1873) = *Pocheina* (Cienk.) Loeblich and Tappan (1961). Stalks consist of superimposed cells, flattened and arranged in one or more tiers, without sheath or membrane; sori globose, unwalled and spores not suspended in slime; habitat wood and bark of trees

   (2) Genus *Guttulinopsis* E. W. Olive (1901). Stalks variable, reduced or conspicuous, composed of compressed cells bounded by external membranes; sori unwalled with spores suspended in slime; habitat dungs and soil

d. Family Fonticulaceae Worley, Raper, and Hohl (1979). Stalks prominent, tapered, becoming hollow as myxamoebae ascend through them to form spores; sori globose, unwalled; myxamoebae small, with filose pseudopodia when swimming; habitat dung

   (1) Genus *Fonticula* Worley, Raper, and Hohl—same as the family

2. Subclass Dictyostelidae (dictyostelids). Myxamoebae with filose pseudopodia and vesicular nuclei with peripherally lobed nucleoli; aggregating cells typically form converging streams; stalks usually cellular and encased by cellulose sheaths (tubes); sori globose to citriform, unwalled; walls of stalk cells and spores contain cellulose

Order Dictyosteliales—same as the subclass

a. Family Dictyosteliaceae Rostafinski (1875). Fructifications (sorocarps) simple or branched, consisting of cellular stalks bearing globose to citriform sori

   (1) Genus *Dictyostelium* Brefeld (1869). Sorocarps simple or irregularly branched; sori globose to citriform, usually terminal but may be sessile

   (2) Genus *Polysphondylium* Brefeld (1884). Sorocarps with branches in whorls; sori globose and terminal on central axis and branches

(3) Genus *Coenonia* van Tieghem (1884). Sorocarps simple or branched, with prominent crampon-like bases; sori gelatinous, borne in terminal cellular cupules. (Known only from original description)

b. Family Acytosteliaceae Raper and Quinlan (1958). Sorocarps delicate, consisting of very thin acellular stalks bearing terminal globose sori

(1) Genus *Acytostelium* Raper (1956), also Raper and Quinlan (1958)—same as the family

B. Class Protosteliomycetes (protostelids). Fructifications mostly unicellular or with 2-4 cells; pseudoplasmodia lacking; very small plasmodia and flagellate cells in some genera. Representative genus: *Protostelium* Olive and Stoianovitch (1960). For full coverage see reviews by L. S. Olive (1967, 1975) and original papers by Olive and Stoianovitch (1960 et seq.)

C. Class Myxomycetes (Acellular or plasmodium-forming slime molds). Vegetative stage multinucleate plasmodia, often large; fructifications in form of sporangia of many types. Representative genus: *Physarum* Persoon (1794). For full coverage see Martin and Alexopoulos, 1969, pp. 1-477, pls. 1-47; also Gray and Alexopoulos, 1968, pp. 1-288

II. Division Eumycota (true fungi—molds and mushrooms). Vegetative stage consists of tube-like hyphae

ACRASIDAE

Acrasis

The genus *Acrasis* was described a century ago but the species upon which it was based still awaits rediscovery. Unfortunately, in describing *A. granulata* van Tieghem (1880) published no illustrations, gave virtually no information about the trophic stage, and included less than one needs to know about its fruiting structures. Still his report contained enough facts to give a fair idea of what the fructifications were like, even if dimensions were given only for the spores. Following convergence (aggregation) the myxamoebae were said to form small clusters, and by sliding one over the other the still independent cells rose up vertically to form cones that became longer and thinner. Meanwhile, in the central axis of each cone cells in a single tier were modified to form a stalk that supported a chain of spores. Such structures commonly occurred singly, but in rich cultures coremium-like clusters of 10 to 12 might be formed, each bearing its chain of spores and the whole suggesting a "penicillus." The basal cell was typically dilated to form a palm-like holdfast, and in larger structures this might be buttressed by other cells applied against it. The numbers of cells in stalks and spore chains varied and, not surprisingly, seemed to be influenced by environmental factors. Cells in the stalks were reported to be 1½-2 times as long as broad and to be bounded by "une membrane de cellulose." The spores were large, variable in size from 10 to 15 μm in diameter, violaceus-brown, and globose with granular walls that contained cellulose.

Of special significance was the following observation: "Chacun des myxamibes épars dans le liquide, quand la crossance a pris fin, entre donc directement et tel quel dans la constitution du fruit où, suivant la place qu'il occupe, il devient, soit une spore, soit une cellule du pied."

For eighty years the genus *Acrasis* remained an enigma; then in 1960 L. S. Olive and Carmen Stoianovitch described an organism as *Acrasis rosea* that conforms with the description of *A. granulata* in many particulars but not in all. While the two slime molds probably belong together, a question of generic identity may arise when and if *A. granulata* is rediscovered. Much of the uncertainty stems from our lack of information about the myxamoebae of *A. granulata*: What type of pseudopodia did they have? Were they lobose as in *A. rosea* or filose as in the dictyostelids? What of the nucleus? Did it have a single, centrally positioned nucleolus as in *A. rosea* or peripherally lobed nucleoli as in the dictyostelids? To what degree did the stalk cells differentiate? If isolated would they revegetate as do those of *A. rosea*? Or as their tiered arrangement and elongate shape might suggest, were they more like the vacuolate, dead stalk cells of a dictyostelid? These are but some of the questions that await answers.

In the meanwhile Olive and Stoianovitch (1960) and Olive (1975) have published comprehensive accounts of the growth and development of their *Acrasis rosea* and have provided us with living cultures as well, and so further characterization of the genus is based upon that organism (Fig. 10-1). The myxamoebae, pinkish in color, are characterized by lobose pseudopodia, commonly range from 20 to 30 μm in diameter, and are usually uninucleate (Fig. 10-1A). The nuclei contain compact, centrally positioned nucleoli. Trophic cells may feed upon bacteria but seem to grow better on yeasts such as *Rhodotorula rubra*. Following vegetative growth the myxamoebae often spread outward from the yeast colony and collect into small mounds (Fig. 10-1B,C) from which arise fruiting structures of differing size and complexity. These vary from simple tiers of slightly modified cells bearing single chains of spores to columns several cells in cross section and bearing spores in many branched and tangled chains (Fig. 10-1D-F). The cells in the stalks, as the spores, can germinate to renew the growth cycle. There is no evidence for an expanded holdfast at the base of the column as reported for *A. granulata*, and one senses that the cells of the stalk are much less strongly differentiated than were those in van Tieghem's species. The smooth-walled, spherical to slightly elliptical spores range from about 10 to 15 μm in diameter and are only a little smaller than the microcysts that are often formed by isolated and unaggregated cells. Walls of the spores and microcysts were reported initially to contain cellulose, but of this the species authors are now less certain.

Acrasis rosea was initially isolated from the dry florets of *Phragmites* and is, apparently, quite common in nature, having been isolated by Olive and Stoianovitch from dead plant parts collected from many countries and

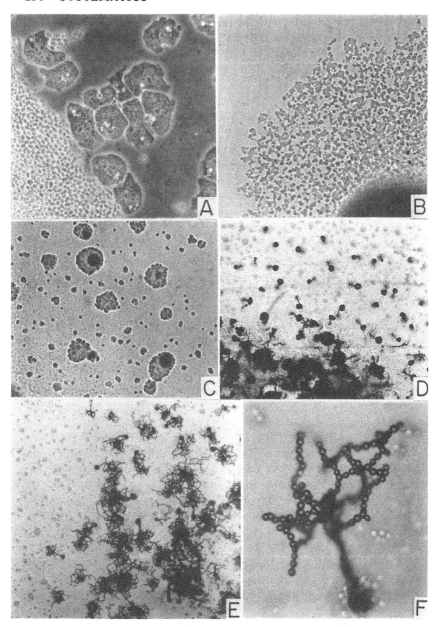

FIG. 10-1. *Acrasis rosea* Olive and Stoianovitch. *A*. Myxamoebae feeding on the yeast *Rhodotorula rubra*. × 300. *B*. Myxamoebae outside a yeast colony following vegetative growth. × 18. *C*. A similar area after two hours; cell aggregates, or pseudoplasmodia, have formed. × 24. *D*. Fruiting structures in process of formation with rounded sorogens from which chains of spores will develop later. × 24. *E*. A similar area after 5 hours; many fruiting structures (sorocarps) are now completely formed. × 12. *F*. A single sorocarp showing detail of branched spore chains. × 200. (From Raper, 1960)

representing many different plant species. Considering the ubiquity of *A. rosea*, and if the two species are closely related, it seems incredible that someone, somewhere has not rediscovered *Acrasis granulata*.

For further information concerning *Acrasis rosea*, the reader may consult Olive and Stoianovitch (1960) and Olive (1975) with regard to growth and morphogenesis, Fuller and Rakatansky (1965) concerning carotenoid pigments, Reinhardt and Mancinelli (1968) for light responses, and Reinhardt (1975) regarding inter- and intrastrain variation.

Copromyxa

Copromyxa protea (Fayod) Zopf (1885) is representative of, if not in fact the sole species of, this genus. First reported as *Guttulina protea* by Fayod in 1883, the generic name was changed to *Copromyxa* by Zopf two years later. As originally described, the organism was said to produce upright fructifications in the form of "spindles, horns, clubs or little knobs" 1-3 mm high and to show no sign of demarcation into stalk and sporogenous regions, the cells throughout consisting of closely packed sorocysts. It was upon this basis that Zopf (1885) removed it from *Guttulina*, for in describing *G. rosea* Cienkowski (1873) had recorded the presence of a stalk and emphasized the specialized character of its cellular components.

As studied in this laboratory, *Copromyxa protea* when grown on dung and selected laboratory substrata produces upright fructifications that range from raised cones and simple columns to richly and irregularly branched structures up to 2.5-3.0 mm high (Fig. 10-2A,B). Columns and branches consist of similarly encysted cells throughout and may be few to several cells thick. The constituent cysts are held together by secreted slime and are often, but not always, quite irregular in form and dimensions (Fig. 10-2D-F). The myxamoebae are of the limax type and commonly range between 25 and 35 μm in major diameter, contain one or more contractile vacuoles, and are usually uninucleate, the nuclei being characterized by single prominent, centrally positioned nucleoli (Fig. 10-2C). Myxamoebe that do not aggregate commonly form cysts on the agar surface that are strikingly similar to those that comprise the columns.

Fayod's original description of this slime mold failed to mention branched structures, and when Nesom and Olive encountered them they presumed their isolates to represent a new species and described it under the name *Copromyxa arborescens* Nesom and Olive (1972). Additionally, they did not observe spherical cysts as reported and illustrated by Fayod (1883). Since that time many more isolates have been studied, and in the course of such investigations Spiegel and Olive (1978) obtained evidence of a sexual stage that became apparent when certain isolates, representing compatible mating types, were brought together. In such paired cultures, and only in these, they observed within a few days plaques of round cysts with

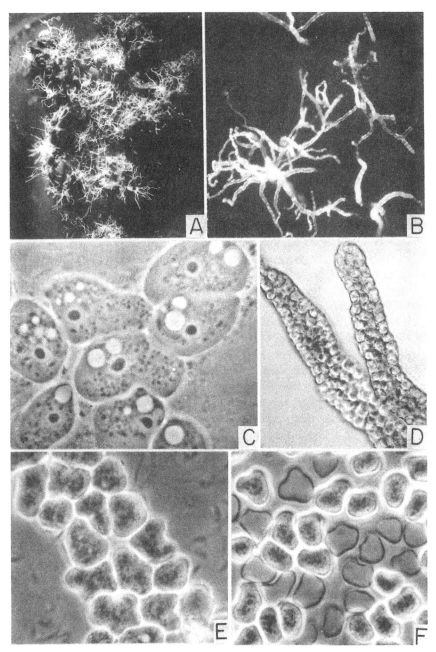

FIG. 10-2. *Copromyxa protea* (Fayod) Zopf. *A*. Profuse development of fruiting structures on sterilized horse dung. × 2. *B*. Irregularly branched fruiting structures formed on a weak glucose-yeast extract agar in association with a mixed indigenous bacterial flora. × 12. *C*. Myxamoebae growing in association with *Escherichia coli* on ¼ strength nutrient agar—note especially the conspicuous nuclei (with prominent central nucleoli) and the large contractile vacuoles. × 850. *D*. Two columns showing characteristic cellular structure and absence of enveloping walls. × 210. *E*. A portion of the same showing the irregular form of the constituent-sorocysts. × 1000. *F*. Sorocysts from column showing empty cyst walls following germination. × 850. (In part from Raper, 1973)

thicker, somewhat roughened and brownish walls, structures to which they applied the name sphaerocysts in order to distinguish them from the irregular sorocysts that constitute the upright columns and the microcysts that form on the substrate surface. Since cysts of the same type had been clearly illustrated by Fayod, Spiegel, and Olive concluded that he and they worked with the same species and so proposed that the name *C. arborescens* N. & O. be regarded as a synonym of *C. protea* (Fayod) Zopf.

The genus *Copromyxa* has received minimal attention during the century since it was first recognized despite its common occurrence on the dungs of such herbivorous animals as horses and cows. In the laboratory it grows well and seems to fruit best in association with a mixed indigenous flora on agar substrates of low nutrient content such as 0.1% glucose, 0.05% yeast extract.

Copromyxella

Copromyxella is a relatively new genus introduced in 1978 by Raper, Worley, and Kurzynski to accommodate four new species of cellular slime molds that are in certain respects suggestive of *Copromyxa*. As in that genus, the fructifications consist wholly of encysted myxamoebae and show no demarcation into areas of supportive and reproductive cells. Fruiting structures are much more delicate, however, and the cellular elements are appreciably smaller than those of *Copromyxa*. The myxamoebae often show a complete absence of contractile vacuoles, albeit the cytoplasm may at times appear almost frothy due to an abundance of very small clear vacuoles of unknown origin and function (Fig. 10-3B). Pseudopodia are lobose and often explosive in their extension, and the myxamoebae tend to become quite elongate when moving freely in liquid substrates (Fig. 10-3C). Cells are generally uninucleate, and the nuclei, as in other acrasids, are characterized by centrally positioned nucleoli. The sorocysts are small, thinwalled, and usually globose to slightly elliptical (Fig. 10-3G) rather than irregular to angular as in *Copromyxa*.

Of the four known species, *Copromyxella filamentosa* R. W. and K. (1979) may be taken as representative of the genus and is illustrated in Figure 10-3. Cell aggregation occurs without stream formation (Fig. 10-3D). Fructifications in this species are singularly delicate and thin, often measuring 1.0-1.5 mm or more in length and only 25-40 μm in diameter near the base to 10-15 μm just below the apex. They may be simple or irregularly branched with midsections usually several cells in thickness and terminal areas often thinning to a single row of superimposed cells (Fig. 10-3F,G). In rich cultures, such as that shown in Fig. 10-3A, the columns often overlap and coalesce so that one sees not single sorocarps but a network of interwoven thread-like structures, from which the species takes its name. In thin cultures, as at the margin of a colony, the character of individual sorocarps may be clearly revealed (Fig. 10-3E).

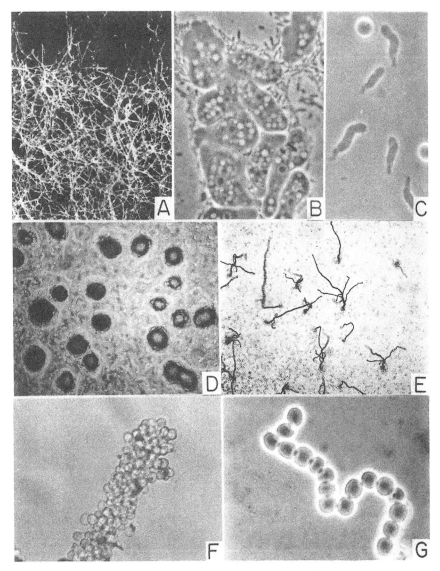

FIG. 10-3. *Copromyxella filamentosa* Raper, Worley, and Kurzynski. *A*. Thin, interlacing sorocarps produced on enriched hay infusion agar in association with *Escherichia coli*; surface lighting. × 14. *B*. Myxamoebae consuming *E. coli* on a very dilute nutrient agar (cells somewhat compressed). × 950. *C*. Myxamoebae in dilute nutrient broth showing extreme limax form when moving freely. × 550. *D*. Mounds of aggregated myxamoebae prior to sorocarp formation. × 80. *E*. Isolated sorocarps formed outside margin of bacterial colony; note branched character and microcysts on agar beneath. × 23. *F*. A portion of a column, somewhat contracted showing cellular composition and absence of external wall or membrane. × 360. *G*. Terminal portion of small branch that consists of a single tier of sorocysts. × 950. (From Raper, Worley, and Kurzynski, 1978)

Other species differ in details. *Copromyxella corralloides* is more consistently branched and typically shows clusters of arms arising from a much thicker base or stalk. It is the only one of the four to be isolated from dung, in this case that of a bald eagle. *C. spicata*, as the name suggests, produces fructifications that normally taper to a blunt point from a somewhat thicker base; it was isolated from a decaying mushroom. *C. silvatica*, like *C. filamentosa*, was isolated from forest soil and differs from that species in habits of growth and in producing less delicate sorocarps.

Guttulina (Pocheina)

Second in age to Brefeld's *Dictyostelium mucoroides* (1869) among the Acrasiomycetes is Cienkowski's *Guttulina rosea*, first reported to a meeting of Russian naturalists at Kazan in 1873. As originally described, the sorocarps consisted of microscopic droplets of rose color (sori) 70 μm in diameter, not ensheated, and borne on stalks of the same length. The cells of the heads (spores) were spherical and slightly roughened whereas those of the stalk were pressed together and wedge-shaped in appearance. The myxamoebae contained a red pigmented plasm and a nucleus, presumably of the acrasid type, as did the stalk cells. Upon germination the released myxamoebae were said to resemble *Amoeba limax* Dujardin.

For many years *Guttulina rosea* was known only from Cienkowski's description based upon specimens he found on rotting wood. During the 1960s, however, Mrs. Nannega-Bremerkamp made numerous collections from the barks of trees in the Netherlands that duplicated quite well the original description of *Guttulina rosea*. Some of the latter specimens were sent to us, and it is from them that three of the pictures contained in Fig. 10-4 were made, the other being from a slide preparation sent to us by Mrs. N-B. Unfortunately, neither she nor we were able to cultivate the slime mold in the laboratory, or to germinate the spores.

Several years ago Olive and Stoianovitch (1974) succeeded in isolating from the bark of trees an organism that they equated with Cienkowski's *G. rosea*. It was characterized by pink sorocarps with stalks up to 100 μm in length bearing spherical to oblong sori of about equal length. While they were not able to cultivate the slime mold, they did succeed in germinating the spores (and stalk cells) with singularly interesting results, for the cells germinated in two contrasting ways. In some cases, the spore wall split and a uninucleate myxamoeba with lobose pseudopodia emerged and moved away; in other cases, germination was preceded by a single nuclear division that was sometimes completed as the protoplast emerged, after which flagella began to appear, a single pair being associated with each nucleus. The binucleate cell subsequently separated into two uninucleate, anteriorly biflagellate cells. Until very recently this was the only

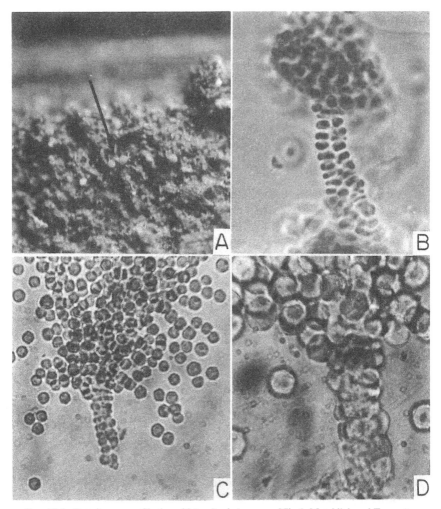

FIG. 10-4. *Guttulina rosea* Cienkowski (= *Pocheina rosea* [Cienk.] Loeblich and Tappan). *A*. Sorocarps (arrow) collected from bark of living tree. × 27. *B*. Basal portion of sorocarp showing "wedge-shaped" stalk cells. × 400. *C*. Sorocarp mounted in water with sorus broken and spores dispersed. × 375. *D*. A portion of the preceding enlarged to show irregular contour of spores. × 950. Specimens and stained preparation (B) contributed by N. E. Nannega-Bremekamp of Doorwert, Holland. (From Raper, 1973)

reported case of flagellate cells being found in any member of the Acrasiomycetes.

In the aforementioned paper, and in Olive's *The Mycetozoans* (1975) also, the organism just described was referred to not as *Guttulina rosea* but as *Pocheina rosea* (Cienk.) Loeblich and Tappan (1961), the change in name having been made because d'Orbigny in 1839 (*fide* L. & T.) had used *Guttulina* as the generic name for a fossil foraminifer.

Within the past year or so Olive and Stoianovitch (personal communication) have isolated three seemingly related slime molds, all with pink to reddish sori. In one of these, collected from oak bark, the trophic stage consists of uninucleate amoeboid cells very similar to those of *Acrasis rosea* together with some multinucleate protoplasts. It forms sorocarps with stalks consisting of single rows of cells that bear apical sori similar to those described by Cienkowski. It fruits in cultures with *Aureobasidium* sp., fails to produce flagellate cells, and is believed to represent a variety of Cienkowski's species. A second collection, obtained quite recently from a dead *Juniper*, consisted of numerous sorocarps, some with three or four lines of wedge-shaped stalk cells, some with two lines as shown in Fig. 10-4B, and a few with one line of flattened cells. This has not yet been grown in culture, but spores and stalk cells have germinated and yielded amoeboid cells. It is similar to the collections from the Netherlands and is believed to represent *Guttulina rosea* as observed and described by Cienkowski.

The third isolate, obtained from the bark of a living short-leaf pine, is being cultivated in association with *Aureobasidium* sp. It fruits abundantly, and either amoeboid or flagellate cells are released when spores and stalk cells germinate. The amoeboid cells are similar to those of the above except that the multinucleate protoplasts are smaller. This third isolate, interestingly enough, is said to be similar to the one upon which the paper "A Cellular Slime Mold with Flagellate Cells" was based (Olive and Stoianovitch, 1974) and in which the generic name *Pocheina* was applied for reasons already given. This has now been described as a new species, *Pocheina flagellata*, by Olive, Stoianovich, and Bennett (Mycologia *75*: 1019-1029, 1983), thus leaving the nonflagellate strains with wedge-shaped stalk cells to represent *Pocheina rosea* (Cienk.) L. & T., or as we prefer, *Guttulina rosea* Cienkowski.

However designated, the aforementioned slime molds are, as Olive has noted, similar to *Acrasis rosea* in many ways: (1) pink to reddish pigmentation, (2) similar myxamoebae, (3) hila on spores but not on stalk cells, and (4) mitochondria with lobose cristae.

Guttulinopsis

The genus *Guttulinopsis* was created by E. W. Olive in 1901 to include three new species which he designated *G. vulgaris*, *G. clavata*, and *G. stipitata*. All were isolated from dungs. Sorocarps as described were usually stalked, but sometimes sessile in *G. vulgaris*, and bore globose to elongate, unwalled sori that consisted of spores suspended in a thin to slightly gelatinous slime. The genus was said to differ from *Guttulina*, which E. W. Olive did not see, in having pseudospores rather than true spores, that is, propagules that upon reactivation simply rehydrated and resumed an amoeboid state. Only the first of the three species could be recognized by Raper, Worley, and Kessler in their recent examination of the genus (1977). *Guttulinopsis vulgaris* is common on the dung of many animals. It appears as sorocarps of varied size and pattern, sometimes showing definite stalks and sori (Fig. 10-5A); or as sessile masses of spores that appear as glistening droplets; or less commonly as grape-like clusters of sori that develop successively over a period of time. The myxamoebae have prominent lobopodia and move with rapid changes in shape and direction, have nuclei (usually single) with centrally positioned nucleoli, and usually contain prominent contractile vacuoles (Fig. 10-5B). The spores may be somewhat irregular in shape and commonly range between 5.0 and 7.5 μm in diameter. Contrary to E. W. Olive's report (1902), thin hyaline spore coats are left behind following spore germination, as L. S. Olive has indicated (1965).

A new and more interesting species, *Guttulinopsis nivea* (Fig. 10-5C-F), described and discussed in some detail by Raper, Worley, and Kessler (1977), has been isolated several times from tropical forest soil in addition to monkey dung, the type habitat. It can be cultivated with excellent sorocarp production on a variety of laboratory substrates in the absence of dung or dung extract in association with Gram-negative bacteria such as *Escherichia coli* and *Klebsiella pneumoniae*. The myxamoebae are smaller than those of *G. vulgaris* but are of the same general pattern. Cell aggregation, as in other Acrasidae, occurs by the regional convergence of individual cells to certain points (Fig. 10-5E) from which sorocarps subsequently arise that are characterized by prominent tapering stalks bearing globose milk-white sori (Fig. 10-5C). The stalk consists of myxamoebae loosely packed together and surrounded by a self-secreted membranous envelope, while the sorus is formed by myxamoebae that apparently ascend through this matrix, collect in a globose droplet at its apex, and differentiate as small globose to subglobose, slightly roughened spores (Fig. 10-5F). As the spores age they may show limited vacuolation just inside the semirigid walls, a condition that disappears upon rehydration.

While we cannot now say what van Tieghem had in hand when he described *Guttulina sessilis* and *Guttulina aurea* in 1880, it is believed that the former might have represented E. W. Olive's ubiquitous *Guttulinopsis vulgaris* (see above). Concerning *Guttulina aurea* we hesitate to offer an opinion, for this was said to resemble closely Cienkowski's *Guttulina rosea* except for its color, and, as already noted, the latter species hardly merits additional uncertainty.

Fonticula

This genus is represented by a single species, *Fonticula alba* Worley, Raper, and Hohl (1979). In outward form and dimensions it bears a striking

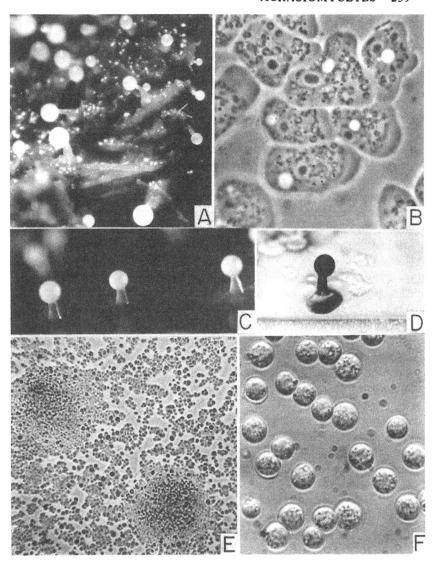

Fig. 10-5. *Guttulinopsis* species. *A-B*. *Guttulinopsis vulgaris* E. W. Olive. *A*. Stalked sorocarps produced on sterilized cow dung in association with a mixed bacterial flora; surface
lighting. × 12. *B*. Myxamoebae growing on dilute nutrient agar in association with *Klebsiella
pneumoniae*; note prominent nuclei, contractile vacuoles, and lobose pseudopodia. × 100.
C-F. *Guttulinopsis nivea* Raper, Worley, and Kessler. *C*. Sorocarps produced in association
with *K. pneumoniae* on low-nutrient agar; surface lighting. × 30. *D*. Single sorocarp photographed by transmitted light to emphasize deltoid base and tapering stalk. × 30. *E*. Two
developing aggregations (pseudoplasmodia); converging cells do not form streams. × 120.
F. Spores in saline following rehydration (incomplete phase). × 950. (From Raper, Worley,
and Kessler, 1977)

resemblance to *Guttulinopsis nivea* (Fig. 10-6C,E), yet it differs from that slime mold in many important characteristics—so much so, in fact, that it doubtfully belongs in the Acrasidae. At the same time it cannot be assigned with the Dictyostelidae, for it differs from the dictyostelids even more. Since we hesitate to create a third subclass (or order) to accommodate a single genus and species, we shall consider it tentatively as representing a fourth family, the Fonticulaceae, within the Acrasidae, recognizing at the same time that its correct position remains in doubt.

The myxamoebae of *Fonticula alba* are unlike those of either the acrasids or the dictyostelids. Small and usually uninucleate, the myxamoebae are irregular in form, range between 9-15 × 6-12 μm in diameter, and show filose pseudopodia when moving freely in liquid; on agar beneath a coverglass they show distinct areas of clear ectoplasm and granular or "frothy" endoplasm (Fig. 10-6A). The round to oval nuclei are readily deformable and may show either interior or peripheral nucleoli, as in the dictyostelids (Deasey, 1982). Cell aggregation occurs without stream formation, but myxamoebae moving toward a particular center often seem to follow preexisting paths. As more and more cells converge, definite mounds are formed (Fig. 10-6B), from which sorocarps of unique pattern develop, first by the formation of a tapered column of cells, followed by the opening of the completed column at its apex, and lastly by the rapid ascent of the cells within to form the terminal, unwalled globose sorus (Fig. 10-6C-E). The spores are slightly elliptical and differ only superficially from the globose microcysts that form on the agar surface from unaggregated myxamoebae (Fig. 10-6F,G).

The slime mold was originally isolated by L. S. Olive from dog dung, but it can be grown quite successfully on nutrient agar containing tryptone 2.5 g, yeast extract 2.5 g, glucose 0.5 g, K_2HPO_4 0.5 g, and agar 20 g/liter water, or diluted to ½ strength, in association with Gram-negative bacteria such as *Escherichia coli* or *Klebsiella pneumoniae*. The optimum range for the slime mold is pH 8.0-8.5, which is somewhat above that for *Guttulinopsis nivea* but not *G. vulgaris* also isolated from dung.

The singular manner of prespore cell ascension in the developing sorocarps is without parallel in any other cellular slime mold. Increased turgor in the closed base and stalk of the sorocarp affords a probable explanation; and recent studies by Deasey and Olive (1981) indicate that the Golgi apparatus of amoeboid cells remaining in the bulbous base continues to elaborate stalk material, and that the mucous surrounding this absorbs moisture from the outside, thereby exerting enough pressure within the stalk to cause the spores to rise and to be expelled. The fact that water is drawn from the outside could account for the concave depressions in the agar surface that one observes when intact sorocarps are removed. In any case, *Fonticula alba* today stands as a unique cellular slime mold that may or may not be intermediate between the more primitive Acrasidae and the more advanced Dictyostelidae that follow.

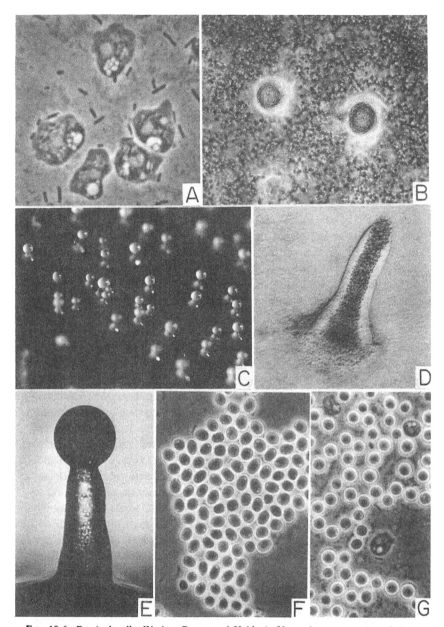

FIG. 10-6. *Fonticula alba* Worley, Raper, and Hohl. *A*. Vegetative myxamoebae photographed under coverglass; note nuclei (nucleoli indistinct), contractile vacuoles, and food vacuoles containing bacteria. × 1020. *B*. Cell aggregations shortly before forming upright columns. × 85. *C*. Mature sorocarps showing globose, milk-white sori; surface lighting. × 20. *D*. Young fruiting column in stained preparation showing thick hyaline wall, cellular content within, and apical opening through which cells ascend as differentiating spores. × 170. *E*. Maturing sorocarp showing spore-forming cells ascending through the stalk to form the sorus. × 135. *F*. Spores; note slightly elliptical shape. × 625. *G*. Microcysts formed on agar surface by individual, unaggregated myxamoebae; note globose shape. × 625. (From Worley, Raper, and Hohl, 1979)

CHAPTER 11

Dictyostelidae

Classification of the Dictyostelidae presents a singular challenge. This results in part from a variety of environmental pressures that influence growth and development, and in part from characteristics of the dictyostelids that are in themselves unique. The latter merit primary consideration, for the systematist is faced not so much with classifying organisms in a true sense as with classifying *communities* of organisms that aggregate, integrate, and differentiate in reasonably constant ways. The separation of growth and morphogenesis in these slime molds cannot be overemphasized. Unlike the situation in higher plants and hyphal fungi, where the two processes proceed simultaneously, in the dictyostelids they are sequential: growth is completed before morphogenesis begins. We cannot at this time classify the dictyostelids upon the basis of differences in the vegetative state, for the myxamoebae appear strikingly similar in all cases. To discern the varied patterns that characterize different species, we are left, then, with the manner in which thousands of amoeboid cells come together, how they intermix, and how, as building units, they ultimately arrange themselves prior to terminal differentiation; together with such accessory structures (e.g., stalk tubes) as they may form to mold the varied patterns that characterize different species. That they achieve reasonable measures of uniformity is remarkable. Until terminal differentiation is near at hand, the collaborating cells are bounded only by their permeable membranes and the slime they secrete; and as such they are constantly subjected to the vicissitudes of the physical environment within which they play out their hereditary but previously unsettled roles.

Subject to certain genetic constraints and singularly sensitive to the external environment, the fruiting structures, or sorocarps, of a given species typically conform to a general if somewhat elastic pattern. Furthermore, there is for each species an upper if often ill-defined limit to sorocarp size, this being determined by the maximum number of cells that can collaborate effectively in building a sorocarp. In the presence of excess cells, the number of sorocarps simply increases with the richness of the culture. On the other hand there is seldom if ever a comparable lower limit, and in older cultures especially minute structures can usually be found.

Sorocarps may be generally solitary, in which case a fructification develops from one aggregation or from a severed part of an inflowing myxamoebal stream. In contrast, sorocarps may be clustered; that is, several sorogens may originate within an aggregation center, not infrequently over a limited period of time, and each produce a sorocarp independently. In still other cases both modes of development may occur, and the investigator is hard pressed to determine which if either merits precedence.

Branching of sorocarps is of two types, here designated primary and secondary. Of these, the former is relatively rare and results from a splitting of the sorogen and the subsequent formation of two essentially independent sorophores with a single base, as may occur in *Dictyostelium rosarium* or as is commonplace in *D. bifurcatum*. The latter, or secondary type of branching, is far more commonplace and assumes a variety of forms, ranging from the whorls of side branches that characterize the genus *Polysphondylium*, through the richly but irregularly branched structures typical of *Dictyostelium aureo-stipes* and certain other species of that genus, to the occasional branch that may be seen in *D. mucoroides* or *D. purpureum*. In all the latter cases branches develop from fractional masses of cells abstricted from rising sorogens, which then may or may not cleave but in whole or as parts develop what amount to diminutive sorocarps anchored to the main axis, just as it in turn is anchored to the substrate. "Branches" in *D. discoideum* originate in a different way; a second pseudoplasmodium climbs part way up an existing stalk and builds an independent sorocarp outward from this support.

Spore patterns afford useful criteria for classification and are diagnostic in a few cases. Spore dimensions are also useful, but these should be determined and recorded soon after a strain is isolated; otherwise, cultures tend to develop diploid components with continued laboratory cultivation. Spores of the genus *Polysphondylium* and those of many species of *Dictyostelium* contain conspicuous polar granules, suggesting a measure of relatedness. Traub and Hohl (1976) have in fact proposed erecting a new genus based upon this character, coupled with additional ones that include the formation of microcysts and a negative correlation for cAMP induced cell aggregation. Taken together, these characteristics are thought to provide tangible evidence of close relationships.

Concerning heritable variations, observable differences exist among strains of any species of dictyostelids for which several natural isolates are known. In some cases this is of minor concern, while in other and generally more cosmopolitan species it may be sufficient to befog or virtually obliterate the boundaries between related species, as for example within the so-called *Dictyostelium mucoroides* complex. One then finds it necessary to select certain isolates as representative and to center the species description on these cultures, indicating if possible the outer limits of variation to be accommodated under the name in question. For example, in what we now regard as *Polysphondylium violaceum* one finds strains that produce sorocarps ranging from fairly erect and symmetrical structures to others that are long and rangy and produce few side branches in poorly defined whorls. Another phenomenon still unexplained, but common to many strains of *Dictyostelium*, is an annoying tendency for cultures to grow less and less satisfactorily in the laboratory following successive transfers or, stated more succinctly, for cultures to deteriorate with time. For this reason it is prudent to conserve cultures for long-term storage by lyophilization or other means as promptly as possible following isolation.

Apart from the aforementioned considerata, which are inherent within the slime molds, a variety of external factors, both nutritional and environmental, may influence profoundly the growth and morphogenesis of the dictyostelids. Their effects may range all across the board—from promoting optimal fructification in one case to distorting the process in another so that we have no sorocarps to classify. Such external factors may include the following: substrate composition (including the bacteria upon which the myxamoebae feed), light, temperature, humidity, the culture vessels, and autoinhibitory gases.

SUBSTRATES. One cannot say which of the listed parameters is the more important, for, speaking generally, there is an interplay between two or more in any particular situation. It can be said, however, that the amount of cell growth, the general well-being of the myxamoebae, and their capacity to aggregate and subsequently interact optimally in fructification clearly depend upon the availability of a suitable nutritive bacterium growing (or resting) upon a favorable substrate. This relationship has been considered at length in the chapter on cultivation, hence needs no amplification here except to say we have chosen a few substrates, and dilutions thereof, as more or less standards for comparative purposes.

LIGHT. Light hastens cell aggregation, and in many species profoundly influences the behavior of developing sorogens by determining not only their orientation but the duration of sorophore elongation and hence of sorocarp dimensions and proportions. In *Dictyostelium discoideum* it directs the movement of the migrating pseudoplasmodium and can trigger sorocarp formation as well. In this, and in other species also, sorocarp formation is markedly accelerated by exposure to light; and the response is especially striking when cultures in the pseudoplasmodial stage are moved from darkness to light.

TEMPERATURE. Cell aggregation and sorocarp formation are accelerated also by limited increases in temperature; concurrently the number of sorocarps is increased and their size decreased. The extent of migration, or sorophore formation in most species, is likewise extended at lowered temperatures, or curtailed if elevated by a few degrees. Sorocarp patterns are also affected, the ratio of stalk to spore cells being somewhat greater in sorocarps produced at lower temperature. For each slime mold there is a relatively narrow temperature range within which it grows, aggregates, and fruits best, and this should be utilized as fully as possible in constructing species descriptions.

HUMIDITY. A film of free surface water is necessary for cell aggregation and pseudoplasmodial migration, while sorocarp formation is favored in some cases, if not in all, at levels of atmospheric moisture near saturation. For example, in *Dictyostelium polycephalum* optimal fruiting occurs at 98% RH (Whittingham and Raper, 1957). Actual values are not available for other species, but we suspect that comparable levels may apply quite commonly.

CULTURE VESSELS. Glass or clear plastic Petri dishes may be used, and of these glass is preferred for reasons already given in Chapter 4.

AUTOINHIBITORY GASES. The phenomenon of sorocarp avoidance has been known since it was reported by Bonner and Dodd in 1962. Equally important was the observation that fruiting in some dictyostelids is greatly improved by adding activated charcoal near sites of slime mold growth (Bonner and Hoffman, 1963). This practice has now been adopted by other investigators, and in our laboratory it is routine procedure for stimulating sorocarp formation in seemingly recalcitrant cultures. The nature of the gas(es) emitted by developing sorocarps in *D. discoideum* is not known, or whether it is the same in different species. We do know that in poorly fruiting or nonfruiting cultures something can be removed or suppressed by activated charcoal that oftentimes enables sorocarp formation to proceed in a more normal manner. We also know that if one hopes to classify the Dictyostelidae successfully, one must have for comparison the best-formed sorocarps it is possible to obtain.

As presently known, the subclass Dictyostelidae contains a single order, the Dictyosteliales, of which the specifications are those of the subclass. Two, and possibly three, families are included. The Dictyosteliaceae and Acytosteliaceae possess many characteristics in common and are closely related, yet easily distinguished one from the other. They differ primarily in the character of their sorophores: those of the former consist of columns of cells contained within sheaths or tubes rich in cellulose; those of the latter consist only of very narrow acellular cellulosic tubes. The uncertainty arises concerning van Tieghem's *Coenonia denticulata* (1884), which is known only from the original description. Stalks were reported to be cellular, to contain cellulose, and to arise from crampon-like bases, all characters that suggest the Dictyosteliaceae. Coupled with these characteristics, however, were gelatinous sori formed in chalice-like cupules with dentate rims borne on stalks with external cells showing small upturned teeth. In no other cellular slime mold are like structures found, and the correct assignment of *Coenonia*, whether in the Dictyosteliaceae or in a third family, must await its rediscovery and careful study.

CHAPTER 12

Dictyosteliaceae: *Dictyostelium*

TECHNICAL DESCRIPTION

Dictyostelium Brefeld, in Abh. Senckenberg. Naturforsch. Ges. *7*: 85-107, Tafs. I-III, figs. 1-36 (1869), and in Untersuch. Gesammtgeb. Mykologie *6*: 1-34 (1884). Also Olive (E. W.), in Proc. Amer. Acad. Arts and Sci. *37*: 338-341 (1901), and in Proc. Boston Soc. Natur. Hist. *30*: 451-513, 68 figs., in pls. 6-8 (1902).

Vegetative stage consists of small uninucleate, free-living amoeboid cells that feed upon bacteria; fruiting stage characterized by the coordinated aggregation of these myxamoebe into pseudoplasmodia, and the subsequent development from them of erect, semi-erect, or partially prostrate sorocarps with cellular sorophores bearing unwalled sori; sorophores unbranched or with branching patternless and usually sparse; sori globose to somewhat citriform, composed of spores suspended in semi-viscous slime; dimensions and patterns of sorocarps vary greatly in different species; macrocysts and/ or microcysts produced in some species.

GENERAL DESCRIPTION AND COMMENTS

Dictyostelium is the oldest and much the largest genus of the Dictyostelidae, embracing well over half of all the described species. Known as such since 1869, it had been observed and even illustrated a few years earlier by Coemans (1863), who identified it erroneously as a nonexistent pycnidial stage of *Rhizopus nigricans*. In the years following Brefeld's description of *D. mucoroides* two additional species, *D. roseum* and *D. lacteum*, were added by van Tieghem in 1880, and one by Marchal in 1885, who redescribed as *D. sphaerocephalum* a form previously isolated in Amsterdam by Oudemans and described by him (1885) as *Hyalostilbum sphaerocephalum*. In his "Monograph of the Acrasieae," E. W. Olive (1902) was able to list seven species of *Dictyostelium*, including three new taxa described by him the previous year, *D. purpureum*, *D. aureum*, and *D. brevicaule*. The genus attracted little attention during the next three decades, and it was not until 1935 that another new species, *D. discoideum* Raper, was described. Since that time interest in the genus has increased dramatically, with the number of reported species and varieties now numbering about forty and derived from global sources. Whereas most of the pre-1902 species were of coprophilous origin, virtually all of those reported since that time have come from forest soils, where they constitute an integral part of the microflora.

Of the many species that now comprise the genus, the type *Dictyostelium mucoroides* is the most commonplace and, as Cavender and others have demonstrated, can be found wherever soils are collected and screened for cellular slime molds. It does not however, produce the largest or the smallest sorocarps, for within the genus *Dictyostelium* may be found species with sorocarps ranging from less than a millimeter in height, as in *D. deminutivum* Anderson, Fennell, and Raper, to several centimeters in length, as in *D. giganteum* Singh and *D. purpureum* Olive when these are cultivated in darkness or in one-sided light. The basic patterns of sorocarps vary greatly in different species also and range from the relatively simple structures found in *D. mucoroides* to the coremiform arrangements of *D. polycephalum* Raper, the clustered sorocarps of *D. polycarpum* Traub, Hohl, and Cavender, and the fruiting structures with multiple sessile sori seen in *D. rosarium* Raper and Cavender. Quite aside from such inherent differences in form, the patterns and dimensions of sorocarps within individual species may vary also, depending largely upon the relative abundance of the vegetative myxamoebae and upon the cultural environment, which can profoundly influence the manner in which such cells coordinate their movements during cell aggregation and subsequently differentiate to effect sorocarp formation.

In the genus *Dictyostelium*, and in *Polysphondylium* as well, we have elected to use for species characterizations substrates of intermediate nutrient content wherever possible, including most often agar media containing 1 g each of lactose and peptone (0.1L-P), the same with added phosphates (0.1L-P, pH 6.0), and hay infusion agar, with or without similar pH adjustment (see Chapter 4). Some of the larger species of *Dictyostelium* will grow quite well on richer substrates, and robust forms such as *D. discoideum*, *D. purpureum*, and *D. giganteum* can thrive and fruit optimally on media of five- or tenfold the nutrient content just cited. On the other hand, many of the more delicate species of the genus cannot grow optimally or fruit satisfactorily on substrates such as 0.1L-P or hay infusion agars, in which cases they are diluted with nonnutrient agar (NNA) to achieve a suitable nutrient level. Additionally, for certain species it is desirable to substitute glucose for lactose or yeast extract for peptone. The former becomes especially important if one needs or elects to shift from such conventional bacterial associates as *Escherichia coli*, *E. coli* B/r, or *Klebsiella pneumoniae* to *Pseudomonas fluorescens* or *Serratia marcescens*. In any case, species characterizations are optimally useful only when certain cultural parameters are controlled and recorded, including the nature of the substrate, the identity of the bacterial associate, the temperature of incubation, and in many cases the nature of the light.

Ideally, species characterizations should be based upon all aspects of a dictyostelid's growth and development. In practice, however, we find that species at this time have to be defined primarily upon the physical characteristics of the fructifications, including sorocarp dimensions under stated cultural conditions; extent and pattern of sorophore branching; pigmentation of the sori and sorophores, if present; pattern and dimensions of spores; presence or absence of microcysts and/or macrocysts; and, in some cases, stalk-free migration of the pseudoplasmodia prior to sorocarp construction. Some of the foregoing merit special consideration.

Among the fungi generally, spores are reasonably constant in pattern, form, and dimensions, and these are often cited as among the more stable characteristics. Within the genus *Dictyostelium* this is partially true. During the 1950's it was realized that with continued laboratory cultivation the type strain of *D. discoideum* (NC-4) tended to shift from haploid to diploid, with a consequent increase in spore lengths from about 7-8 to 10-12 μm (see Chapter 6), while a limited proportion appeared reniform rather than oblong with obtuse ends (Fig. 6-6). Whereas chromosome counts are known for only four or five species, it is not uncommon to find spores of mixed dimensions in slide mounts from the sori of many species, and it is assumed that such variation probably represents differences in ploidy also.

Concerning some other differences there can be no question: the spores of *D. rosarium* and of *D. lacteum* are spherical, whereas those of *D. polycephalum* germinate not by the extension of a lateral fissure but by the near circumferential rupture of the spore wall at midpoint of the ellipsoid spore. The spores of many dictyostelia, including most of the larger species, appear to be homogeneous throughout, while equally ellipsoid spores of many other species, like those of *Polysphondylium*, contain conspicuous inclusions that normally occupy polar positions and are referred to as "polar granules." Little is known about their origin or possible significance in the life cycles of the dictyostelids that produce them. Nonetheless, they are very useful as a taxonomic criterion, for their presence or absence can be determined quickly as a rule with the light microscope. Furthermore, spores with polar granules usually correlate with certain cultural characteristics including a lack of chemotactic response to cAMP and the formation of microcysts by unaggregated myxamoebae. Ultrastructurally the polar granules consist of clusters of small electron-transparent vesicles embedded or held together in an electron-dense matrix (Traub and Hohl, 1976). It has been suggested that they may contain trehalose (Müller and Hohl, 1973; Gregg and Badman, 1970) or lipid droplets (Schaap, van der Molen, and Konijn, 1981b). Additionally, spores that have polar granules do not show mitochondrial profiles prominently lined with ribosomes, whereas spores lacking these granules do show such profiles (H. Hohl, personal communication).

Most members of the genus range from white to uncolored, but some species are strikingly pigmented, such as *D. purpureum*, with sori deeply purple and sorophores lightly but similarly tinted, and *D. mexicanum*, with sori and sorophores in golden to lemon-yellow shades. In other species pigmentation is less pronounced and in some cases variable. This is especially true among the crampon-based species and in *D. aureum*, where the sori are lemon-yellow and the stalks uncolored, while in *D. aureostipes* quite the opposite is true. Even in *D. discoideum*, the most studied species, the sorus is generally white to cream colored, but under some conditions, still not fully understood, the sori are bright yellow.

The myxamoebae of many cellular slime molds are capable of encysting as individual cells, thus entering a transient dormant stage that is more resistant than trophic cells but less so than spores. This is true of many if not a majority of species presently assigned to the genus *Dictyostelium*. As noted already, this capability seems to correlate generally with the presence of spores that contain polar granules and an inability of the myxamoebae to aggregate in response to cAMP, as shown by Traub and Hohl (1976).

Seven species of *Dictyostelium* are now known to form macrocysts, and we know not how many more might reveal this capability if sufficient numbers were examined or, more importantly, if investigators could isolate sufficient strains and pair these in all seemingly logical combinations. Apparently illustrated by Brefeld and termed "dwarf sporangia," macrocysts were not recognized as such until 1937 in a strain of *D. mucoroides*, NC-12, that is now lost from our collection. They were next encountered in strains of *D. minutum* (Raper, 1951) but were not accorded careful study until the mid-fifties, when Blaskovics and Raper (1957) turned their attention to macrocyst development in *D. mucoroides*. Intensive studies have since been conducted by Filosa, O'Day, and their coworkers in Canada, by Nickerson, Erdos, Wallace, Chang, and Raper in Wisconsin, by Francis and associates in Delaware, and by Robson and Williams in Australia. Out of their work has come the still inadequately exploited fact that macrocysts represent the true sexual stage of the dictyostelids. In some species, such as *D. mucoroides* and *D. minutum*, macrocysts arise from self-compatible strains; in others, such as *D. giganteum* and *D. rosarium*, macrocysts have been found only after pairing compatible mating types; while in still others, including *D. discoideum* and *D. purpureum*, both routes to cyst formation are known, the mating-type route being far more prevalent in the former species and probably in the latter as well.

The *Dictyostelium mucoroides* Complex

As isolated from nature, Dictyostelia with milk-white sori and simple or irregularly branched sorophores far outnumber all others, and it must have been a culture of this general type upon which Brefeld (1869) based his genus and species, *Dictyostelium mucoroides*. No type culture exists, however, and little more can be said. Measurements were given only for the spores, 4.0 × 2.4 μm; and some additional information is afforded by his scaled drawings. Using this approach, mature sorocarps are found to be about 1.0-1.6 mm high and to consist of unbranched sorophores 10-17 μm wide bearing globose sori about 100-160 μm in diameter. Shown also are some rounded, cellular structures 25-35 μm in diameter that we interpret as representing macrocysts (Fig. 1-1).

Particular difficulty arises when one attempts to distinguish between *D. mucoroides* and two other taxa with sori equally white and obviously closely related. Sorocarps in these are generally somewhat larger or the sorophores more often branched. One of the two, *D. sphaerocephalum* (Oud.) Sacc. et March., is quite old, dating from 1885; the other, *D. giganteum* Singh, was described more recently (1947a). E. W. Olive (1902) questioned whether Oudeman's species should be separated from *D. mucoroides*, and Raper in 1951 voiced a similar doubt based upon cultures observed up to that time. More recently, Traub (1977) has questioned whether Singh's *D. giganteum* should be separated from *D. sphaerocephalum*, while Hagiwara (1983) believes *D. sphaerocephalum* (Oud.) Sacc. et March is a synonym for *D. mucoroides* Brefeld (*sensu stricto*).

What one encounters is a great number of slime molds with many characteristics in common, yet differing substantially among themselves in sorocarp form and dimensions, and in developmental behavior as well. For the overall assemblage it has become common to use the term "*Dictyostelium mucoroides* complex," a sobriquet first introduced by Cormier, Whittingham, and Raper in an oral presentation at the Annual Meeting of the Botanical Society of America in 1956 and later committed to print by Bonner in 1959. It is an informative and useful designation, applicable in certain situations but inadequate for our present purpose. Fortunately, in the course of a comprehensive study of cultures belonging to this complex, Anne Marie Norberg, a former doctoral student, was able to recognize types of Dictyostelia that we believe can be meaningfully associated with each of the aforementioned historic names, freely admitting that cases of overlap commonly occur. Concerning *D. giganteum* there can be no question, for she had in her possession, as we still do, Singh's type culture, sent to us by him and since conserved in lyophilized form. Concerning *D. mucoroides*, there are isolates that develop relatively short sorophores when cultivated in full light, and of these many produce macrocysts homothallically, as we have reason to believe was true of Brefeld's culture. We are somewhat less confident of *D. sphaerocephalum*, but it was reported to be larger than *D. mucoroides*, to have generally heavy stalks, and to be irregularly branched. These three species are central to the concept of a *D. mucoroides* complex; they do not, however, preclude inclusion of such species as *D. intermedium* Cavender or even *D. aureum* Olive, which, like the above, produce neither microcysts nor spores with polar granules.

Heterosphondylium

The genus *Dictyostelium* as we now envision it contains many species that differ substantially from those just considered. In fact many of them share certain characteristics with the genus *Polysphondylium*, namely: the presence of spores with polar granules, an absence of chemotactic response to cAMP, and the frequent formation of microcysts by individual myxamoebae. These correlations were first recognized and reported in 1976 by Franz Traub and Hans Hohl in a landmark paper entitled "A New Concept for the Taxonomy of the Family Dictyosteliaceae (Cellular Slime Molds)." The following year (1977) Traub introduced but did not formally describe a new genus, *Heterosphondylium*, to include such forms and to occupy a position intermediate between Brefeld's two genera, *Dictyostelium* and *Polysphondylium*. Species comprising the new genus would differ from the latter primarily in the absence of branches arranged in definite whorls. Additionally, many, but not all, species assignable to *Heterosphondylium* are richly (but irregularly) branched, often rather delicate, and generally favored by substrates of reduced nutrient content.

The importance of Traub and Hohl's observations cannot be questioned, and strong arguments for the creation of an intermediate genus can be advanced. Furthermore, recognition of a separate genus such as that suggested may in time be warranted when more is known about the chemotactic agents responsible for cell aggregation and about pattern regulation in the slime molds that would comprise it. For the present, it is our considered belief that an effective nomenclature will be better served by continuing to recognize as *Dictyostelium* those dictyostelids with unbranched or irregularly branched sorocarps.

Keys to Species of *Dictyostelium*

Although *Heterosphondylium* is not recognized as a separate genus, a striking characteristic of the suggested taxon, namely the presence versus absence of polar spore granules, does provide a convenient and generally reliable basis upon which the genus *Dictyostelium* can be subdivided. For this reason the character is accorded primary importance in the dichotomous key that follows. Having established that the spores appear homogeneous and devoid of polar granules, or inhomogeneous due to their presence, the investigator can then use a series of supplemental characteristics to identify species. The key is admittedly artificial and is not intended to convey implications of phylogenetic relationships. Species with several characters in common fall together naturally; a very few, such as *D. minutum* and

D. vinaceo-fuscum, fail to fit comfortably within the subdivisions to which they are assigned; while some like *D. lacteum* and *D. rosarium* are placed together upon the basis of a very obvious but doubtfully kin-related character, their globose spores. Hopefully, a more natural arrangement can be achieved as additional species are discovered, and as information continues to accumulate concerning species now in laboratory culture.

A semi-synoptic key that lists species showing some particular or unusually striking characteristic(s) is also included for the avowed purpose of helping an investigator identify unclassified isolates as quickly as possible. The characteristics included in this key are not weighted; hence the order of listing implies no suggestion of taxonomic hierarchy.

DICHOTOMOUS KEY

I. Spores ellipsoid[1] or capsule-shaped
  A. Spores appear homogeneous, lack prominent granules
    1. Sori white to cream-colored, globose to citriform; sorophores unpigmented with bases rounded, clavate, or expanded
      a. Sorophores unbranched or sparsely so, apices plain; pseudoplasmodia rarely show stalk-free migration
        1'. Sorocarps mostly 2-5 mm in overhead light, often 1 cm or more in unidirectional light; sori commonly 125-200 μm.
          a'. Spores from collapsed sori do not germinate immediately .... *D. mucoroides* Brefeld
          b'. Spores from collapsed sori germinate immediately; myxamoebae aggregate and form small sorocarps .... *D. mucoroides* v. *stoloniferum* Cavender & Raper
        2'. Sorocarps 5-10 mm in overhead light, often 3-6 cm in unidirectional light; sori 125-250 μm *D. giganteum* Singh
        3'. Sorocarps robust, 1.0-1.5 cm high with very large sori, often up to 500 μm or more, erect or semi-erect, arising from expanded conical bases. (Optimum temp. 15-17°C) .... *D. septentrionalis* Cavender
        4'. Sorocarps poorly formed; stalks thick and warty; sori misshapen .... *D. irregularis* Nelson, Olive, & Stoianovitch

---

[1] The terms "ellipsoid" or "elliptical," sometimes with modifiers, are used for the shapes of most spores. Stearn (1966, p. 352) lists *oblongus apicibus obtusis* as a more precise designation for such objects having nearly parallel sides and rounded ends.

  b. Sorophores commonly branched, with apices
     modified or plain; pseudoplasmodia may or may
     not migrate

    1'. Sorocarps 2-4 mm when erect, 0.5-1.5 cm or
       more when prostrate; cupule-like residues at
       bases of sori; stalk-free migration common at
       15-20°C . . . . *D. sphaerocephalum*
                                               (Oud.) Sacc. et March.

    2'. Sorocarps small, erect, mostly 0.5-0.85 mm;
       sorophore apices plain; pseudoplasmodia do
       not migrate . . . . *D. minutum* Raper

2. Sori off-white to ivory or yellow; sorocarps unbranched; stalk-free migration alternative to direct
  fructification

    a'. Sorocarps commonly 3-5 mm high; sori white
       to old ivory; stalk-free migration common;
       sorophore bases often deltoid . . . . *D. intermedium*
                                               Cavender

    b'. Sorocarps mostly 3-8 mm high; sori white to
       lemon yellow; stalk-free migration frequent . . *D. dimigraformum*
                                               Cavender

    c'. Sorocarps mostly 1.5-3.8 mm, erect; soro-
       phores strongly tapered with expanded basal
       disks; sori white to yellow; pseudoplasmodia
       typically migrate . . . . *D. discoideum* Raper

3. Sori pigmented in different shades; sorocarps
  sparsely branched with sorophore bases not distinctive

    a'. Sori bright lemon yellow, 150-250 μm; soro-
       carps solitary or clustered, up to 1 cm or
       more; sorophores in yellow shades . . . . *D. aureum* v. *aureum*
                                         Cavender, Worley, & Raper

    b'. Sori yellow to pale yellow, commonly larger
       than 250 μm; sorophores often yellowish . . . . *D. aureum* v. *luteolum*
                                         Cavender, Worley, & Raper

    c'. Sori in tan to dull brown shades; sorophores
       anchored by expanded aprons of congealed
       slime . . . . *D. brunneum* Kawabe

    d'. Sori in purple shades, usually dark; sorocarps
       large, often 1 cm or more, strongly photo-
       tropic; sorophores may show purple tints . . . . *D. purpureum* Olive

    e'. Spore mass spherical, reported to be bright
       red with spores as in *D. mucoroides* but
       larger 8 × 4 μm. (Known only from original
       description) . . . . *D. roseum* van Tieghem

B. Spores contain prominent granules, usually bipolar

1. Sori hyaline, white, or cream colored; sorophore
  bases rounded, clavate, or somewhat enlarged

a. Sorophores colorless or only faintly pigmented

1'. Sorocarps separate, sorophores not adherent

  a'. Sorocarps usually less than 1 mm high, rarely branched; spores usually less than 5.0 μm in length

    1". Cell aggregations with streams lacking or limited

      a". Sorocarps 0.2-0.6 mm high, solitary or clustered; young sorogens claviform ...... *D. deminutivum* Anderson, Fennell, & Raper

      b". Sorocarps 0.25-1.0 mm high, clustered; young sorogens vermiform.............. *D. multi-stipes* Cavender

    2". Cell aggregations with streams well-developed; sorocarps 0.24 to 1.2 mm .... *D. microsporum* Hagiwara

  b'. Sorocarps commonly 1-4 mm high; spores usually exceed 5.0 μm in length

    1". Myxamoebae predatory on other dictyostelids; sorophores unbranched, thin, 0.7-1.5 mm with cells in single tiers except near bases; sori 40-100 μm .... *D. caveatum* Waddell, Raper, & Rahn sp. nov.

    2". Myxamoebae not reported predatory on other dictyostelids; sorophores mostly 2-4 mm, occasionally branched, cells often in single tiers; sori 100-250 μm............. *D. fasciculatum* Traub, Hohl, & Cavender

  c'. Sorocarps delicate, 1-5 mm high, typically branched

    1". Branches numerous, pattern irregular

      a". Sorophores hyaline, thin; cells mostly in a single tier; sori 50-100 μm.............. *D. tenue* Cavender, Raper, & Norberg

      b". Sorophores hyaline to faintly yellow; cells often in single tier above; sori 75-150 μm *D. delicatum* Hagiwara

    2". Branches limited in number, patterns distinctive

      a". Sorophore branches in bifurcate pattern ... *D. bifurcatum* Cavender

      b". Sorophore branches in monochasium-like pattern .... *D. monochasioides* Hagiwara

2'. Sorocarps in coremium-like clusters; sorophores adherent in basal areas or more

  a'. Sorocarps arise from thin, elongate, migrating pseudoplasmodia; upright sorophores adherent *D. polycephalum* Raper

  b'. Sorocarps arise directly from cell aggregations; migrating pseudoplasmodia lacking; sorophores adherent at bases .... *D. polycarpum* Traub, Hohl, & Cavender

b. Sorophores pigmented in various shades
  1'. Sorophores in yellow shades; pigment fading
    in bright light; bases clavate
    a'. Sorophores comparatively thin, flexuous;
      branching prominent . . . . *D. aureo-stipes*
      Cavender, Raper, & Norberg
    b'. Sorophores stout, erect; branching profuse . . . *D. aureo-stipes* v. *helvetium*
      Cavender, Raper, & Norberg
  2'. Sorophores in bluish shades; bases digitate;
    branches lacking or limited . . . . *D. coeruleo-stipes* Raper &
      Fennell
2. Sori pigmented in different shades; sorophore
  bases variously modified
  a. Sorophore bases weakly to strongly digitate; sori
    in gray, olive, vinaceous, or fuscous shades
    1'. Sori bluish gray to brownish lavender, terminal
      and sessile on sorophores; bases weakly
      digitate . . . . *D. laterosorum* Cavender
    2'. Sori yellowish, grayish, or brownish olive;
      bases crampon-like . . . . *D. rhizopodium* Raper &
        Fennell
    3'. Sori in lavender, vinaceous to fuscous shades;
      bases digitate
      a'. Sori in lavender, vinaceous, or light smoke
        shades . . . . *D. lavandulum* Raper &
          Fennell
      b'. Sori in deep vinaceous gray to dark smoke
        shades. (Polar granules questionable) . . . . *D. vinaceo-fuscum* Raper &
          Fennell
  b. Sorophore bases commonly discoid; apices ex-
    panded and flattened above
    1'. Sori and sorophores in bright yellow to golden
      shades . . . . *D. mexicanum* Cavender,
        Worley, & Raper
II. Spores globose or nearly so
  A. Sorocarps comparatively large with sessile sori
    borne on ascending sorophores . . . . *D. rosarium* Raper &
      Cavender
  B. Sorocarps very small with terminal sori only
    1. Sorogens, one to many, arise separately from
      small aggregations . . . . *D. lacteum* van Tieghem
    2. Sorogens arise from clustered papillae on mounds
      of aggregated cells . . . . *D. lacteum* v. *papilloideum*
        Cavender

Semi-synoptic Key[2] or Helpful Hints

Spores globose, subglobose, or oval .... *D. rosarium, D. lacteum, D. lacteum* var. *papilloideum*

Sori sessile and terminal .... *D. rosarium, D. laterosorum*

Sori pigmented—several shades:

Sori yellow....................... *D. aureum* var. *aureum, D. mexicanum*

Sori white to yellow .... *D. discoideum, D. dimigraformum, D. aureum* var. *luteolum*

Sori purple, bluish gray, or vinaceous *D. purpureum, D. laterosorum, D. lavandulum, D. vinaceo-fuscum*

Sori grayish or brownish olive .... *D. rhizopodium*

Sori tan to dull brown .... *D. brunneum*

Sori pink to rose .... *D. roseum*

Stalk bases disk-like .... *D. discoideum, D. mexicanum*

Stalk bases more or less digitate........ *D. rhizopodium, D. lavandulum, D. laterosorum, D. coeruleo-stipes, D. microsporum, D. vinaceo-fuscum*

Stalks pigmented in different shades:

Stalks in yellow shades .... *D. aureo-stipes, D. aureostipes* var. *helvetium, D. mexicanum, D. dimigraformum*

Stalks in blue or purple shades .... *D. coeruleo-stipes, D. vinaceo-fuscum, D. purpureum*

Interrupted stalk formation .... *D. giganteum, D. purpureum, D. sphaerocephalum, D. mucoroides* (occasionally)

Stalks usually composed of single tiers of cells........................... *D. lacteum, D. l.* var. *papilloideum, D. deminutivum, D. multi-stipes, D. microsporum, D. caveatum, D. tenue*

Sorocarps adherent in clusters.......... *D. polycephalum, D. polycarpum*

Stalk-free migration of pseudoplasmodia *D. discoideum, D. polycephalum, D. dimigraformum, D. intermedium, D. sphaerocephalum* (at lower temperatures)

Cell aggregation typically without streams *D. minutum, D. deminutivum, D. multi-stipes, D. lacteum, D. caveatum*

Optimum growth temperature below 20°C *D. septentrionalis, D. polycarpum*

Macrocyst formation reported:

Self-compatible strains known .... *D. mucoroides, D. minutum*

Self-incompatible strains known ...... *D. giganteum, D. mexicanum, D. rosarium*

---

[2] The term "semi-synoptic" is used because species not showing a particular characteristic (always a majority) are not indicated.

Self-compatible and self-incompatible
strains known . . . . *D. discoideum, D. purpureum*
cAMP reported as chemoattractant in cell
aggregation . . . . *D. discoideum, D. purpureum, D. rosarium, D. mucoroides, D. giganteum, D. sphaerocephalum, D. aureum*
Myxamoebae prey upon cells of other
dictyostelids . . . . *D. caveatum*

SPECIES DESCRIPTIONS

Dictyostelium mucoroides Brefeld, in Abh. Senckenberg. Naturforsch. Ges. *7*: 85-107, Tafs. I-III (1869). Also Olive in Proc. Amer. Acad. Arts Sci. *37*: 338-339 (1901), and in Proc. Boston Soc. Natur. Hist. *30*: 504-505, 45 figs., in pls. 6-8 (1902); Harper in Bull. Torrey Bot. Club. *53*: 229-268, figs. 1-33 (1926). *Synonyms: Dictyostelium brevicaule* Olive, in Proc. Amer. Acad. Arts Sci. *37*: 340 (1901), and in Proc. Boston Soc. Natur. Hist. *30*: 506, pl. 8, fig. 108 (1902). *Dictyostelium firmibasis* Hagiwara, in Bull. Natl. Sci. Mus. (Tokyo), *14*: 356, 358-359, fig. 4 (1971).

Cultivated upon substrates such as 0.1L-P, 0.1L-P (pH 6.0) and hay infusion agars in association with *Escherichia coli* or *E. coli* B/r at 22-25°C. Sorocarps usually solitary, erect, semi-erect, inclined, or partially prostrate, plain or sparingly branched, commonly 2-5 mm high but often more, especially in one-sided light, phototropic. Sorophores uncolored, appearing white in reflected light (Fig. 12-1A), arising from clavate to definitely expanded and sometimes flattened bases, commonly 18-45 μm in diameter near the base, unevenly tapered and narrowing to 8-15 μm at the sorus (Fig. 12-1F), apices plain or only slightly enlarged. Sori usually globose, sometimes citriform, milk-white when formed to cream-colored in age, commonly 125-200 μm in diameter but often less and sometimes more, frequently coalescing; spores ellipsoid, variable in size, in many strains 4.5-6.0 × 2.5-3.5 μm, in others 5-8 × 3.5-4.5 μm, the latter sometimes reniform, homogenous throughout with no polar granules (Fig. 12-1B). Cell aggregations radiate in pattern, variable in size, commonly 1-3 mm in diameter (Fig. 12-1D) and fruiting separately, sometimes larger and consisting of relatively long convergent streams (Fig. 12-1E) that may become nodular and segmented, each segment then producing a sorocarp Vegetative myxamoebae (Fig. 12-1C) ranging from about 12 to 20 × 10 to 15 μm, with vesicular nuclei, usually single contractile vacuoles and multiple food vacuoles evident. Macrocysts produced in some strains (Fig. 12-1G), not in others, usually rounded oval or occasionally angular, 35-

50 μm in diameter; microcysts not observed. Aggregating myxamoebae respond positively to cyclic AMP.

HABITATS. Dungs of various animals, leaf mould, forest soil, decaying mushrooms and vegetable products undergoing decomposition. Cosmopolitan in distribution.

No type culture exists, and because of wide variation among cultures that bear the name *Dictyostelium mucoroides,* past and present, we hesitate to designate a neotype. We can, however, list certain cultures that we believe represent the species in a broad sense, including:

Strain S-28b, isolated by KBR in 1951 from soil of a mixed pine-and-hardwood forest, Holly Hill, South Carolina. This macrocyst-forming culture was carefully investigated by Blaskovics and Raper (1957)

Strain DM-7, isolated by Michael Filosa (1962) as a superior macrocyst-producing culture and brought to this laboratry by Ann Weinkauff Nickerson. It was subsequently used for intensive study of cyst formation and germination by Nickerson and Raper (1973a,b)

Strain DC-46, isolated in 1970 by David Castener from forest soil collected at Roaring River, Missouri. Macrocysts not observed

Strain SE-4, isolated by Franz Traub from forest soil in Switzerland. Macrocysts not observed

Dictyostelium mucoroides occupies a very special place among the cellular slime molds. It was the first species to be described, hence represents the type of the genus and of the family Dictyosteliaceae; furthermore, it is encountered more frequently than any other dictyostelid, and because of such ubiquity was the species examined by virtually all eary investigators. Of these George Potts (1902) merits special attention, for he was the first to appreciate and describe the very important role of the culture environment in modulating cell aggregation and sorocarp construction— effects that are in our day as pervasive as they were in his. Like him, we still find sorocarp patterns and dimensions strongly influenced, if not in fact controlled, by such external parameters as light, temperature, relative humidity and the composition of the nutrient base.

Sorocarps of *Dictyostelium mucoroides* are visualized ideally as solitary, erect, or semi-erect, and of intermediate dimensions; and very often this is true when the dictyostelid is implanted as a mixed suspension of spores (or myxamoebae) and nutritive bacteria with subsequent incubation under overhead light. Under conditions of richer and more localized growth sorocarps may be more or less clustered, and in one-sided illumination sorophores may reach 2 cm or more in length and lie prostrate on the agar surface except for a terminal segment of a few millimeters. In most cases the sorophore is continuous from the site of aggregation to the sorus, usually

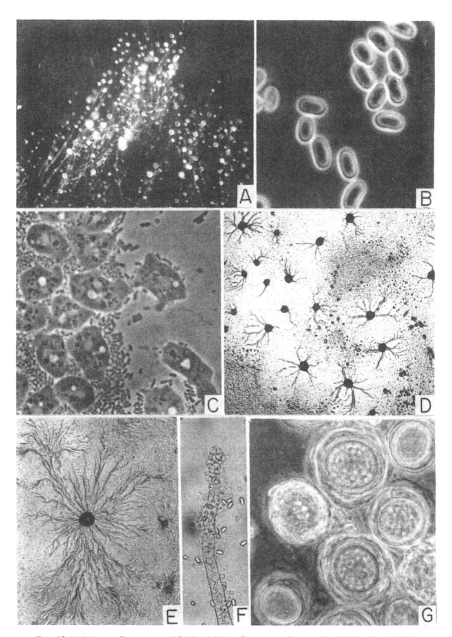

FIG. 12-1. *Dictyostelium mucoroides* Brefeld. *A*. Sorocarps photographed with reflected light; note milk-white character and that some sori in densest region of the colony have coalesced. × 5. *B*. Spores showing homogeneous content and absence of polar granules. × 900. *C*. Myxamoebae at colony edge feeding on bacteria (*E. coli*). × 90. *D*. Numerous small aggregations formed in plate inoculated with mixed suspension of bacteria and slime mold spores. × 9. *E*. Larger and more isolated aggregation with long converging streams. × 13.5. *F*. Terminus of a sorophore; note very slight enlargement and near absence of adherent debris. × 230. *G*. Macrocysts in various stages of development. × 400. See also Fig. 1-1, Brefeld's illustrations (1869), and Figs. 9-5 and 9-8, formation and germination of macrocysts, respectively

somewhat thickened at and above the zone of transition from a horizontal to a vertical orientation. Very rarely one sees examples of primary branching in this species, whereas the formation of unevenly spaced lateral branches is not uncommon; neither is the formation of occasional small sorocarps alongside prostrate sorophores, such having been formed by small masses of cells that lagged behind the advancing sorogens. One wishes that species such as *D. mucoroides* could be defined in more precise quantitative terms. But this cannot be done, and the reader should recognize that the measurements given for this and other species are not absolute but subject to variation. Furthermore, values listed for one species often overlap in part those listed for a closely related taxon. Nowhere else among the cellular slime molds is interstrain variation more troublesome than among the dictyostelia with white sori borne on simple or irregularly branched stalks of intermediate size.

In addition to variations in form and dimensions of sorocarps in single strains imposed by the culture environment, there are inherent differences among strains as well. This we believe to be the basis of Olive's species *D. brevicaule*, first described briefly in 1901 and repeated verbatim in his "Monograph" of the following year, at which time he included an illustration of a single sorocarp. He had two isolates, one each from sheep and goat dung, that were said to differ from the "long, luxuriant, frequently flexuous fructifications of *D. mucoroides* and *D. sphaerocephalum*" in the "possession of short, rather rigid stalks (1-3 mm), bearing sori of comparatively large size." When we consider the culture methods used by Brefeld, and by Olive, and project their reports against our observation of scores of cultures over many years, we find insufficient cause for further recognition of *D. brevicaule* as a valid taxon.

Based upon his published description and illustrations, and upon our examination of the type culture, which Hagiwara kindly sent to us, we also regard *Dictyostelium firmibasis* Hagiwara (1971) as a synonym of *D. mucoroides* Brefeld. As the name implies, emphasis was placed upon the expanded and somewhat flattened sorophore bases, but such are not uncommon and were in fact depicted by Brefeld for stalks in *D. mucoroides* (see his fig. 22). Furthermore, in the same report Hagiwara illustrates rather similar sorophore bases for *D. sphaerocephalum*. Spores of his type culture tend to be narrowly ellipsoid, while the overall pattern and dimensions of sorocarps fit quite well those of *D. mucoroides*.

Special attention is accorded several strains received in August 1966 from Ian A. Worley as isolates from soils of New Zealand. These produce primary sorocarps that are indistinguishable from those of *D. mucoroides* (sorophores 2.5-4.5 mm high bearing globose, white sori 100-200 μm in diameter); then large numbers of small sorocarps arise near the bases of the larger structures. They rarely exceed 1.0-1.5 mm in height or bear sori larger than 50-75 μm. Spores from the primary sorocarps and from the diminutive ones that encircle them are similar in form and dimensions and are devoid of polar granules. Two of the isolates were retained as numbers IW-9 and IW-10.

The presence of macrocysts is a fairly common and conspicuous feature of many strains of *Dictyostelium mucoroides*, and it was in this species that such cysts were first recognized as specialized structures quite separate from the spore-bearing sorocarps (Raper, 1951). It was in this species also that their singular patterns of origin and development were first elucidated by Filosa and Dengler (1972), Nickerson and Raper (1973a,b), and Erdos et al. (1973a,b); and it was in *D. mucoroides* where their sexual character was first suspected (Clark, Francis, and Eisenberg, 1973) and subsequently confirmed (Macinnis and Francis, 1974). As of the present, all known macrocyst-forming strains of *D. mucoroides* are self-compatible, but whether this is universally true is open to doubt, for in other species of which we have information contrasting mating types are generally required, even where occasional self-compatible isolates are known to occur, as in *D. discoideum* and *Polysphondylium violaceum*.

In no case have we observed individual myxamoebae of *Dictyostelium mucoroides* to encyst, forming microcysts. Cysts formed by individual cells were reported and illustrated by Brefeld in his original description of *D. mucoroides* (1869, his figs. 6,7). These, however, were of a size and pattern (heavy roughened walls) to indicate not encysted myxamoebae of a *Dictyostelium* but the cysts of a free-living soil amoeba such as *Hartmanella* sp. That he had such amoeboid organisms in his culture is not surprising, for they are commonly present in the natural habitats of cellular slime molds.

Dictyostelium mucoroides Bref. var. *stoloniferum* Cavender and Raper, in Amer. J. Bot. 55: 510-511, figs. 7-12 (1968).

When cultivated upon 0.1L-P agar in association with *Escherichia coli* at 23-25°C, primary sorocarps resemble those of the species in general form and appearance: sorophores are elongate with bases rounded and only slightly enlarged, and, being phototropic, they develop toward the light. Once formed, the sori enlarge by absorption of water and not infrequently fall to the agar surface. Unlike the species, where spores fail to germinate unless dispersed (Russell and Bonner, 1960), spores of the variety germinate immediately and en masse upon touching the agar; the newly emerged myxamoebae aggregate without regrowth (Fig. 12-2D); and the resulting pseudoplasmodia produce few to several sorocarps in situ within 6 to 8 hours (Fig. 12-2A-C). These smaller secondary structures are erect with comparatively short, gradually tapering sorophores with expanded and sometimes flattened bases (Fig. 12-2E,F) and show little or no response to light. They do, however, exhibit a marked avoidance reaction during

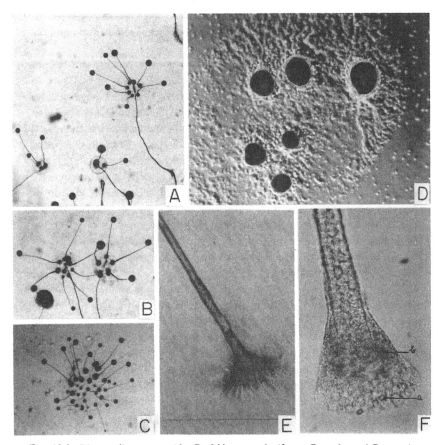

Fig. 12-2. *Dictyostelium mucoroides* Brefeld var. *stoloniferum* Cavender and Raper. *A*. Margin of colony showing "stoloniferous habit." Mature sori fell to agar surface at three sites and secondary sorocarps arose following spore germination and aggregation of emerged myxamoebae. × 10. *B*. Other examples of the same; note striking divergence of secondary sorocarps. × 10. *C*. Cluster of small sorocarps 10 hours after spores were deposited on agar surface. × 10. *D*. Aggregation of myxamoebae 3.5 hours after spores were deposited on 0.1L-P agar. × 50. *E*. Base of a secondary sorocarp. × 170. *F*. Detail of sorophore base; note that the stalk rests upon a flattened cushion of strongly vacuolate cells (a) that extends upward (b) to form a supporting collar. × 270. (From Cavender and Raper, 1968)

development (Fig. 12-2A-C), presumably due to gaseous inhibition as reported for *D. discoideum* and other species (Bonner and Dodd, 1962a).

The varietal name reflects the stolon-like aspect created by the fallen primary and emergent secondary sorocarps.

HABITATS. Surface soil of tropical rain forests, Costa Rica.

HOLOTYPE. Strain Tu-II-6, isolated by J. C. Cavender in 1962 from soil collected at Turrialba, Costa Rica. A second and similar strain, FO-II-1, was isolated in the same year from soil collected at Fortuna, Costa Rica.

In the years since 1968 an occasional strain with characteristics approaching those of the cotypes has been encountered, particularly among cultures isolated from soils collected by Cavender in East Africa and sent to us in 1968. A pattern of development superficially resembling that of *D. m.* var. *stoloniferum* is sometimes seen in *D. giganteum* and other species, but with this difference: when a nascent sorus falls to the surface, the already formed spores spread outward but do not germinate, while the still centrally placed amoeboid cells reaggregate and produce secondary fructifications.

Concerning this freely germinating variety, Professor Bonner tells an interesting story: A senior student, Alison Moore, collected spores and replanted them for six successive "culture generations," all grown without food. The volume of the cells decreased by about half with each recultivation, and the final spores measured only a little more than 3 μm in length on average. Remarkable also is the fact that despite such reduction in cell size, the sorocarps were always normally proportioned.

Dictyostelium giganteum Singh, in J. Gen. Microbiol. *1*: 11-21, text fig. 1, pl, 1, figs. 1 & 2 (1947). Also Norberg in Ph.D. Thesis, University of Wisconsin, Madison, pp. 73-80, figs. 4A-F (1971).

Cultivated on substrates such as 0.1L-P, 0.1L-P (pH 6.0), and hay infusion agars in association with *Escherichia coli* or *E. coli* B/r at 22-25°C. Sorocarps large and variable, mostly solitary, usually semi-erect or inclined under overhead light and prostrate in unidirectional lateral light, strongly phototropic (Fig. 12-3E), 4-10 mm or more high when semi-erect and up to 5-7 cm or more in length when primarily procumbent (Fig. 12-3A). Sorophores hyaline, white, or faintly colored, plain or sparsely branched, sinuous and at times helical, sometimes discontinuous, variable in width, tapered in semi-erect sorocarps and terminal portions of procumbent structures, usually ranging from 20-45 μm in diameter near the base and narrowing to 8-15 μm at the sori; sorophore bases rounded but rarely enlarged, apical regions narrowed and plain, the longest sorophores often terminating without sorus formation. Sori terminal, white, globose to subglobose (Fig. 12-3A), commonly ranging from 125-250 μm in diameter, occasionally larger and often smaller. Spores ellipsoid, mostly 5.0-6.5 × 3.0-3.5 μm but not uncommonly 7.5 × 4.0 μm and occasionally more, the latter often reniform, spore content appearing homogenous, without polar granules (Fig. 12-3B). Cell aggregations initially radiate but becoming irregular in pattern as neighboring centers compete for unaggregated myxamoebae, commonly 2-5 mm in diameter (Fig. 12-3D), sometimes much larger, usually forming single sorogens into which cells continue to stream after sorophore formation has begun. Sorogens at first long and narrow, becoming progressively shorter and thicker as development progresses. Vegetative myxamoebae variable in form, commonly 12-20 × 8-14 μm, with vesicular nuclei and one or more contractile vacuoles (Fig. 12-3C). Microcysts not observed; macrocysts produced when appropriate mating types are paired (Fig. 12-3F).

As stated by Norberg (1971, p. 73), "The particular distinguishing characteristics of these organisms are best observed in cultures grown with *E. coli* on 0.1L-P agar at 23°C in one-sided illumination."

HABITATS. Soils of different type, including forest, prairie, and crop land, also animal dungs, decaying mushrooms, and rotting vegetation. Common in North America and isolated also in England, France, and the Netherlands; probably worldwide in distribution.

HOLOTYPE. Strain "Singh," isolated by B. N. Singh in 1943 from an actively decomposing compost heap of straw and sludge, Rothamsted Experimental Farm, Herts, England.

Other representative strains examined in this and earlier studies include:

Strain S-2, isolated by KBR in 1934 from a decaying mushroom, Cambridge, Massachusetts. Before Singh described *Dictyostelium giganteum*, this strain was regarded as *D. mucoroides* and was reported as such by Raper and Thom (1941) in their study of interspecific mixtures.

Strain WS-125, isolated by KBR in 1954 from soil of a deciduous forest, University of Wisconsin Arboretum, Madison, Wisconsin.

Strain H-109, isolated by Theo. M. Konijn in 1957 from soil collected in a Cemetery, Amsterdam, Netherlands.

Strain DC-14, isolated by David Castener in 1970 from rotting wood, Kansas City, Missouri.

Strain WS-589, isolated by John Sutherland in 1973 from soil of a wet mesic prairie, Whitewater, Wisconsin. This was designated as mating type A1 and formed macrocysts when paired with mating types A2, A3, or A4 (Erdos et al., 1975).

Strain WS-606, isolated by John Sutherland in 1973 from soil of a mesic prairie, Racine, Wisconsin. This was designated mating type A2 and formed macrocysts when paired with mating types A1, A3, and A4 (Erdos, et al., 1975). See Chapter on Macrocysts.

Dictyostelium giganteum Singh is one of the more abundant dictyostelids in nature and has surely been isolated and studied far more often than the name has appeared in published reports. Prior to 1947 it would have been diagnosed as *D. mucoroides*, just as was done by Raper in 1941 when

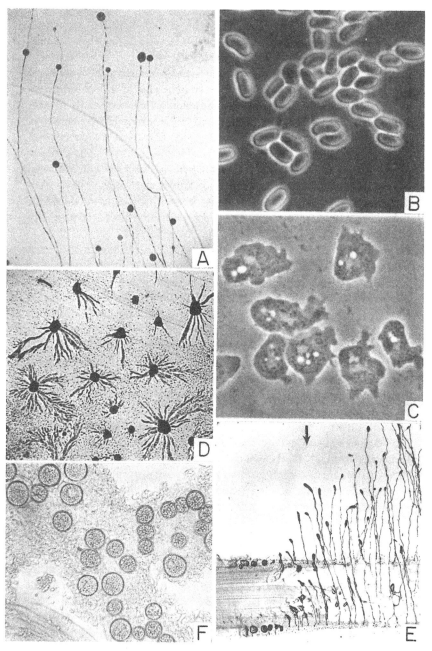

FIG. 12-3. *Dictyostelium giganteum* Singh. *A.* Sorocarps with long sorophores characteristic of the species, strongly phototropic. × 5.5. *B.* Spores—note homogeneous content and absence of polar granules. × 900. *C.* Vegetative myxamoebae photographed just outside a bacterial colony. × 900. *D.* Numerous small aggregates in plate inoculated with mixed suspension of bacteria and slime mold spores. × 10. *E.* Developing sorocarps showing strong and uniform phototropic response; direction of lateral light indicated by arrow. × 5.5. *F.* Macrocysts produced by pairing strains WS-589 (mating type A1) and WS-606 (mating type A2) on 0.1L-P agar with incubation in the dark. × 150

describing *D. minutum*, and by Raper and Thom (also 1941) when studying specific intermixtures among four different dictyostelids. It is withal a large and vigorous species that grows well and is easily handled in the laboratory. Most strains of *D. giganteum* can be grown quite successfully on substrates of substantial nutrient content; those cited above are not necessarily optimal except for comparison with other species. For example, in studies on interspecific mixtures Raper and Thom (1941) employed a wide range of substrates, including lactose-peptone and dextrose-peptone agars containing nutrients up to 1%. Levine's Eosin Methylene Blue Agar (Difco) diluted to ½ to ⅓ nutrient strength is especially useful and yields colored fructifications besides. As the reader might suspect, sorocarps produced on such richer media are more robust, have thicker sorophores, and generally bear larger sori.

The species as now envisioned is subject to appreciable variation, not only among different strains but also within single strains, depending especially upon the substrates employed and the prevailing physical environment. We shall first consider the environment. When a strain like WS-125 is grown with *E. coli* upon a buffered lactose-peptone agar such as 0.1L-P (pH 6.0) and incubated at 23°C directly beneath two 15 w fluorescent lamps held at 18-20 inches above the Petri dishes, the sorocarps are generally erect or semi-erect, seldom more than 6-10 mm high, and bear terminal sori commonly in the range of 150-250 μm in diameter. If similar cultures are incubated so that light enters only through the side of the dish, sorogens are formed as above but build toward the light and, because of their weight, soon fall to the agar surface. The sorophores are then lengthened more and more as they move toward the edge of the culture dish, often fruiting on its vertical wall. This type of response can be somewhat enhanced by using unbuffered 0.1L-P agar, and the effect can be especially dramatic if plates are incubated in a black box with light entering through a narrow slit ca. 2 mm wide, thus focusing the structures toward a point. The sori of such elongate sorocarps are always small and sometimes lacking altogether, for so many cells are used in stalk formation that none remain to form spores. Under less extreme circumstances the sorogen may build a stalk horizontally for a few centimeters, then pause, gradually assume a vertical orientation, and complete its development essentially perpendicular to the substrate, bearing a sorus of somewhat reduced size (Fig. 8-17B). In other cases the sorogen may build in an inclined direction until some surface cells have become transformed into spores and then fall to the agar surface, whereupon the spores spread outward on the agar surface while the central, still amoeboid cells reaggregate and produce one or more sorocarps not unlike but smaller than those first described.

In marked contrast to the above, the sorogens and/or pseudoplasmodia of some strains appear to be relatively insensitive to light, and additionally tend to migrate with the formation of discontinuous stalks in a seemingly random manner before halting to form upright sorocarps, if indeed this ever occurs. Strain WS-589 affords an extreme example of such behavior, and doubtfully would be of special interest were it not for the fact that it constitutes the most studied mating strain of the A_1 type (Erdos et al., 1975). On the other hand, Strain WS-606, which represents type A_2, and with which WS-589 mates readily to form macrocysts, is quite different in pattern and in phototropic response, since it essentially duplicates WS-125 in both regards.

Some strains provisionally assigned to *D. giganteum* differ in still other ways. Foremost among these are cultures such as WS-142 that produce very long but quite thin stalks when incubated in one-sided light. Sorophores often consist of a single tier of superimposed (linear) cells, are rarely branched, and bear small sori seldom larger than 125 μm in diameter. Spores are ellipsoid, appear homogenous within, and contain no polar granules. Clearly they belong in the *D. mucoroides* complex and appear to fit *D. giganteum* more nearly than any other species. It is of some interest, if only coincidental, that another large and very strongly phototropic species, *Dictyostelium purpureum*, also includes strains that are developmentally similar, albeit richly pigmented.

Dictyostelium septentrionalis Cavender, in Can. J. Bot. *56*: 1326-1332, figs. 1-10 (1978).

Cultivated on substrates such as 0.1L-P (pH 6) or 0.1% G, 0.05% Yxt agars in association with *Escherichia coli* at 15-17°C. Sorocarps typically erect, robust in habit, solitary, seldom branched (Fig. 12-4F). Sorophores unusually large, commonly 0.8-1.2 cm in length but ranging from 0.5 to 1.5 cm, generally arising from expanded conical bases 150-375 μm in diameter (Fig. 12-4E,H), conspicuously tapered, ranging from 60-110 μm wide near the base to about 25 μm at the top (Fig. 12-4F-J). Sori globose to slightly elongate, milk-white to cream colored, 250-650 μm in diameter (Fig. 12-4F). Spores broadly elliptical, mostly 6.5-8.0 × 4.0-4.8 μm, not uncommonly 10 × 5 μm and rarely more, content seemingly homogeneous, without polar granules (Fig. 12-4A). Cell aggregations radiate, usually large with converging streams commonly several millimeters in length, typically producing single sorogens and sorocarps. Vegetative myxamoebae usually rounded when feeding, about 12 μm in diameter, elongate when moving and ranging from 13 to 20 × 9 to 12 μm. Microcysts and macrocysts not reported or observed.

HABITAT. Soil and humus of spruce and hemlock forests, coastal area of southeastern Alaska.

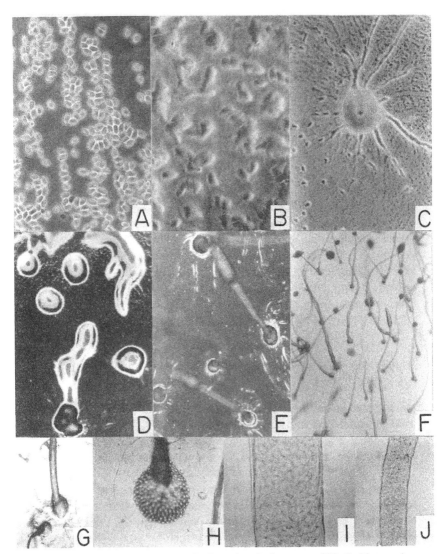

Fig. 12-4. *Dictyostelium septentrionalis* Cavender. *A*. Spores. × 285. *B*. Myxamoebae beginning to clump prior to aggregation. × 180. *C*. Small aggregation. × 75. *D*. Late aggregation prior to sorogen formation. × 18. *E*. Rising sorogens as seen from above. × 18. *F*. Sorocarps as seen from above; note conical bases and tapered sorophores. × 8. *G,H*. Sorocarp bases, × 18 and 75, respectively. *I*. Portion of large sorophore near conical base. × 180. *J*. Portion of similar sorophore near the sorus. × 180. (Figure assembly from J. C. Cavender as used in the species description, 1968)

HOLOTYPE. Strain AK-2, isolated in 1974 by J. C. Cavender from humus collected in a Sitka spruce and western hemlock forest, Tongass National Forest, Juneau, Alaska.

Two characteristics in particular serve to distinguish this species: (1) its low temperature optimum and (2) its unusually robust sorocarps (Fig. 12-4F,I). Both are of special interest, for in no other species is the recommended growth temperature as low as 15-17°C, and in no other species are the sorophores as thick or the sori as large as some of those found in *D. septentrionalis*.

The low temperature required for growth and development of *D. septentrionalis* most probably represent a physiological adaptation to its natural habitat, for Cavender (1978) has reported the mean daily temperature in the forested areas from which it was isolated to be 13°C in July. In the laboratory cultures grow well over a range from about 12 to 20°C but fail to develop optimally at 20°C and fail to fruit at higher temperatures. At 20°C sorocarps appear stunted, sorophores are irregularly thickened and sometimes branched, and the sori, commonly misshapen, may contain poorly formed spores. Such structures stand in sharp contrast to the sorocarps formed at lower temperatures, where sorophores show a progressive taper from conical base to globose sorus and withal appear to be well proportioned, suggesting in some measure the shorter sorocarps of *D. discoideum*. As in that species, each aggregation typically produces one sorocarp; there is, however, no suggestion of a migrating pseudoplasmodium.

Cultures of *D. septentrionalis* may be grown in association with *E. coli* or *E. coli* B/r on unbuffered 0.1L-P agar, but the addition of phosphate buffers to pH 6.0 is clearly advantageous. There is limited evidence also that fruiting is improved by using relatively dry rather than freshly poured plates. Agar media containing glucose (0.1%) and yeast extract (0.05%) support excellent myxamoebal growth and satisfactory fruiting, but sorophores tend to become longer and thinner so that the sorogens often fall to the agar surface as in *D. giganteum* or *D. purpureum* before completing sorocarp formation. When mounted in water the mature sori disperse, as do those of other species, but a substantial body of spores usually adheres to the sorophore, and in some cases there is the suggestion of a persistent collar at the base of the sorus. Such structures are generally present in strains that we believe to belong to *D. sphaerocephalum*.

Dictyostelium irregularis Nelson, Olive, and Stoianovitch, in Amer. J. Bot. *54*: 354-358, figs. 1-26 (1967).

Cultivated on Difco cornmeal-dextrose agar with 0.1% yeast extract, or lactose-peptone and hay infusion agars in association with *Escherichia coli* or *Klebsiella pneumoniae* at 23°C. Sorocarps simple or compound, very irregular in pattern and dimensions, 140-965 μm tall, with irregular, ridged, and warty stalks that are mostly 21-138 μm broad; but diminutive sorocarps with single rows of stalk cells also formed. Sori globular, oval, or elongate, 32-112 × 42-250 μm. Spores hyaline, typically elliptical, but occasionally nearly globose or somewhat irregular, 4.8-9.8 × 3.5-7 μm without polar granules. Cell aggregations abnormal in appearance, consisting of thick and often poorly defined inflowing streams giving rise to one of several sorogens that may fruit in situ or migrate for some distance before sorocarp formation is attempted. Myxamoebae comparatively large but typical of the genus. Macrocysts and microcysts not formed.

HABITAT. Decaying ear of corn (maize) from experimental plot, campus of University of Hawaii, Honolulu, Hawaii.

HOLOTYPE. Strain H-68, isolated by L. S. Olive in March 1963.

Based upon the foregoing description abstracted from Nelson et al. (1967) and upon our observations of the type as submitted to us, together with supplemental comments by the species' authors and their very informative illustrations, we would be inclined to regard this as a depauperate member of the *D. mucoroides* complex that is unable to "get its act together." Alternatively, it might be considered as a more primitive member of the genus, as Nelson et al. suggest. In support of this view they offer its greater plasticity, irregular appearance, and the variability in its sorocarp development. In our view it should probably be regarded as a naturally occurring, developmentally deficient mutant of *D. mucoroides* or possibly *D. sphaerocephalum*. Whereas the "fructifications" are in some instances reminiscent of the sorocarps sometimes seen in the genus *Guttulinopsis*, the true relationship of the slime mold is unquestionably with *Dictyostelium*. This is amply confirmed by the pattern of its spores, the vacuolate pattern of its poorly ordered stalk cells, and the filose pseudopodia and vesicular nuclei (with peripheral nucleoli) of its myxamoebae.

It is of some interest that J. C. Cavender has isolated, reported (1980), and submitted to us a comparable but nonsporulating culture that develops a marked purple pigmentation strongly suggestive of *D. purpureum*. This too might be interpreted in one of two ways: either that an occasional developmentally deficient mutant may arise in a common, recognized species, or that there exist in nature more primitive and less developmentally competent dictyostelids that present isolation methods rarely recover.

Dictyostelium sphaerocephalum (Oud.) Sacc. et March., in El. Marchal. "Champignons coprophiles de Belgique," Bull. Soc. Roy. Botan. Belg.

24: 74, Pl, III, figs. 1-4 (1885). Also Norberg in Ph.D. Thesis, University of Wisconsin, Madison, pp. 34-57, 93-101, fig. 7A-I (1971). *Synonym: Hyalostilbum sphaerocephalum* Oudemans, in Aanwisten voor de Flora Mycologica van Nederland, *9-10*: 39, Pl, IV, fig. 4a-d (1885).

Cultivated upon substrates such as 0.1L-P, 0.1L-P (pH 6.0), and hay infusion agars in association with *Escherichia coli* or *E. coli* B/r at 22-25°C. Sorocarps white, commonly branched, usually solitary, weakly phototropic, erect or semi-erect, 2-4 mm high if developed at sites of aggregation (Fig. 12-5A), or procumbent with terminal portions rising from the substrate in arcs to bear the sori (Fig. 12-5F), often robust in appearance. Sorophores variable in form and dimensions, generally stout and of uneven taper in erect structures, with bases often poorly formed and relatively thin but widening to 30-45 μm at midpoint and narrowing to 15-25 μm in the sori, frequently interrupted (Fig. 12-5F,G) and of lesser diameter when procumbent, the segments enclosed in a continuous slime envelope produced by the advancing sorogen/pseudoplasmodium. Sori terminal on sorophores and branches (if present), milk-white and globose or nearly so (Fig. 12-5A,F), often appearing recurved (Fig. 12-5H), comparatively large, commonly 175-250 μm but up to 300-350 μm in diameter, and in dense cultures coalescing to produce even larger spore masses; sori not completely removed by needle contact, and both terminal and subterminal masses of debris and occasional spores remain adherent to sorophore apices (Fig. 12-5H,I). Spores ellipsoid, mostly 5.5-7.0 × 3.0-3.5 μm but sometimes smaller and occasionally up to 7.5 × 4.0 μm, without polar granules (Fig. 12-5B). Cell aggregations radiate (Fig. 12-5D), commonly 1.5-2.5 mm in diameter under overhead light, larger in diffuse light or darkness, usually producing single sorogens and sorocarps, with pseudoplasmodia often migrating freely or with stalk formation discontinuous (Fig. 12-5E,G) leading to aerial construction of spore-bearing structures. Vegetative myxamoebae commonly 12-20 × 10-12 μm (Fig. 12-5C), with single vesicular nuclei, food vacuoles, and one or sometimes two contractile vacuoles evident. Macrocysts reported (Hagiwara, 1971) but not observed in our laboratory; microcysts not reported or observed.

HABITATS. Soils of different types, including forest, grassland, and garden; dungs of various animals, and different kinds of decomposing vegetation. Apparently cosmopolitan in distribution.

No type culture exists, and we hesitate to designate a neotype. From among many isolates examined, however, the following strains may be listed as representative of the species as currently envisioned:

Strain WS-69, isolated in 1953 by Joan Cormier Blaskovics from soil

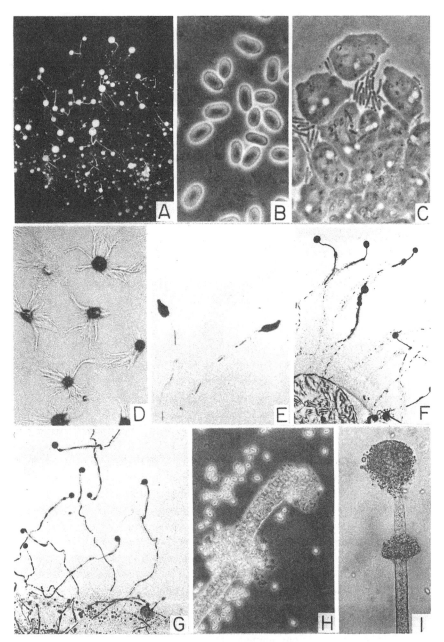

Fig. 12-5. *Dictyostelium sphaerocephalum* (Oud.) Sacc. et March. *A.* Sorocarps developed in situ beneath overhead light; note the relatively short sorophores and large white sori. × 6. *B.* Spores showing homogeneous cell content and absence of polar granules. × 900. *C.* Vegetative myxamoebae at margin of bacterial colony. × 900. *D.* Cell aggregations in culture inoculated with mixed suspension of bacteria and spores. × 6. *E.* Two migrating pseudoplasmodia with stalk segments in their wakes. × 14. *F.* Sorocarps with inclined or vertical stalks developed from migrating pseudoplasmodia. × 6. *G.* Variation of the same— stalks are nearly continuous, and upright sorophores are very short. Migrating structures are generally nonphototactic. × 5. *H.* Terminus of an upright sorophore slightly recurved showing upper and lower masses of adherent cellular debris following removal of the sorus. × 230. *I.* A more striking example of the same phenomenon which is generally characteristic of the species. × 230

of a pine planting, University of Wisconsin Arboretum, Madison, Wisconsin.

Strain GR-11, isolated by Franz Traub, University of Zurich, from forest soil in Switzerland and sent to us as representative of the unpublished provisional species "*Dictyostelium robustum*" (Traub and Hohl, 1976).

WS-130, isolated by KBR in 1954 from prairie soil collected in the University of Wisconsin Arboretum, Madison, Wisconsin.

Strain DC-34, isolated by David Castener in 1970 from forest (?) soil collected at Squaw Creek, Missouri.

Strain WS-696, isolated by John Landolt, Shepherdstown, Pennsylvania, and sent to us for diagnosis in September 1980.

Dictyostelium sphaerocephalum as envisioned and redefined here is marked by three distinguishing characteristics in particular:

(1) The frequent migration of pseudoplasmodia prior to assuming a semi-erect orientation with the formation of an upright sorocarp, or the semi-erect terminal portion of a largely procumbent structure. Such migration may be stalkless in whole or in part, discontinuous with sections of stalk interspersed with stalkless areas, or with the formation of a continuous stalk from the site of aggregation to the terminal sorus. In any case, the upright stalk or portion thereof is usually characterized by an uneven taper as it rises arc-like from the substrate. It is commonly thin and irregular at the point of origin, usually thicker as it rises from the substrate, often reaching a maximum width at about midpoint as it builds upward, then narrowing again toward the sorus. Generally, the upright sorophores are comparatively stout, sometimes quite short, and bear relatively large sori that in some strains may be recurved. Pseudoplasmodial migration increases markedly as incubation temperatures are lowered to 15-18°C.

(2) The inconsistent but frequent branching of the sorophores, which may assume a variety of different forms. This may range from multiple sorophores arising from a common base (as illustrated by Oudemans, 1885), through different levels of aerial branching, to the formation of numerous side branches or small, secondary sorocarps along the length of a prostrate sorophore. In each of the last two cases the structures arise from small masses of cells left behind by the advancing pseudoplasmodium, and their subsequent development as branches or as satellite sorocarps depends upon whether they remain attached to the sorophore or separated from it. Except for their small dimensions, branches and sorocarps so formed mimic the primary sorophore and terminal sorus in general configuration.

(3) The adherence of residual debris and some spores to the ends of sorophores upon removal of the sori by needle contact. Such material may take the form of a simple brush-like enlargement, as was illustrated by Marchal (1885). More often, however, it is subdivided into two parts (Fig. 12-5H,I), one terminal and more or less rounded above, the other (at which

was the base of the sorus) simulating in appearance the calyx of a flower or the bottom of a broken cupule. This singular two-parted residue was first recognized and photographed by Traub and Hohl during their survey of the cellular slime molds of Switzerland; and it was for a time considered to be a distinguishing characteristic of a provisional but unpublished new species of *Dictyostelium, D. robustum* (1976). Significantly, it has appeared with remarkable consistency in virtually all the strains that we would assign to *D. sphaerocephalum* upon the bases of other criteria, both cultural and microscopic.

Individual strains vary considerably among themselves. Some of these, such as WS-69, when grown under overhead light at 22-23°C produce mostly semi-erect and generally stout sorocarps with comparatively thick stalks and large globose sori (Fig. 12-5A). Strain GR-11 develops in much the same manner but the sorocarps tend to develop outward from the bacterial streak wherein they arise and often fall to the agar surface, the sori collapsing as this occurs. While not proven, such behavior could represent an unusually sensitive expression of gaseous inhibition. In strain WS-130 the pseudoplasmodia show an increased tendency to migrate well beyond their sites of origin, often forming continuous or interrupted stalks for several centimeters before reorienting to produce quite short, upright terminal sections that bear the sori. Branching and/or the formation of small sorocarps along the prostrate stalk is commonplace. Strain WS-696 shows a greater degree of aerial branching than most members of the species, and some of the branches may be of substantial size. There is, however, no consistent pattern discernible.

It is our belief, as already stated, that *Dictyostelium brevicaule* Olive (1901, 1902) should be regarded as synonymous with Brefeld's *D. mucoroides*. Nevertheless, strains such as WS-69 when cultivated upon nutrient-poor substrates do at times develop sorocarps of the form and dimensions that Olive reported. That this is true only strengthens our belief that the name probably reflects conditions of culture more than any inherent structural characteristic of a dictyostelid.

Dictyostelium minutum Raper, in Mycologia *33*: 633-649, figs. 1-4 (1941).

Cultivated on 0.1L-P/2, 0.1G-P/2, and dilute hay infusion agars in association with *Escherichia coli* at 22-25°C. Sorocarps small, erect, solitary or clustered, commonly 0.5-0.85 mm in height, occasionally larger, often smaller, frequently branched (Fig. 12-6), not phototropic. Sorophores colorless, 10-20 μm in diameter at the base, tapering gradually to 2.5-5.0 at the sorus, terminal regions flexuous, consisting of single tiers of cells except in basal regions (Fig. 12-6). Sori colorless to milk-white, citriform to rounded-apiculate (Fig. 12-6A-H), commonly 75-125 μm in diameter, occasionally larger. Spores elliptical, hyaline, mostly 5.0-6.0 × 3.0-3.5

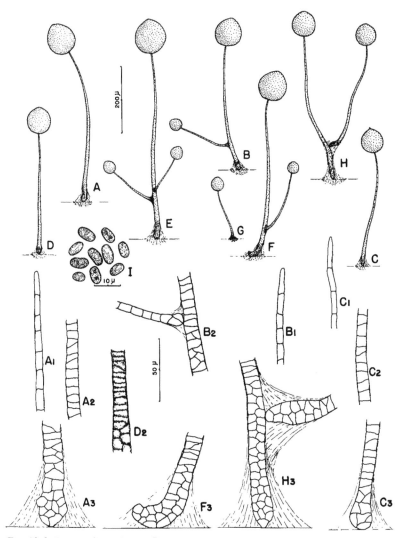

Fɪɢ. 12-6. *Dictyostelium minutum* Raper, mature sorocarps. *A-H*. Camera lucida sketches of typical sorocarps showing relative proportions. A_1, A_2, and A_3, cellular structure of terminal, central, and basal portions of sorocarp *A*. B_1, terminal portion of sorocarp *B*; B_2, structure and anchorage of branch in same sorocarp. C_1, C_2, and C_3, apical, central, and basal portions of sorocarp *C*. D_2, detail of cellular structure near sorophore base, sorocarp *D*. F_3, recurved base of sorocarp *F*. H_3, detail of branch anchorage in sorocarp *H*. *I*. Spores. Scale variable, as indicated. (From Raper, 1941b)

µm, without polar granules (Fig. 12-7B). Cell aggregations of two types: small, rounded mounds without inflowing streams (Fig. 12-7D), and aggregations with interconnected and often seemingly undirected streams (Fig. 12-7E), the former fruiting directly, the latter following emergence of secondary centers. Sorogens uncolored, elongate to clavate. Myxamoebae commonly 12-18 × 7-10 µm, irregular or broadly triangular in

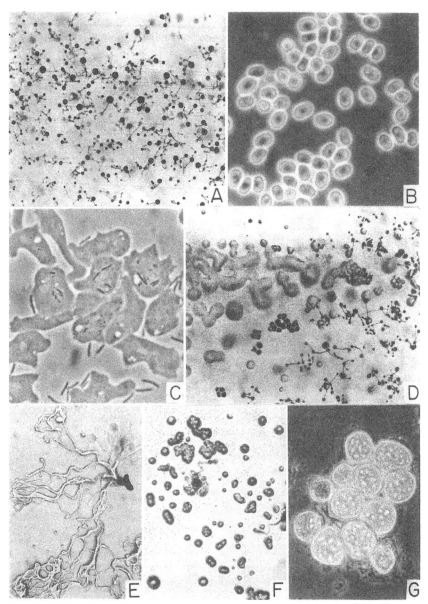

FIG. 12-7. *Dictyostelium minutum* Raper. *A*. Sorocarps developed on 0.1% peptone agar and here shown in silhouette. Some larger "sori" represent examples of coalescence. × 13.5. *B*. Spores showing variations in form and an absence of evident polar granules. × 880. *C*. Vegetative myxamoebae at edge of bacterial colony. × 880. D. Portion of a colony on hay infusion agar: upper left, aggregations forming without myxamoebal streams; center diagonal, emergent sorogens; lower right, mature sorocarps. × 6.5. *E*. Postaggregative "streams" of myxamoebae (see text). × 33. *F*. Small aggregations destined to form from one to several macrocysts. × 33. *G*. A small cluster of macrocysts showing undigested myxamoebae (endocytes) within. × 335. (From Raper, 1941b, in substantial part)

the direction of movement (Fig. 12-7C). Microcysts may form in older cultures; macrocysts produced abundantly in some strains, rare or absent in others (Fig. 12-7F,G).

HABITAT. Surface soil and leaf mould of deciduous and coniferous forests, probably worldwide in distribution. Common in North America and reported from Europe, Africa, and Japan.

HOLOTYPE. Strain V-3, isolated by Raper in 1937 from soil of a deciduous forest near Vienna, Virginia. Current description based upon strain V-3, WS-581 from forest soil, Bergen, Minnesota, and many other isolates.

Of the many dictyostelids now known, *D. minutum* is one of the most cosmopolitan species, for it is found in most forest soils where there is an appreciable accumulation of decomposing leaf litter. In numbers it is exceeded only by *D. mucoroides, Polysphondylium violaceum* and *P. pallidum*. Because of its small dimensions it is not so easily recovered from primary isolation plates as the larger species, but it has one advantage: it develops as relatively compact, heavily sporulating clones. Once obtained in two-membered culture with *E. coli* or other bacterial host, it can be grown on a wide variety of low-nutrient substrates.

The current characterization of the species is based upon the original description (Raper, 1941b), supplemented in some particulars by information more recently obtained. Macrocysts represent a case in point: they were not actually reported until a decade later (Raper, 1951) and then in strains of *D. minutum* other than the type culture. In fact, they are rarely produced by strain V-3, and when present are generally solitary and few in number. In contrast, some strains such as WS-581 produce macrocysts in great numbers, sometimes at the expense of sorocarp development. In still other strains they have not been observed. As in strains of *D. mucoroides* and *D. discoideum*, macrocyst formation is especially good on 0.1L-P agar (nutrients at half strength in this case) and negligible or lacking on the same substrate buffered to pH 6.0 with phosphates, even as sorocarp formation is enhanced. Where known to occur, and where tests have been made, the macrocyst-forming strains have proved to be homothallic, but whether this is always true is an open question. Both homothallic and heterothallic strains are known to occur in *D. purpureum, P. violaceum,* and *D. discoideum* and *P. pallidum*, so why not in *D. minutum*? The answer lies in further study.

Cell aggregation in *Dictyostelium minutum* differs from most dictyostelids in that it follows two contrasting patterns. These were noted and illustrated by Raper in 1941(b), but their significance was not comprehended. The subject was further studied and clarified by Gerisch in two papers some twenty years later (1963e and 1964a), the first of which included the paragraph that follows (freely translated from German):

> As a rule the collective amoebae of the genus *Dictyostelium* join to form cell strands at the onset of aggregation. . . . This is different with *Dictyostelium minutum* Raper. In this species, after the consumption of the food bacteria, there appear round elevations in a homogeneous layer of single cells. These elevations soon increase in size due to an influx of neighboring cells. These centers are the source of a strong attractive force. The cells, which move in from the environment in a *directed* fashion, are nevertheless neither elongated nor connected to one another. The cell mass of the center later gives rise to fruiting bodies, which include only part of the amoebae when the centers are large, while the excess amoebae emigrate. Only during this latter *dispersion phase* do the amoebae come together in strands, which can increase their size by the influx of amoebae that are still free and that later on condense to secondary centers, which in turn give rise to fruiting bodies.

When the elevated centers (rounded mounds) are quite small, they usually produce solitary sorocarps, or if somewhat larger become papillate, then form multiple sorogens and sorocarps, commonly from two to six per mound. Under other conditions they fruit incompletely and behave in the manner Gerisch has described. Although streams of emigrating cells are, as a rule, associated with specific fruiting centers (Fig. 12-6E), at other times and places appreciable networks of interconnecting streams seem to form independently of any apparent fruiting phenomena.

In comparative cultures *D. minutum* bears a superficial resemblance to *D. multi-stipes*. However, the two species can be distinguished as follows. The sorocarps and sori of *D. minutum* average somewhat larger, bases of the sorophores are somewhat thicker, the spores show no polar granules, and fewer sorocarps arise from an average primary aggregation. Additionally, streams of myxamoebae in *D. minutum*, when present, signal the dispersal of myxamoebae from existing centers more often than the convergence of cells to form new aggregations.

As with many cultures long maintained in the laboratory, strain V-3, the type of *D. minutum*, no longer reproduces in all particulars the species description as first published (Raper, 1941b). This is especially true of the spores, which are now appreciably larger and frequently reniform or recurved in pattern rather than short and elliptical. It is suspected that this change reflects an increase in ploidy, as has been shown to occur in *D. discoideum* and other species. Whatever the cause, we have chosen not to widen the species diagnosis to suit the "type" in its present state; rather, we have drawn more heavily upon strain WS-581 and other recent isolates that meet the original prescription and fit more nearly the measurements of strains as these are recovered from nature.

Two very informative and beautifully illustrated studies of *D. minutum* have been published recently by Schaap, van der Molen, and Konijn. The first of these (1981a), based upon light, phase, and electron microscopy, contrasts the developmental cycle of this species with that of *D. discoideum*. The second (1981b) is focused upon the vacuolar apparatus of *D. minutum* and records that in the late aggregation phase all cells develop heavily coated vacuoles said to be equivalent to the prespore vacuoles found in prespore cells of *D. discoideum*. In *D. minutum*, however, these are destroyed in the stalk cells and fuse with the plasma membrane of prespore cells during spore formation.

Dictyostelium intermedium Cavender, in Amer. J. Bot. *63*: 63-66, figs. 23-35 (1976).

Cultivated on 0.1L-P or 0.1L-P (pH 6.0) in association with *Escherichia coli* at 22-25°C. Sorocarps erect or semi-erect (Fig. 12-8A), unbranched, variable in size and proportions, commonly 3-6 mm in length but often less and sometimes more, strongly phototropic and/or phototactic, pseudoplasmodia migrating freely without stalk formation in darkness or weak unidirectional light (Fig. 12-8F) and often fruiting 1-5 cm from sites of origin. Sorophores commonly 25-50 μm or more in diameter near the base and tapering to 5-10 μm above, bases enlarged, sometimes approaching discoid, more commonly deltoid (Fig. 12-8H) or foot-shaped. Sori globose to citriform, mostly 150-300 μm in diameter, white to off-white, sometimes old ivory in age. Spores elliptical to reniform, commonly 7.0-8.5 × 3-4 μm (Fig. 12-8B) but sometimes up to 10-12 × 4.5 μm, without polar granules. Cell aggregations as in other large species of the genus, basically radial in pattern (Fig. 12-8D), with prominent inflowing streams up to 3-4 mm in length, such streams later breaking apart and segments fruiting individually (Fig. 12-8I). Myxamoebae not distinctive, commonly 15-25 μm in major diameter as viewed on agar surfaces (Fig. 12-8C). Neither macrocysts nor microcysts observed or reported.

HABITAT. Humus and leaf mould from tropical rain forest, Peutjang Island, Java, Indonesia.

HOLOTYPE. Strain PJ-11, isolated by J. C. Cavender from soil collected during the summer of 1970 and sent to us later that year.

This species, known only as the type culture, is clearly distinctive and, as the name suggests, appears to be intermediate between two-well-established taxa, *D. mucoroides* and *D. discoideum*. Some of its characteristics are common to both of these slime molds while others are shared with one or the other individually. Like those species it is vigorous and readily

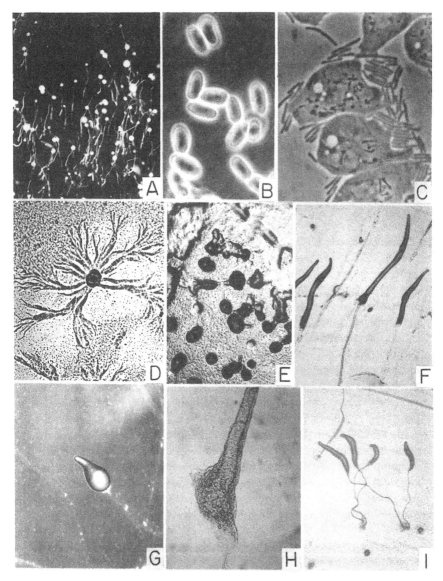

Fig. 12-8. *Dictyostelium intermedium* Cavender. *A*. Sorocarps formed on 0.1L-P agar (pH 6.0) in unidirectional light. × 5. *B*. Spores; note absence of polar granules. × 900. *C*. Vegetative myxamoebae feeding on bacteria; cells are somewhat flattened by coverglass. × 900. *D*. Cell aggregation on 0.1L-P agar at 46 hours. × 9.5. *E*. Cell aggregation in a rich culture on EMB/3 agar; pseudoplasmodia form by blocking out masses of myxamoebae. × 9.5. *F*. Migrating pseudoplasmodia moving toward light at upper right. × 13.5. *G*. Pseudoplasmodium beginning to assume a vertical orientation. × 20. *H*. Base of a sorocarp formed from a migrating pseudoplasmodium; note the residual cell mass that surrounds the bottom of the sorophore. × 375. *I*. Four sorogens forming sorocarps directly from the sites of cell aggregation. × 13.5. (*G* from J. C. Cavender, 1976a)

cultivated, and it can be grown successfully on substrates of substantially increased nutrient content. When cultures are exposed to overhead light, fructification follows the pattern of *D. mucoroides*: sorocarps arise from the sites of cell aggregation, are generally erect or semi-erect, and bear globose to citriform sori that are white to cream-colored. In form and dimensions they could hardly be distinguished from the sorocarps of a large *D. mucoroides*.

When grown in darkness or in weak illumination from one side the process of fructification in *D. intermedium* mimics that of *D. discoideum*. The myxamoebae form an upright column as aggregation progresses and upon its completion the column bends over, contacts the agar surface, and migrates freely for some distance before stopping to form a sorocarp. This pattern of behavior can be beautifully demonstrated by placing culture plates in a black box into which light enters through a narrow slit 2 mm in width. The migrating pseudoplasmodia in this case move directly toward the slit. Migrating pseudoplasmodia of *D. intermedium* are strikingly similar to those of *D. discoideum* and, as in that species, vary somewhat in form and dimensions depending upon the cultural environment. Sorocarps formed from migrating pseudoplasmodia may in some cases appear to have disk-like bases, but these usually consist of residual vacuolated cells loosely arranged around and against the somewhat expanded bases of the sorophores rather than integral parts of the fructifications. Far more common are somewhat expanded bases that are flattened below (Fig. 12-8H) or somewhat foot-shaped, the toe (often elongate) representing the site where sorophore formation began and the heel the point where vertical orientation occurred. In any of these cases the general pattern of the fructification is more that of an isolated sorocarp of *D. mucoroides* than of *D. discoideum*, since the sorophores are thinner, less consistently erect, and lack the degree of taper that characterize the latter.

In describing the species, Cavender (1976a) reported the spores to be ellipsoid and to measure 5.5-7.5×3.3-4.2 μm. Although working with the same strain (type), we now find most of the spores to fall outside this range and in some cases to be substantially larger, up to 10-12×4.5 μm. Such large spores may assume various shapes but are commonly reniform, S-shaped, or recurved, even suggesting the pattern of a boomerang. We suspect that they may represent some degree of ploidal increase, while the spores measured by Cavender soon after strain isolation could have been haploid—a change known to occur in *D. discoideum* with continued cultivation.

Dictyostelium intermedium bears a strong resemblance to *D. dimigraformum*, isolated from tropical forest soil from Trinidad, West Indies. That species differs, however, in that it produces sori of an intense lemon-yellow color, its migrating pseudoplasmodia may form sorophores as they move across the agar surface in response to light, and the sorogens do not leave conical masses of vacuolate cells or form flattened or foot-shaped sorophore bases when culminating.

Insofar as we know, interspecific grafts between migrating pseudoplasmodia of *D. intermedium* and those of *D. discoideum* or *D. dimigraformum* have not been attempted.

Dictyostelium dimigraformum Cavender, in J. Gen. Microbiol. *62:* 113-117, pls. 1 & 2 (1970).

Cultivated on low-nutrient agars such as 0.1L-P or 0.1G-P in association with *Escherichia coli* or *Klebsiella pneumoniae* at 20-25°C. Sorocarps usually solitary, erect or inclined (Fig. 12-9A), unbranched, arising from the site of cell aggregation or at a distance following stalk-free migration of the pseudoplasmodia (Fig. 12-9D-H), strongly phototropic and/or phototactic, variable in size and dimensions. Sorophores sinuous, unpigmented to slightly yellow, mostly 3-10 mm in length but sometimes more, with bases spatulate or rounded and slightly enlarged, sorophores gradually tapering from 25-60 μm near the base to half this diameter or less at the sori. Sori subglobose to citriform, off-white to lemon-yellow, ranging from 100 to 400 but mostly 200-350 μm in diameter. Spores elliptical to reniform, variable in size, usually 8.8-10.0 × 3.0-4.5 μm (Fig. 12-9B) but sometimes larger, without polar granules. Cell aggregations at first small, rather diffuse, with centers poorly defined, rarely exceeding 2 mm in diameter, later merging as dominant centers emerge from which radiate prominent converging streams (Fig. 12-9C), usually producing single unpigmented sorogens or migrating pseudoplasmodia. Myxamoebae not distinctive, generally rounded when feeding, about 10-12 μm in diameter, becoming elongate up to 20 × 10 μm when moving actively. Neither macrocysts nor microcysts have been reported.

HABITAT. Leaf mould and surface soil from a moist tropical forest, Trinidad, West Indies.

HOLOTYPE. Strain AR-5b, isolated by J. C. Cavender from soil collected in the summer of 1968.

Although the species *Dictyostelium dimigraformum* is known only by the type which appeared as a clone in one of Cavender's isolation plates, it is nonetheless a very distinctive taxon marked by its large lemon-yellow sori and, more particularly, by its alternative patterns of fruiting. When cultivated at 25°C in full overhead light, cell aggregation normally leads directly to sorogen formation, followed without interruption by the construction of erect or semi-erect sorocarps a few millimeters in height. If grown at a slightly lower temperature with diffuse but uneven lighting,

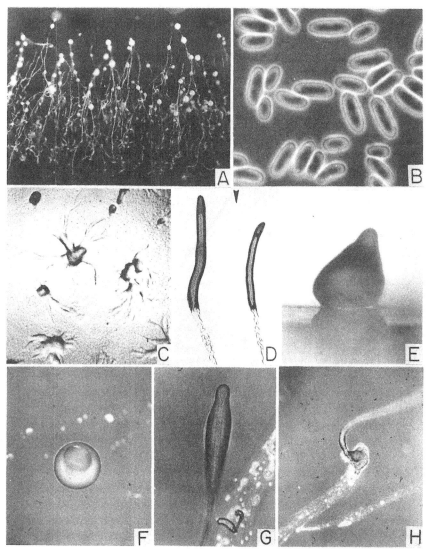

Fig. 12-9. *Dictyostelium dimigraformum* Cavender. *A.* Sorocarps produced on 0.1L-P agar in unidirectional light. Surface lighting. × 6. *B.* Spores from a similar culture. Note variations in size. × 810. *C.* Cell aggregations. × 18. *D.* Two migrating pseudoplasmodia moving toward a light source (arrow). × 20. *E.* Young fruiting structure in which orientation has shifted from horizontal to nearly vertical. × 60. *F.* A similar stage photographed from above. × 35. *G.* A sorogen moving on the agar surface; note the internal stalk with funnel-like terminus. × 54. *H.* The base of an upright sorocarp constructed after a pseudoplasmodium ceased migrating. × 35. (*C* and *E-H* from J. C. Cavender)

many of the sorogens will in their phototropic response touch the agar surface and thereafter migrate for some distance while forming a stalk before again assuming an erect posture to complete the fruiting process (Fig. 12-9G). If grown at 20-22°C in one-sided light, as in a black box with light entering through a narrow slit, most of the aggregations give rise to erect columns of cells which then bend toward the light and upon touching the agar migrate freely toward the source (Fig. 12-9D). Under such conditions, and in the absence of any fruiting, one would be hard pressed to distinguish such freely migrating structures from the migrating pseudoplasmodia of *D. discoideum*. Like the latter they are colorless by nature and when produced under nutritionally comparable conditions are of approximately the same dimensions. To illustrate, on 0.1L-P/2 agar with *E. coli* at 22°C the migrating pseudoplasmodia of strain AR-5b range from 0.4 to 1.2 mm long × 50 to 200 μm wide.

The migrating pseudoplasmodia, and to a lesser degree the yellow color of its sori, suggest a possible relationship with *D. discoideum*. This, however, is thought to be more illusory than real, for although the sori of *D. discoideum* may at times appear yellow, there is no suggestion of a basal disk in the sorocarps of *D. dimigraformum*, even those formed following extended migration. Instead, as the migrating body ceases forward movement it begins to lay down a stalk, quite thin at first and gradually enlarging to form an elongate horizontal to upturned base of a diameter slightly greater than that of the erect sorophore into which it merges imperceptibly. Stalk-free migration by pseudoplasmodia may vary from a few millimeters to one or more centimeters, and the extent of such wandering is strongly influenced not only by light but by the composition of the substrate and the relative abundance of free water at the agar surface. For example, stalk-free migration is appreciably greater on freshly poured than on older plates. It is also greater on unbuffered than on PO_4-buffered agars, and on 0.1L-P/2 agar is greater than on full strength 0.1L-P—all of which fits quite nicely with what Slifkin and Bonner reported for *D. discoideum* many years ago (1952). Additionally, stalk-free migration and migration whilst forming a stalk, from which behavior the species takes its name, are interchangeable to a very substantial degree, and the same pseudoplasmodium may alternate between the two modes as it moves across the agar, leaving behind a collapsed sheath of slime that it secretes continuously. Whereas no studies have been reported concerning the wavelengths effective in guiding pseudoplasmodial migration in *D. dimigraformum*, it is suspected that these would closely approximate those reported by Francis (1964) for attracting comparable structures in *D. discoideum* and *D. purpureum*. We are indebted to Cavender (1970) for the interesting observation that sorocarp formation by dark-"grown" migrating pseudoplasmodia can be initiated immediately by exposing them to light.

Other species that produce yellow sori include *Dictyostelium aureum* and *D. mexicanum*, but there is little to suggest a close relationship with *D. dimigraformum*.

Dictyostelium discoideum Raper, in J. Agr. Res. *50*: 135-147, pls. 1-3 (1935).

Cultivated on substrates such as 0.1L-P, 0.1L-P (pH 6.0), and hay infusion agars in association with *Escherichia coli, E. coli* B/r, or *Klebsiella pneumoniae* at 20-23°C. Sorocarps erect, unbranched, consisting of strongly tapered sorophores that arise from flattened disks and bear at their apices unwalled globose to citriform sori, variable in size but nearly constant in form (Figs. 12-11C,E), commonly 1.5-3.8 mm high but often less and sometimes more. Sorophores colorless to very light tan, generally rigid below, ranging from 30-65 μm in diameter near their bases, narrowing to 10-15 μm at the sori, the upper ⅓ usually flexuous. Sori globose, rounded-apiculate or citriform, usually uncolored when formed but often becoming pale to fairly bright yellow depending upon conditions of culture, in size generally proportional to the fructification as a whole and commonly ranging from 200 to 350 μm in diameter, often less and rarely more. Spores narrowly elliptical, hyaline, with seemingly homogeneous content and no polar granules (Fig. 12-10A), typically 6.5-8.0 × 2.5-3.5 μm, but often much larger in strains long cultivated. Cell aggregations typically radiate in pattern (Figs. 12-10C,D), variable in size from relatively small, 2-3 mm in diameter, to quite large, 1 cm or more, strongly influenced by culture conditions; streams converge to form upright columns of cells that may fruit directly but typically bend over, touch the agar surface, and in the form of strongly phototactic, cartridge-shaped, multicellular bodies migrate freely (Fig. 12-10E) for some distance before building sorocarps (Fig. 12-11A,D). Vegetative myxamoebae irregularly rounded to roughly triangular in shape (Fig. 12-10B), commonly 13-16 × 9-11 μm in diameter but ranging from 10 to 20 × 8 to 12 μm, each containing a single vesicular nucleus with nucleolus showing peripheral lobes, one or two contractile vacuoles, and numerous food vacuoles containing ingested bacteria. Macrocysts are formed by pairs of compatible mating types, rarely by self-compatible strains (Figs. 12-11G,H); microcysts have not been observed. The acrasin is cyclic AMP.

HABITATS. Surface soil and leaf litter of forests, particularly deciduous forests of temperate North America; reported also from India and Japan.

HOLOTYPE. Strain NC-4, isolated by KBR in 1933 from partially decomposed leaves taken from a hardwood forest in Little Butts Gap (Fig. 2-2), Craggy Mountains, western North Carolina. This strain has been studied more intensively than any other.

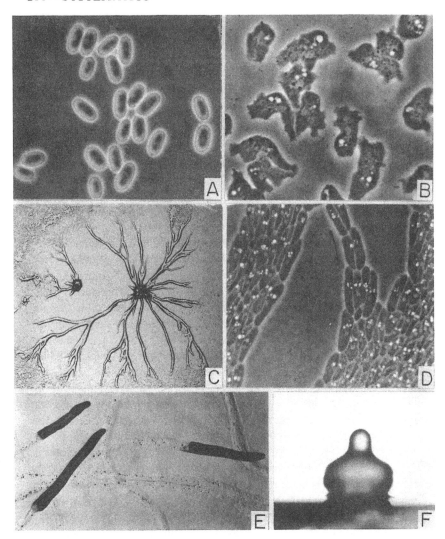

Fig. 12-10. *Dictyostelium discoideum* Raper. *A.* Spores showing characteristic form and absence of polar granules. × 900. *B.* Preaggregative myxamoebae—note inconstant shape, lack of uniform orientation, irregular pseudopodia, nuclei (appear as light gray areas), and contractile vacuoles. × 750. *C.* A well-advanced developing aggregation: the smaller organization at left resulted from severance of a large inflowing stream. × 12. *D.* Detail of myxamoebae in two converging streams. × 365. *E.* Migrating pseudoplasmodia moving across an agar surface; note the trails of slime left in their wakes. × 16.5. *F.* Early stage in sorocarp formation: the mass of myxamoebae has assumed a vertical orientation, and stalk formation is just beginning. × 35

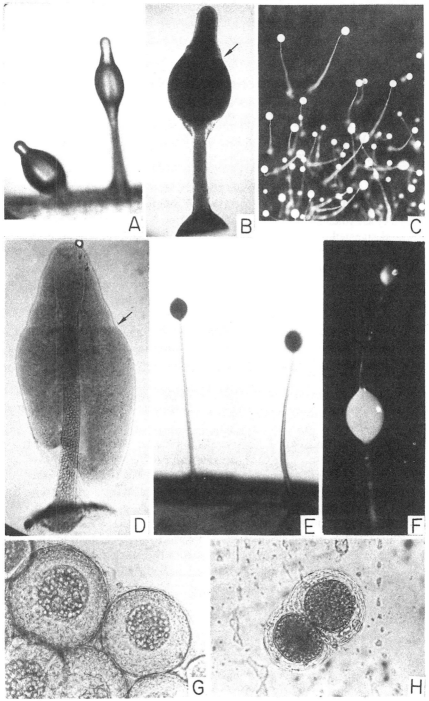

FIG. 12-11. *Dictyostelium discoideum* Raper. *A.* Two young sorocarps: at left, main body of sorogen is just separating from cells which will form the basal disk; at right, sorocarp formation is substantially more advanced. × 50. *B.* Young sorocarp fixed to reveal demarcation between prespore and prestalk forming parts (arrow). × 95. *C.* Mature sorocarps photographed with reflected light. × 5. *D.* Developing sorocarp fixed and stained to show stalk-tube deposition in apical region. Arrow shows line of demarcation between prestalk and prespore cells. × 130. *E.* Two fully developed sorocarps viewed from the side. × 20. *F.* Large sorocarp where the main sorus has formed midway up the stalk. × 30. *G.* Young macrocysts showing enlarging giant (or phagocytic) cells. × 240. *H.* Two macrocysts at endocyte stage—the giant cell has engulfed all peripheral myxamoebae and is now digesting them. × 240. *G*, strain AC-4, self-compatible; *H*, strains WS-10 × WS-582, mating types

Of many strains isolated and studied over the past half-century, the following may be cited as representative of the species, and in most cases are of some special interest as well.

Strain V-12, isolated by KBR in 1937 from soil and leaf litter collected in a deciduous forest near Vienna, Virginia. This strain was sent to Gunther Gerisch in 1957 and became the source of substrain V-12 M/2, much studied by him and his associates. It was paired with NC-4 by Erdos, Raper, and Vogen in 1973 and found to produce macrocysts, whereupon it was designated mating type A_2, while NC-4 was designated A_1. This and parallel studies by Clark, Francis, and Eisenberg, also 1973, clearly established heterothallism as the dominant, but not exclusive, mode of sexual reproduction in *D. discoideum*.

Strain WS-10, isolated by KBR in 1953 from surface soil and humus of a mixed oak forest in the University Arboretum, Madison, Wisconsin. This strain, representing mating type A_1, together with strain WS-582, representing mating type A_2, was examined in great detail by M. A. Wallace in a study of genetic exchanges in the macrocysts of *D. discoideum*, the work being reported by Wallace and Raper in 1979.

Strain WS-582, isolated in 1971 by John B. Sutherland from soil collected in Parfrey's Glen, Baraboo, Wisconsin. (See preceding account.)

Strain AC-4, isolated by J. C. Cavender in 1964 from forest soil collected near Acayucan, Vera Cruz, Mexico. This is the most productive of two known self-compatible macrocyst-producing strains of *D. discoideum*. It was included in Nickerson's studies of macrocyst formation and germination among the dictyostelids, as reported by Nickerson and Raper in 1973.

Strain, A3, a mutant strain of NC-4 isolated by W. F. Loomis in 1971 following treatment of haploid cells with N methyl N′nitro N nitrosoguanidine that grows axenically in selected media, i.e., in the absence of any bacteria. It has been much used in his laboratory and elsewhere for biochemical and physiological studies. If so desired, it can be grown in association with *Klebsiella pneumoniae* on agar substrates where it fruits satisfactorily, albeit the sorocarps average somewhat smaller than those of the parent strain.

The species description given above is based upon substrates of relatively low nutrient content for comparative purposes. However, it should be remembered that *D. discoideum*, like *D. purpureum* and certain other large dictyostelids, can be and usually is cultivated upon richer media, representing in some cases a tenfold nutrient increase (see Chapter 4). Under such circumstances, not only is the number of sorocarps greatly increased but the sorocarps average somewhat larger and may at times reach heights of 4.5 mm, have basal disks 550-650 μm wide, and bear sori up to 450-500 μm in diameter.

Many bacteria other than those listed above may be used as nutrient for the myxamoebae of *D. discoideum* if proper attention is given to substrate composition and the physical environment; some of these serve special purposes admirably. For example, because of their large size, cells of *Bacillus megaterium* are very good for demonstrating the ingestion and digestion of bacteria (Fig. 6-4); and *Serratia marcescens* is especially suited for "staining" migrating pseudoplasmodia when making grafts, or for confirming that the anterior fractions of such structures form the stalks of completed sorocarps (Fig. 8-7).

At the risk of redundancy, we note again that the spores of *D. discoideum* are, in our experience, uniformly small (6.5-8.0 × 2.5-3.5 μm) and presumed to be haploid as strains are isolated from nature but show a progressive tendency to shift to diploid when long maintained in continuous culture. Unfortunately, nearly all our old stock cultures of strains NC-4 and V-12 now contain a majority of large spores (10-12 × 3.5-4.5 μm). For many studies, therefore, it is necessary to reclone and carefully screen resulting populations to ensure that one is working with a uniform cell population.

Structurally the sorocarps are quite distinctive. They arise from flattened, broadly conical disks that are cut off from the rising sorogens, hence formed subsequent to the bases of the sorophore which they surround and support in positions essentially perpendicular to the surface, whether this be horizontal, vertical, or inverted. Additionally, the sorophores are more strongly tapered than in other species. They may show a gradual reduction in width from base to sorus, or they may be of essentially uniform diameter to approximately ⅓ their ultimate length, above which diameters decrease markedy. More important, of whatever dimensions, if produced under *optimum cultural conditions* the proportions of the stalks, and of these to the supporting disks and the supported sori, remain much the same (Raper, 1941a). This near uniformity of sorocarp pattern (Figs. 8-9 and 8-16), together with the freely migrating pseudoplasmodia, account, we believe, in large measure for the singular interest *D. discoideum* has elicited from developmental biologists throughout the world.

Stalk-free migrating pseudoplasmodia are not unique to *D. discoideum*, for they occur from time to time in other species and under certain conditions in several large dictyostelids, including *D. purpureum* and members of the *D. mucoroides* complex. In *D. intermedium* and *D. dimigraformum* they are generally present, and in *D. polycephalum* they constitute an integral and perhaps essential part of the developmental cycle. In no other species, however, can one predict with comparable certainty the behavior or ultimate fate of constituent cells as one can for an undisturbed migrating pseudoplasmodium of this species.

Returning briefly to sorocarp form, we have for many years been amazed by the near uniformity of pattern and dimensions among scores of strains of *D. discoideum* as these were isolated from natural sources. Still a few distinctive strains have been encountered, and of these perhaps the most striking is one sent to us in 1981 by David Waddell of Princeton University as an isolate from bat guano out of a cave in Arkansas. In almost any strain of *D. discoideum*, and under conditions not fully defined, one may see an occasional sorocarp where the bulk of the spores are found not in a terminal sorus but in a large citriform body at about ⅓ the total height of the stalk (Fig. 12-11F). In Waddell's strain virtually all the sorocarps are of this type. What seems to happen is this. As the sorophore lengthens and as spores begin to differentiate at its periphery, the continuity of the slime envelope that has previously surrounded the whole cell mass is interrupted in a circumferential zone. Once this occurs, due to the weight of the prespore cell mass and/or the low viscosity of the intercellular slime, most of the prespore cells remain in place and form spores whilst the prestalk cells, being already committed, continue to build a sorophore of the dimensions preordained for a normal sorocarp. In strains such as the type NC-4, spores taken from the low sorus yield wholly normal cultures with no detectable decrease in normal sorocarps. In Waddell's strain spores yield cultures of the aberrant type whether taken from a small terminal sorus or from a much larger low sorus. We suspect that this strain represents a naturally occurring mutant characterized by intercellular slime of less than normal viscosity, perhaps induced by long-continued growth in a wholly saturated atmosphere.

Two other subjects have been discussed at length in earlier chapters, hence need no elaboration here. Still it is meet to record that it was with *D. discoideum* that Konijn, Bonner, and others first established the chemotactic role of 3′, 5′, -cyclic AMP in cell aggregation in a dictyostelid (see Chapter 7); and it was in this species, along with *D. mucoroides*, that the recombination of characters was first demonstrated by Francis, Erdos, Wallace, and others, showing beyond conjecture that macrocysts represent the true sexual stage of the dictyostelids (see Chapter 9). It was in *D. discoideum* also that indisputable evidence of the cell-free nature of the sorophore sheath, or tube, was first obtained by Raper and Fennell, a fact that is strikingly illustrated in Figure 12-12. The tube clearly consists of a continuous sheet of cellulose surrounding but in no way attached to the inherently fragile column of compressed vacuolate cells within (see Chapter 8).

Dictyostelium aureum Olive var. *aureum* Cavender, Worley, and Raper, in Amer. J. Bot. *68*: 375-376, figs. 1-6 (1981). *Dictyostelium aureum* E. W. Olive, in Proc. Amer. Acad. Arts Sci. *37*: 340-341 (1901); also Proc. Boston Soc. Natur. Hist. *30*: 507, p. 6, figs. 63 & 64 (1902).

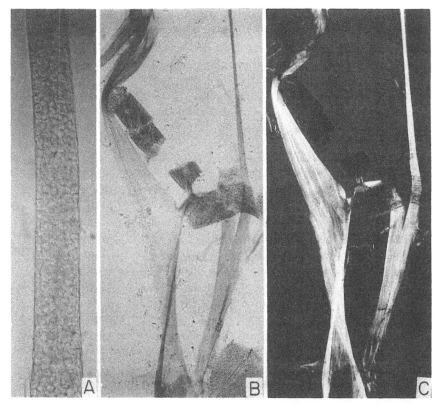

Fɪɢ. 12-12. *Dictyostelium discoideum* Raper. *A*. Portion of a very large sorophore showing internal parenchyma-like cellular structure and the uninterrupted nature of its thin external "wall." × 225. *B*. A similar sorophore stained with chloroiodide of zinc and crushed to show the continuous sheet-like character of the sorophore sheath or tube, here torn lengthwise, and the column of cells, now broken, that previously filled it. × 150. *C*. The same photographed with polarized light showing strong birefringence in the cell-free sorophore sheath. Stalk cells show little or no birefringence due to a lack of consistent orientation and the thickness of the column. × 150. (Rearranged from Raper and Fennell, 1952)

Cultivated on low-nutrient substrates such as 0.1L-P (pH 5.5 or 6.0) or 0.1% G, 0.05% Yxt agars in association with *Escherichia coli* at 22-25°C. Sorocarps solitary or clustered, erect, semi-erect, or inclined toward light, sinuous, becoming tangled as numbers increase (Fig. 12-13C), unbranched or irregularly branched. Sorophores generally long and thin, commonly 0.3-1.0 cm in length, sometimes more, with taper imperceptible, often ranging from 15 to 25 μm in diameter at midpoint, yellow to yellowish in color with two or more stalks often adherent. Sori globose to citriform, commonly 150-250 μm in diameter (Fig. 12-13C), sometimes larger and often smaller, bright lemon-yellow to muted yellow in color. Spores elliptical or slightly reniform, mostly 5.0-6.5 × 3.0-3.5 μm (Fig. 12-13D) but variable from 4.0 × 3.0 to 8.0-8.5 × 3.5-4.0 μm, spore contents homogeneous, without polar granules. Cell aggregations vary, small and relatively compact in some strains (Fig. 12-13A), radiate and more open in others (Fig. 12-13B), aggregations and sorogens colorless. Vegetative myxamoebae typical of the genus, commonly 15-18 × 12-15 μm when feeding (Fig. 12-13E). Macrocysts and microcysts not reported.

HABITAT. Soils from semi-deciduous forests of Trinidad and Guyana, South America, and a salt marsh in Ohio.

NEOTYPE. Strain SL-1, isolated in the summer of 1969 by J. C. Cavender from forest soil collected near Simla, Trinidad. A second culture, strain RI-1, basically like the proceding, was isolated by JCC in 1978 from soil of a salt marsh at Ritma, Ohio.

Dictyostelium aureum as described by E. W. Olive (1901) was based upon a single isolate communicated by Professor Thaxter. The source was mouse dung from Puerto Rico, and the slime mold was characterized by its light to golden yellow sori, its irregularly lobed and nodulated myx-amoebae, and its slow and distinctive manner of growth. Fruiting structures were said to be 1.5-4.0 mm high and to bear oval to inequilateral spores 5-8 × 2.5-3.0 μm. The type culture was not conserved, and the species was not reported again until 1968, when Cavender and Raper listed it among their isolations from tropical and subtropical America. The cultures then in hand, however, possessed several features not mentioned by Olive and doubts arose concerning their identity.

Yellow-pigmented dictyostelia were later isolated from soils of tropical forests in Trinidad and Guyana, and from a salt marsh in Ohio, that were believed to approximate the slime mold upon which Olive based his description. Other investigators in other places isolated yellow-pigmented strains as well—Traub (1972) from soil of a deciduous forest in Switzerland and Benson and Mahoney (1977) from chaparral in southern California, respectively. The slime molds in each case were reported as *D. aureum*. Having accumulated a considerable body of material by the mid-seventies, a comparative study was undertaken by Cavender, Worley, and Raper. Three new taxons were recognized (1981): *D. aureum* var. *aureum*, be-lieved to conform with Olive's original concept; *D. aureum* var. *luteolum*, characterized by somewhat reduced pigmentation in sori and sorophores; and *D. mexicanum*, typified by smaller fructifications, more intense pig-mentation, and quite distinctive sorophores.

Although we cannot know precisely what Olive had when he described *D. aureum*, we believe that its sorocarps, except for color, must have resembled those of the ubiquitous *D. mucoroides*; otherwise differences would have been noted. Upon this premise we would assume that his slime mold was basically similar to strains SL-1 and RI-1 as portrayed above.

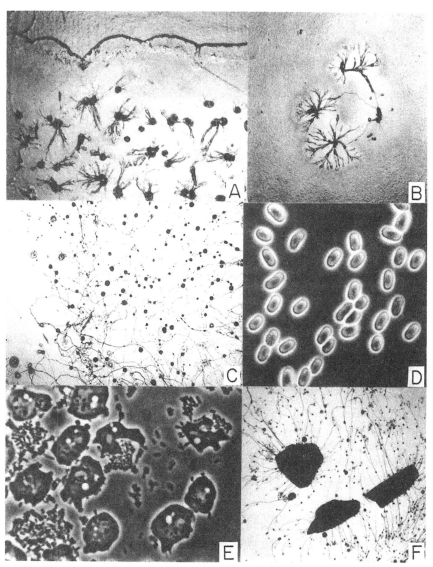

FIG. 12-13. *Dictyostelium aureum* Olive var. *aureum* Cavender, Worley, and Raper. *A*. Strain SL-1 showing characteristic patterns of cell aggregations. × 6.2. *B*. Aggregation in strain RI-1; three sorogens from original center (at right) fell to the agar surface and initiated satellite aggregations. × 6. *C*. Mature sorocarps consisting of long, thin, sinuous sorophores bearing globose sori. × 6.2. *D*. Spores of characteristic form lacking polar granules. × 780. *E*. Myxamoebae feeding at edge of bacterial colony. × 780. *F*. Culture to which activated charcoal was added at time of cell aggregation; note the strong positive response of sorocarps. × 4.5. (From Cavender, Worley, and Raper, 1981)

Sorocarps in these strains do attain lengths greater than those reported by Olive. Such discrepancies, however, may be accounted for by differences in substrates and the unidentified bacteria that accompanied his cultures. Emphasis was given to the "irregularly lobed and nodulated myxamoebae," which he regarded as unique for his *D. aureum*. Thus it is of special interest that we have encountered in *D. a.* var. *aureum* C. W. & R., and more especially in the second variety, *D. a.* var. *luteolum* C. W. & R., great numbers of myxamoebae that fit quite well his description and illustrations (Olive, 1902, figs. 63 and 64). Cavender, Worley, and Raper (1981) wrote of such cells in this way:

> Plates of 0.1L-P agar streaked with mixed suspensions of slime mold spores and *E. coli* may at times present a striking and somewhat puzzling picture. Large, dendritic aggregations often develop at 2-3 days (up to 1.0-1.5 cm side to side), and at the sites of their extension the streams may be broad and fan-like. These "fans" extend into dense fields of post-vegetative myxamoebae that are convoluted in pattern and under a low-power binocular suggest massed microcysts in the way they reflect light. The cells are in fact very irregular, do not lie flat on the agar, and because of rounded surfaces create the impression of densely crowded microcysts. The cells are still amoeboid, however, and with time slowly enter into and enlarge the aggregations already present, a process that may continue for a day or more. In the meantime the proximal portions of the streams become discontinuous, after which the resulting segments give rise to separate sorogens and sorocarps.

As in many other species, cell aggregation and fructification are greatly enhanced by the addition of activated charcoal, and when this is added in the form of small chips, developing sorogens move directly toward the particles while the still maturing sori collapse against the faces of the chips (Fig. 12-13F).

Dictyostelium aureum Olive var. *luteolum* Cavender, Worley, and Raper, in Amer. J. Bot. *68*: 376-377, figs. 7-12 (1981).

Cultivated on low-nutrient substrates such as 0.1L-P (pH 5.5 or 6.0) in association with *Escherichia coli* at 22-25°C. Sorocarps much as in *D. a.* var. *aureum* except the yellow pigmentation of sori is generally less, sorophores are more flexuous and may be uncolored, sori average somewhat larger, spores are more oval in outline and may approach subglobose, and myxamoebae appear slightly larger. Cell aggregations are dendritic in pattern, usually larger, and long inflowing streams regularly break into segments before forming sorogens and sorocarps. Pseudoplasmodia and sorogens are uncolored. Sori tend to darken somewhat with age, and pigmentation is enhanced on acid substrates (pH 5.5).

HOLOTYPE. Strain WS-629 isolated in 1974 by Manetta Benson and Daniel P. Mahoney from soil collected beneath mixed chaparral near Lake Wolford, San Diego County, California.

Of many other strains examined, two may be cited: WS-630, also from Benson and Mahoney as an isolate from soil collected in Southern California; and NW-1, isolated by Franz Traub from soil of a deciduous forest in Switzerland.

Optimum growth and development occur at 22-23°C. At 20°C both growth and pigment production are reduced appreciably. There is little growth at 26°C and none at 28°C.

Aggregations in the variety *luteolum*, as in var. *aureum*, are often initiated by sorogens that, for lack of rigidity in the sorophores, drop to an agar surface covered by a lawn of preaggregative myxamoebae. Whereas this is a natural phenomenon, we believe it is quite comparable to that seen in *D. discoideum* when migrating pseudoplasmodia are moved into a similar field of unaggregated myxamoebae (Bonner, 1949)—in fact, it could result from the same acrasin. When tested by Chris Town and Ann Worley, postvegetative myxamoebae aggregated in response to added cAMP in both *D. aureum* var. *aureum* (strain SL-1) and var. *luteolum* (strain WS-629). The response, however, was not as strong as in *D. discoideum* (strain NC-4) included as control.

Dictyostelium brunneum Kawabe, in Trans. Mycol. Soc. Japan *23*: 91-94, figs. 1 & 2 (1982).

Cultivated on substrates such as 0.1L-P or 0.1G-P agars in association with *Escherichia coli* or *Klebsiella pneumoniae* at 22-23°C. Sorocarps erect or semi-erect, commonly 3-8 mm high under overhead light, sometimes more, often sinuous and interwoven, strongly phototropic in unidirectional light, up to 3.0-3.5 cm in length with sori often small. Sorophores usually unbranched or with branches limited and spaced irregularly, commonly about 25-40 μm in diameter at midsection, occasionally more, lightly pigmented in tan shades, with bases more or less clavate, typically surrounded by prominent aprons of congealed slime. Sori globose to citriform, off-white to cream-colored when young, becoming tan to dull brown with age, commonly 200-300 μm in diameter but sometimes more and often less. Spores ellipsoid to cylindrical, without polar granules, mostly 7-8 × 3.0-3.5 μm but up to 10 × 4 μm, the larger spores often reniform. Cell aggregations radiate in pattern, variable in dimensions, in spread plates commonly 1.5-2.5 mm in diameter, evenly spaced, and each producing a single sorocarp; in streak cultures asymmetrical, often 3-5 mm in diameter, with prominent inflowing streams that may become discontinuous to form multiple sorocarps. Sorogens colorless, cartridge-shaped, up to 2 mm in length by 200-250 μm in diameter. Vegetative myxamoebae as in other species but with nuclei relatively inconspicuous, food vacuoles prominent, and contractile vacuoles commonly numbering two or more. Macrocysts and microcysts not reported or observed.

HABITAT. Soil of forested area along the upper reaches of the Ohi River, southern Japanese Alps.

HOLOTYPE. Kurumi-308 (WS-700). Isolated in 1977 by Kawabe from surface soil, Kurumi-sawa, Shizuoka, Japan. Type sent to us in January 1983.

In some respects *Dictyostelium brunneum* resembles species of the *D. mucoroides* complex, and it produces sorocarps of the same general patterns and dimensions as *D. giganteum* Singh. Additionally, the size and distribution of developing cell aggregations bear a striking resemblance to those of that species. *D. brunneum* differs markedly, however, in developing a brownish to brown pigmentation in its mature sori and to a lesser degree in its sorophores. Such pigmentation is at first limited or lacking, becomes increasingly evident as cultures age, and is typically pronounced within 7-10 days. Kawabe recalled (1982) that Sussman and Sussman (1963) reported a brown-pigmented mutant in *D. discoideum* but he correctly dismissed this as evidence of close kinship, since there is no freely migrating stage. He could have added, also, that there is no cellular basal disk and the sorocarps are quite devoid of constancy in overall proportions. Brownish-olive sori are sometimes seen in *D. rhizopodium*, but the presence of crampon-like sorophore bases and spores with polar granules clearly separate that species from *D. brunneum*.

Kawabe also cited "dilated sorophore bases" as diagnostic of the species, and it may be said that the isolated sorocarps arise from broad and relatively prominent aprons of congealed slime. Such aprons, however, contain no cells and hence represent only the slime residue left on the agar surface by the ascending sorogens. Similar deposits are often seen in other species but not with the prominence and clarity displayed in *D. brunneum*; for this reason they are of diagnostic value.

Developing sorocarps of *D. brunneum* are strongly phototropic, and upon falling to the agar surface may move for appreciable distances while producing stalks continuously as they advance. Such behavior is especially striking in cultures illuminated from one side. While in no sense exact, there is generally an inverse relationship between stalk length and sorus size.

In describing *D. brunneum*, Kawabe employed pregrown cells of *Klebsiella pneumoniae* deposited on agar plates as a nutrient source. We have assumed that he used nonnutrient agar and have found cultures of that type to be satisfactory but not better than those where the bacteria and slime mold were grown concurrently on substrates such as 0.1L-P and 0.1G-P agars. Unexpectedly, we observed sorocarp formation to be better on unbuffered 0.1L-P agar than on the same medium buffered to pH 6.0 with phosphates. In the latter case the numbers of aggregations and sorocarps were substantially increased while their dimensions were correspondingly reduced. Additionally, the sorocarps were relatively fragile and often tangled. For reasons still obscure aggregations that developed subsequent to the primary crop were substantially larger, as were the sorocarps that arose from them.

Kawabe reported strain Kurumi-308 to fruit normally at 20°C but not at 28°C. We have used and can recommend an intermediate level of 22-23°C.

Dictyostelium purpureum E. W. Olive, in Proc. Amer. Acad. Arts Sci. *37*: 340 (1901), and E. W. Olive, "Monograph of the Acrasieae," in Proc. Boston Soc. Natur. Hist. *30*: 451-513, pl. 6, figs. 84 & 87, and pl. 7, figs. 98-104 (1902). *See also*: Whittingham and Raper, Proc. Natl. Acad. Sci. (USA) *46*: 642-649, figs. 1-11 (1960); and Hagiwara, Bull. Natl. Sci. Museum *14:* 355-356, fig. 2 (1971).

Cultivated on low-nutrient agars such as 0.1L-P or 0.1G-0.05 Yxt in association with *Escherichia coli, E. coli* B/r, or *Klebsiella pneumoniae* at 22-24°C. Sorocarps usually solitary, at first erect or semi-erect, then inclined, strongly phototropic (Fig. 12-14A). Sorophores with bases not enlarged, robust, variable in length, often reaching several centimeters in unidirectional light, commonly 20-40 μm below (Fig. 12-14H) and tapering to half that diameter within the sorus, usually sinuous, sometimes tangled in richer cultures, colorless or pigmented in light purple shades. Sori dark vinaceous purple to almost black when mature, globose to citriform, commonly 200-350 μm in diameter but ranging from 125 to 450 μm and occasionally more. Spores narrowly elliptical or capsule-shaped, mostly 5.0-7.2 × 2.5-3.2 μm (Fig. 12-14B) but ranging up to 8 μm or more in length, variable in different strains and within the same isolate, sometimes showing evidence of limited polar granulation. Cell aggregations commonly large, up to 1 cm or more in diameter, may develop first as broad sheets of inflowing cells, then as well-defined converging streams (Fig. 12-14D), the latter often becoming severed as they advance, each segment forming a separate sorogen; centers of aggregation arise either as fixed focal points (Fig. 12-14D) or emerge from the rims of vortex-like sites of cell aggregation (Fig. 7-9B); sorogens develop from primary centers or from rounded segments of severed streams, erect and columnar (Fig. 12-14F). Myxamoebae much as in other dictyostelideae (Fig. 12-14C), about 12-18 × 10-15 μm when feeding on an agar surface, becoming more

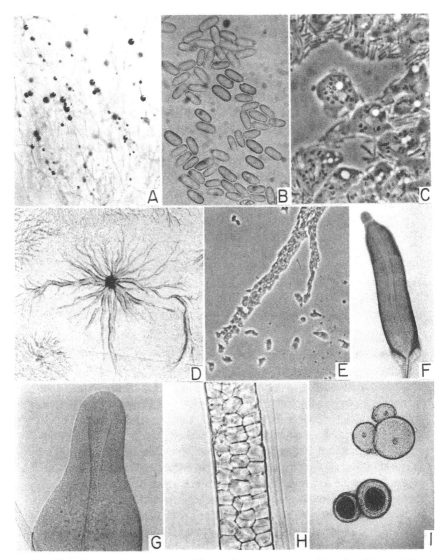

FIG. 12-14. *Dictyostelium purpureum* Olive. *A.* Sorocarps as seen with surface light. × 4.7. *B.* Spores with finely granular content. × 715. *C.* Myxamoebae growing in association with *Escherichia coli*; note cell organelles: nuclei, contractile vacuoles, and food vacuoles containing bacteria. × 715. *D.* Cell aggregation with prominent streams of converging myxamoebae. × 10. *E.* Distal end of an inflowing stream, enlarged. × 185. *F.* Sorogen moving across agar in response to light; sorophore extends through body and into apical tip where it is formed. × 35. *G.* Enlarged view of similar tip showing characteristic funnel-like terminus of the cellulose stalk tube. × 55. *H.* Section of a mature stalk showing detail of its cellular structure. × 495. *I.* Macrocysts of two ages: above, young macrocysts with giant (or phagocytic) cells still small; below, cysts in which giant cells have engulfed a substantial portion of the peripheral cells. × 75. (In part from Whittingham and Raper, 1960)

elongate to 20 μm or more when moving actively. Macrocysts produced in some strains (Fig. 12-14I) but not in others, homothallic or heterothallic; microcysts not observed or reported. The acrasin of *D. purpureum* is cAMP, as in *D. discoideum* or *D. mucoroides*.

HABITATS. Dungs of many animals, deciduous forest soil, and decomposing leaf litter. Common in North America and reported from India, Liberia, and Japan. It was not found by the writer during six months of searching in France, Netherlands, and England but was isolated by Dr. Franz Traub from soil collected in Majorca.

Representative cultures include #3645, isolated from forest soil in Illinois; V-15, from forest soil in Virginia; WS-321, from the rhizosphere of a plant in India; and Liberia-24, isolated from soil sent to us from Monrovia, Liberia.

The cultures upon which E. W. Olive based the species are no longer extant, but if a neotype is to be cited this could be strain #3645, since it has probably been studied more carefully than any other now in our possession (see Whittingham and Raper 1956, 1960).

The size and character of aggregations and sorocarps as given above apply to substrates of the nutrient level cited. However, most isolates of *D. purpureum* will grow and fruit quite well on richer media with a consequent increase in the number of sorocarps, accompanied in many cases by a limited increase in sorocarp dimensions. For information concerning such media the reader is referred to Chapter 4.

Dictyostelium purpureum as a species must be considered in very broad terms, since individual isolates vary appreciably from one another, while still other isolates with purple sori, not presently included within the species as described, are believed to be allied with it. For example, strain D-6, unchanged in culture for nearly forty years, is noteworthy because of its thinner sorophores that consist of single tiers of cells throughout most of their length. These may reach 4 or 5 centimeters in one-sided light and bear sori that average substantially smaller than those noted above. At the same time the relationship to *D. purpureum* of certain other strains with purple or dark vinaceous sori must be questioned. Among the latter may be cited strain CPRP, isolated from Panamanian soil in 1958, and PT-6, of Mexican origin, cited by Cavender and Raper (1968, p. 510) as "a variant of *D. purpureum*." Both have small, often clustered sorocarps that seldom measure more than 2-3 mm in height and bear dark purple sori that range from about 50 to 150 μm in diameter. There is no suggestion of a digitate sorophore base; hence both are excluded from *D. vinaceo-fuscum*, the other described species with dark vinaceous to purplish-black sori (Raper and Fennell, 1967). It is of interest that the spores in each of these strains are quite large, up to 10 μm in length, and often contain polar to subpolar clusters of small granules.

Whereas most investigations on cellular slime molds are and have been conducted with *Dictyostelium discoideum* for obvious reasons, *D. purpureum* has been employed effectively in a number of studies. Prominent among these was verification of Brefeld's initial contention that the differentiated stalk cells in *Dictyostelium* were nonviable (Whittingham and Raper, 1960). Equally dramatic was the discovery that maturing sori of *D. purpureum* remained milk-white in the presence of certain concentrations of volatilized phenol (Whittingham and Raper, 1956). In early experiments phenol in aqueous solution was added at one side of a Petri plate and the effect noted as recorded in Fig. 12-15. This did not, however, distinguished between an effect due to poisoning the substrate and thus influencing cell growth (as occurred near the site of application) and one of inhibiting pigment production within maturing spores. The latter was shown to be the correct explanation when small disks applied to the inner surface of plate lids were saturated with differing amounts of known concentrations of phenol: cultures treated with 10^{-2} M phenol exhibited sorocarps with white sori, the area occupied by these increasing proportionately with the volume of solution added. Cultures similarly treated with 10^{-3} M phenol showed no sorocarps with white sori. Other compounds producing comparable results included *m*-cresol, *p*-cresol, and resorcinol. Spores taken from milk-white sori were viable and yielded clones with normally pigmented spores. It is interesting and presumed significant that all compounds that inhibited pigment synthesis in *D. purpureum* induced the same reaction in *Polysphondylium violaceum*.

While never exploited to their full potential, it was found that mutants producing milk-white sori could be produced by U-V irradiation. Such mutants were studied briefly in mixtures with purple-pigmented wild types soon after Wilson and Ross (1957) erroneously reported myxamoebal fusions yielding zygotes to be a prerequisite for cell aggregation. No suggestions of recombinant strains were obtained and the program was soon discontinued. However, now that we know the sexual stage resides within the macrocysts, it would seem that spore pigmentation and lack of it might be used as contrasting markers should one select *D. purpureum* as an object for genetic experimentation.

Macrocysts (Fig. 12-14I) are known to be produced homothallically by some strains of *D. purpureum* (Nickerson and Raper, 1973a,b) and heterothallically by others (Clark, Francis, and Eisenberg, 1973), but the possible frequency of their occurrence has not been adequately explored; neither have the conditions that might favor their germination. It is known that darkness, free surface moisture, and moderate incubation temperatures found to be optimal for *D. mucoroides* strain DM-7 and *D. discoideum* also promote maximal cyst formation in *D. purpureum*. For strain Za-3b,

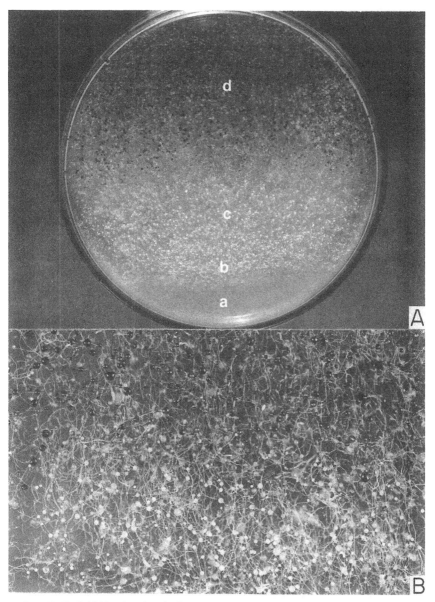

FIG. 12-15. Inhibition by phenol of sorus pigmentation in *Dictyostelium purpureum*. *A*. Culture to which 0.4 ml of 10^{-1} M phenol was added. No pseudoplasmodia have formed at site of application (a); adjacent to this pseudoplasmodia have aborted (b); further removed a broad band of sorocarps with milk-white sori have developed (c); and beyond this sorocarps with dark purple sori have formed (d). *B*. Enlarged view of the transition zone between sorocarps with milk-white and normal, dark purple sori. × 3.4. (From Whittingham and Raper, 1956)

the most thoroughly investigated macrocyst-producing strain of the latter species, the absence of phosphates is especially important. Insofar as known, the development of macrocysts in *D. purpureum* follows the same morphogenetic course as in *D. discoideum*, and, perhaps significantly, the mature cysts commonly show an unusually thick outer wall or covering, as in the latter species. Although little studied to date, such information as we have (Nickerson and Raper, 1973b) indicates that cyst maturation is very slow and the ability to germinate increases with age to attain a limited level in about 50 days.

Dictyostelium roseum van Tieghem, in Bull, Soc. Bot. France, *27*: 317 (1880). Also E. W. Olive, in Proc. Amer. Acad. Arts Sci. *37*: 339 (1901), and in "Monograph of the Acrasieae," Proc. Boston Soc. Natur. Hist. *30*: 451-513 (1902).

The species is known only from van Tieghem's description, here presented as a translation from the French: "The spherical spore mass that terminates the stalk is colored bright red. The spores are oval-elongate, as those of *D. mucoroides*, but a little larger, measuring on average 8 × 4 μ. Found several times on various dungs, especially on rabbit dung in company with *Pilobolus microsporus*." France.

What van Tieghem had in hand a century ago is not known, and one may question whether *D. roseum* should be recognized as a possible valid species. It was in fact challenged long ago by Pinoy, who surmised that van Tieghem may have described strains of *D. mucoroides* that were growing in association with bacteria such as "*Microbacillus*(?) *prodigiosus*" (1903) or "*Bacillus kiele*" (1907) and had absorbed a red bacterial pigment. Possibly so. In our experience, however, the myxamoebae of strains identifiable as *D. mucoroides* rarely retain the red pigment prodigiosin when cultivated with *Serratia marcescens* (= *Bacillus prodigiosus*). Van Tieghem notes that the species was found several times, so presumably he was not describing a single isolate. Additionally, we have little reason for questioning his report—the existence of another new species, *D. lacteum*, published at the same time, was for many years questioned until Raper in 1950 found it to be quite as van Tieghem had described it. Finally, if we have species with purple sori, as we do, and others with yellow sori, as we have, there would seem to be no compelling reason why a species with red or rose-colored sori should not exist. So, for the present, we shall retain the taxon and keep looking for the specimens.

Dictyostelium deminutivum Anderson, Fennell, and Raper, in Mycologia *60*: 49-64, figs. 1-20 (1968).

Cultivated on dilute glucose-peptone agars such as 0.1G-P/2 and 0.1G-

P/4 in association with *Pseudomonas fluorescens* at 24-25°C. Sorocarps diminutive (Fig. 12-16A), delicate, solitary or clustered, commonly 0.2-0.6 mm in height, occasionally 0.7 mm, rarely branched. Sorophores colorless, very thin, 5.0-8.5 μm in diameter at the base and gradually tapering to 1.5-3.0 μm above (Figs. 12-16H,I), consisting of single tiers of cells, bearing small, terminal sori, colorless to milk-white (Fig. 12-16A). Sori typically globose, commonly 25-50 μm in diameter, seldom more. Spores narrowly reniform or capsule-shaped (Fig. 12-16B), mostly 3.5-5.0 × 1.5-2.0 μm, occasionally 6.0 × 2.5 μm, with polar granules prominent. Cell aggregations of two types, irregular mounds formed by the influx of cells separately or in small clusters (Fig. 12-16D), and occasionally small radiate pseudoplasmodia formed by convergence of loosely organized "streams" (Fig. 12-16F). Myxamoebae vary in form and dimensions, commonly about 8 × 10 μm when feeding, but irregular and elongate when actively moving (Fig. 12-16C). Macrocysts not observed or reported; microcysts commonly formed by unaggregated myxamoebae, sometimes dominant in the culture (Fig. 12-16J).

HABITAT. Soil and leaf mould of humid tropical forests, Poza Rica, Mexico.

HOLOTYPE. Strain PR-8, isolated by J. C. Cavender in 1961 from leaf mold collected in Mexico.

Although *D. deminutivum* can be cultivated in association with *Escherichia coli* on low-nutrient lactose-peptone agars, we have found that it grows equally well and fruits better in association with *Pseudomonas fluorescens*, a common soil bacterium, when grown on substrates containing glucose and peptone. Of such combinations the most useful are 0.05%G-P (0.1G-P/2) and 0.025%G-P (0.1G-P/4) agars, and of these the latter is generally best for observing cell growth and aggregation. The pH of this medium is about 4.5-5.0 prior to inoculation and rises to about 5.2-5.5 following growth of the bacteria and slime mold, which is comparatively low but seems to be optimal for this dictyostelid. Additional measures may be taken to enhance cell aggregation and sorocarp formation, namely: the addition of activated charcoal to the culture plate as the time for aggregation approaches or, even better, the substitution of an unglazed porcelain cover (Coors) for the glass lid of the Petri dish. The explanation of this beneficial effect is not known but could result from the absorption or escape of some inimical gas produced during the fruiting process. At the same time a limited reduction in atmospheric moisture has not been ruled out. However accomplished, the successful cultivation and study of *D. deminutivum*, the smallest of all the *Dictyostelia*, can be a rewarding experience. While considered to be different from them, it is apparently

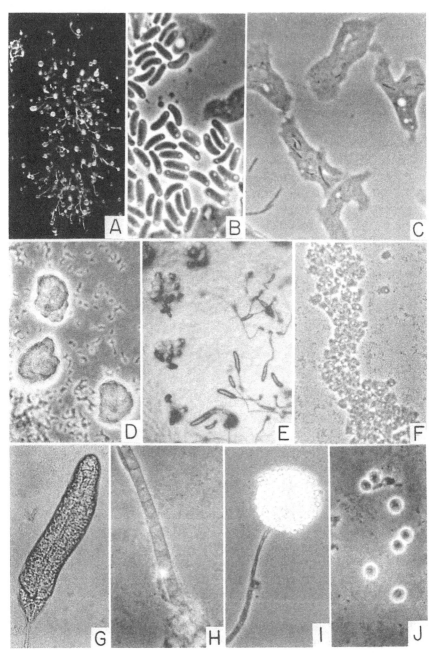

Fig. 12-16. *Dictyostelium deminutivum* Anderson, Fennell, and Raper. *A.* Sorocarps as seen with reflected light. × 17.5. *B.* Spores; note their elongate shape and prominent polar granules. × 1260. *C.* Vegetative myxamoebae in active movement on an agar surface. × 1225. *D.* Small aggregations formed without stream formation: cells converge individually. × 100. *E.* An area showing nascent sorogens and sorogens in process of sorocarp formation. × 70. *F.* Detail of "stream" in relatively large aggregation; note that inflowing cells are not elongate or oriented as in most radial aggregates. × 350. *G.* A sorogen in process of sorocarp formation; note the funnel-like terminus of the lengthening stalk. × 245. *H.* Basal area of typical sorophore. × 385. *I.* Terminal area of sorocarp with collapsed sorus. × 185. *J.* Microcysts formed by unaggregated myxamoebae. × 430. (*B-J* from Anderson, Fennell, and Raper, 1968)

closely related to *D. microsporum* Hagiwara (1978) and *D. multi-stipes* Cavender (1976).

Cell aggregation may assume either of two types. In the first of these the myxamoebae converge individually or in small groups to form a small rounded mound. This subsequently cleaves into a number of sections, each of which gives rise to single sorogens (rarely two) and, in due course, sorocarps. Unlike the sorogens in *D. multi-stipes*, which are comparatively long and very thin (vermiform), those of *D. deminutivum* are as a rule, and from the outset, relatively compact (Fig. 12-16E,G); yet they construct sorocarps with sorophores even thinner than those of *D. multi-stipes*. The type of aggregation and pattern of fructification just described are most frequent and account for virtually all the completed sorocarps. Aggregations of the second type are more suggestive of those formed by most dictyostelids. They are basically radial in pattern, usually small, and only rarely develop effective inflowing streams. Rather, a center of limited potential emerges, and toward this myxamoebae in the immediate environs are attracted and move in a desultory manner, being neither elongate and uniformly oriented nor closely applied one against the other (Fig. 12-16F). Yet they are attracted sufficiently to present a discernible wheel-like pattern. On the one hand aggregations may be quite small (ca. 200 μm) and show clearly defined centers, or they may be more expansive (up to 3 ± mm) with numerous "streams" more or less focused on a large common but centerless area. In patterns of the latter type we have seen evidence that the prior presence of developing sorocarps could have stimulated aggregation of the type.

Insofar as we know the chemotactic substance(s) in *D. deminutivum* have not been investigated.

Under conditions still undefined, but clearly unfavorable for completion of the normal developmental cycle, myxamoebae in large numbers may encyst individually to form microcysts (Fig. 12-16J). As in other microcyst-forming species, such cysts are encased in relatively firm cellulose walls, and when returned to conditions favoring growth, they germinate to release amoeboid cells that reinitiate the vegetative stage.

Dictyostelium multi-stipes Cavender, in Amer. J. Bot. *63*: 63, figs. 12-22 (1976).

Cultivated on substrates of low nutrient content such as 0.1L-P/2 and 0.05% lactose, 0.025% yeast extract agars in association with *Escherichia coli* at 22-25°C. Sorocarps small and delicate, erect or semi-erect, not phototropic, sometimes developing singly but usually in clusters (Fig. 12-17A), the number varying with the richness of the culture. Sorophores consisting of single tiers of cells (bases often two cells wide), mostly 0.25-0.65 mm in length but ranging from 0.15 to 1.0 mm; in diameter, 8-12 μm at the base and tapering to 2-3 μm in the sorus. Sori globose to subglobose, 25-65 μm in diameter, rarely more, white to hyaline. Spores comparatively short, ellipsoid to slightly reniform, mostly 4.0-5.5 × 2-3 μm, occasionally larger to 6.5-7.5 × 3-3.5 μm, with polar granules fairly prominent and usually bipolar (Fig. 12-17B). Cell aggregations small, rounded, or flattened without converging streams (Fig. 12-17D), mostly 250-500 μm in diameter but ranging from 100 to 800 μm, smallest aggregates producing single sorocarps, larger aggregates becoming segmented with each segment producing few to several sorogens (Fig. 12-17G); sorogens delicate, at first vermiform (Fig. 12-17E,F) later becoming club-shaped as sorophores lengthen. Vegetative myxamoebae not distinctive, mostly 12-18 × 9-12 μm when viewed at colony edge (Fig. 12-17C). Macrocysts and microcysts not reported or observed.

HABITAT. Humus and leaf mould of tropical rain forest, Udjung-Kulon, Java, Indonesia.

HOLOTYPE. Strain UK-26b, isolated in the summer of 1970 by J. C. Cavender from soil collected in Udjung-Kulon, Java, and sent to us early in 1971.

As the name implies, the species is characterized by the multiple sorocarps that arise from each pseudoplasmodium, or fraction thereof if it has undergone segmentation. This, together with the very small dimensions of its cell aggregations and sorocarps, serves to distinguish the species. In some ways *D. multi-stipes* clearly resembles *D. deminutivum* Anderson et al. (1968), but it differs from that species in producing relatively long vermiform sorogens, a greater number of sorogens per cell aggregation, and somewhat larger sorocarps. It also resembles rather closely *D. microsporum* Hagiwara (1978) but develops pseudoplasmodia and sorocarps that are somewhat smaller than those of that species. It is of interest, however, that Hagiwara recorded and illustrated the occurrence of "a few comparatively large spores exceeding 3 μm in width," for we have encountered a similar size variation among the spores of *D. multi-stipes*.

Cell aggregation in *D. multi-stipes* occurs without stream formation except in very rare cases (Fig. 12-17F). The myxamoebae converge as individuals to form small rounded clumps or somewhat larger low, flattened mounds (Fig. 12-17D). Myxamoebae are usually drawn from areas about three times the diameter of the mounds, and sorogens begin to emerge soon after the cells have assembled. Although not fixed, there is a general relationship between the size of the aggregation and the number of sorocarps that develop from it. For example, a clump 100 μm in diameter or less will normally produce a single sorocarp, as Cavender (1976a) has shown; if 150-175 μm wide it will probably produce 4-6 sorocarps; if ca. 250 μm the number rises to 10-15; and if 500 μm a dozen or more sorogens will usually form at the center of the aggregation, leaving temporarily a

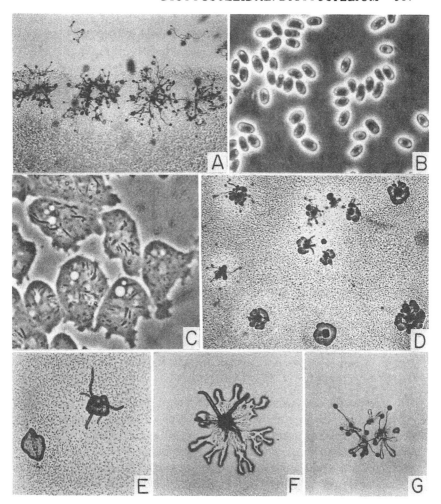

Fig. 12-17. *Dictyostelium multi-stipes* Cavender. *A*. Clusters of sorocarps produced on 0.1L-P/2 (pH 6.0) and shown in silhouette; each cluster developed from a primary aggregation that subsequently subdivided. × 13.5. *B*. Spores characterized by relatively small dimensions and prominent polar granules. × 910. *C*. Vegetative myxamoebae at colony margin as seen under a coverglass; cells are somewhat flattened. × 910. *D*. Field in which several aggregations formed and are now subdividing preparatory to fruiting. × 13.5. *E*. Two young aggregations, one of which is now developing sorogens. × 30. *F*. A larger aggregation with developing sorogens and an unusual pattern of internal streaming. × 30. *G*. A cluster of sorocarps that have developed from four unequal segments of a primary aggregation. × 20. (*E-G* from Cavender, 1967a)

raised outer rim. Later this may subdivide into segments, each of which produces a few sorocarps, or their cells may move back to the central area before completing the fruiting process. Under optimal culture conditions it is not unusual to find clusters of 20-25 or even more sorocarps developing from a single primary aggregation.

Microcysts have not been observed, but unaggregated myxamoebae in great numbers may penetrate the agar surface, become irregularly lobed, and reflect light in a manner that superficially resembles massed microcysts when plates are examined at low magnifications. The numbers of such contorted myxamoebae are greater on mildly acid than on neutral substrates. There is some evidence that these myxamoebae can later aggregate and produce sorocarps.

There is in *D. multi-stipes*, as in other dictyostelids, a maximum number of myxamoebae that can communicate effectively in forming an aggregation, and a smaller number that can collaborate effectively in constructing a sorocarp. The numbers here are quite small when compared with almost all other species. For this reason it is essential that substrates of low-nutrient content be used when cultivating this slime mold. In addition to the agar media listed above, low levels of glucose and peptone (0.05%) or of yeast extract (0.025%) are quite satisfactory, as is agar based upon a dilute hay infusion. Such a medium, along with nonnutrient agar, was recommended by Cavender (1976a) as a base upon which to spread a mixture of dictyostelid spores and pregrown *E. coli*. Such cultures have certain advantages but require additional steps in preparation, namely growing and harvesting the bacteria.

Dictyostelium microsporum Hagiwara, in Bull. Natl. Sci. Museum Ser. B (Bot.), *4*: 27-32, figs. 1A-E & 2A-I (1978).

Cultures of *D. microsporum* have not been available for observation in our laboratory. However, Hagiwara's description and illustrations are quite clear and the species merits recognition as he has described it. For this reason Hagiwara's characterization of the species is presented herewith in essentially unaltered form except for the insertion of figure references.

Cultivated upon pregrown cells of *Escherichia coli* deposited on nonnutrient agar and incubated at room temperature. Sorocarps white, erect or inclined, usually unbranched (Fig. 12-18A) but occasionally bearing one or more lateral branches, developing in loosely integrated groups, not phototropic. Sorophores delicate, consisting of single tiers of cells except in basal parts, sinuous (Fig. 12-19A), usually 0.14-1.2 mm in length but sometimes up to 1.5 mm, tapering gradually from 3-15 μm in diameter near the base to 1-5 μm in diameter near the tip (Figs. 12-18C), sorophore bases usually clavate (Fig. 12-18E) but sometimes conically expanded or weakly digitate (Figs. 12-18G,H). Sori globose (Fig. 12-19A), usually 25-100 μm in diameter but sometimes up to 130 μm, consisting of a large number of small spores and a few comparatively large spores exceeding 3 μm in width (Figs. 12-18B and 12-19C). Spores hyaline, ellipsoid, smooth, mostly 3.4-5.1 × 1.7-2.8 μm, having polar granules (Fig. 12-19B). cell aggregations consist of well-developed radiate streams 0.5-2.0

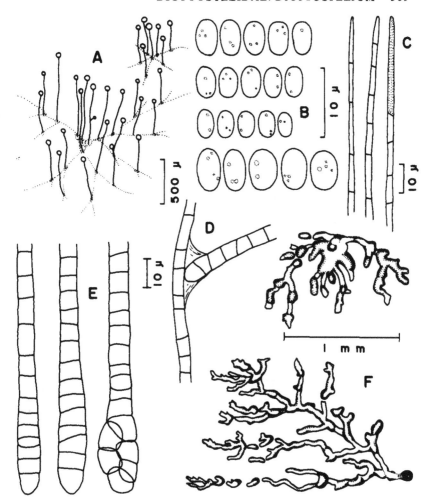

FIG. 12-18. *Dictyostelium microsporum* Hagiwara. *A*. Mature sorocarps formed following breakup of convergent myxamoebal streams; note very small dimensions. *B*. Spores showing the two size classes cited in the species description. *C*. Terminal portions of sorophores showing the thin, elongate cells of which they are composed. *D*. Portion of a rare branched sorophore. *E*. Sorophore bases; note that sorophores consist of a single tier of cells. *F*. Cell aggregations prior to general fragmentation preparatory to sorocarp formation. (Hagiwara's fig.1, 1978)

mm in diameter that break up into several smaller aggregations prior to sorocarp formation (Fig. 12-19F). Microcysts and macrocysts not reported.

HABITATS. Surface soil and leaf mould of beech (*Fagus crenata*) and subalpine forests, Japan.

TYPE CULTURE. Hagiwara's strain 143 isolated in 1971 from soil col-

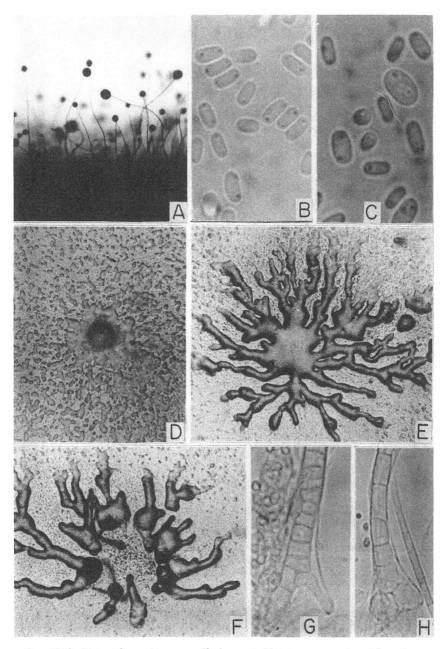

Fig. 12-19. *Dictyostelium microsporum* Hagiwara. *A*. Mature sorocarps viewed from the side. × 40. *B*. Spores showing polar granules. × 2000. *C*. Spores showing variation in size cited in species description. × 2000. *D*. Early stage in cell aggregation. × 60. *E*. Later stage in aggregation. × 60. *F*. Fully developed aggregate with sorogen arising at the center. × 60. *G*. Weakly digitate base of a sorophore. × 800. *H*. Conically expanded sorophore base. × 800. (Photographs courtesy of H. Hagiwara as used in his fig. 2, 1978)

lected at an altitude of about 1000 meters. Mt. Daisen, Tottori Prefecture, Japan. Species description drawn to include other isolates obtained from soils collected in different areas and at altitudes up to 1600 meters.

Superficially, *D. microsporum* bears a striking resemblance to *D. deminutivum* Anderson, Fennell, and Raper (1968) but differs markedly in its pattern of pseudoplasmodium formation. Whereas aggregations assume the form of simple mounds in the latter and show only poorly developed inflowing streams of myxamoebae, in *D. microsporum* the development of radiate aggregations 0.5-2.9 mm in diameter is quite characteristic, albeit these regularly break up into smaller aggregations which fruit separately and in situ along the paths previously occupied by the converging streams (Fig. 12-18A). Additionally, the completed sorocarps average somewhat larger than those of *D. deminutivum*, and the sorophore bases may be conically expanded as in *D. monochasioides* or weakly digitate (Figs. 12-19G,H) as in the smaller sorocarps of the crampon-based dictyostelia.

Spores of the type strain, #143, were reported to range from 3.0 to 6.1 μm in length; and among these were two distinct modes with regard to width, those 3.0 μm wide and less and those ranging from 3.0 to 4.0 μm in width (Figs. 12-18B and 12-19C). In one sorus of over 2200 spores about 6% were of the larger mode, while the remainder were of the lesser width. This relationship was also observed in other strains; and in clonal cultures from two of these (including the type) normal sorocarps containing both large and small spores were produced whether derived from large or small spores. The separation thus appeared to be hereditary.

Dictyostelium caveatum Waddell, Raper, et Rahn, sp. nov.; ref. Waddell in Nature *298*: 464-466, 1982.

Cultum ad 18-20 C in "0.1% lactose-0.1% peptone" cum *Escherichia coli: sorocarpi* parvi, delicati, gregarii, recti vel semirecti, saepe implicati; sorophori tenues, fragiles, typice unicellulares in crassitudine, basibus maximarum structurarum exceptis, plerumque 0.7-1.5 mm in longitudinem, fere 8-15 μm in diametro in inferiore parte et 2-4 μm in superiore parte; *sori* globosi, hyalini vel albi, plerumque 40-100 μm in diametro; *sporae* ellipsoideae, capsulaformes, typice 5.0-6.5 × 2.5-3.5 μm, aliquando majores, conspicuas polaresque granulas continentes; *aggregationes* exiguae, 150-300 μm in diametro; cellulae congregant sed nullum flumen formant; *soroferae structurae* paucae vel multae in singulis aggregationibus, clavatae, plerumque 100-200 × 25-50 μm; vegetae *myxamoebae* variformes, 15-25 × 10-15 μm in agaro, filiformia pseudopodia continentes in liquido substrato; *microcysti* et *macrocysti* absunt.

D. caveatum aliorum dictyostelidorum cellulas edent et eorum fructificationem obturant; talis obstructio possibilis est inter preaggregationem et sorocarporum formationem.

HABITAT. In stercore vespertilionis, Blanchard Springs Cavern, Arkansas.

HOLOTYPUS. Waddell's cultura B4-3 (= Wisconsin cultura WS-695), in mense Septembre anni 1980 a David R. Waddell, Princeton University, Princeton, N.J., isolata.

Cultivated on low-nutrient substrates such as 0.1L-P or hay infusion agars in association with *Escherichia coli* or *E. coli* B/r at 18-20°C. Sorocarps small, delicate, typically clustered (Fig. 12-20A), erect or semierect, often tangled, very weakly phototropic, arise from small and often persistent cell mounds. Sorophores thin and fragile, typically one cell thick (Fig. 12-20F), except for basal portions of largest structures, commonly 0.7-1.5 mm high, sometimes more, ranging from 8 to 15 μm in diameter at the base and narrowing to 2-4 μm at the sorus. Sori globose (Fig. 12-20A), hyaline to translucent-white, commonly 40-100 μm diameter, sometimes less but seldom more. Spores ellipsoid, capsule-shaped, mostly 5.0-6.5 × 2.5-3.5 μm, occasionally 7.0-7.5 μm, with prominent granules, usually but not consistently bipolar (Fig. 12-20B). Cell aggregations small, mound-like (Fig. 12-20D), 150-300 μm in diameter, typically develop without stream formation, produce few to many sorogens depending on size; sorogens clavate, without conspicuous tips, mostly 100-200 × 25-50 μm. Vegetative myxamoebae commonly 15-25 × 10-15 μm when moving on agar (Fig. 12-20C), uninucleate, with contractile and food vacuoles prominent and ecto- and endoplasm clearly distinct, the latter projecting filose pseudopodia in liquid substrates. Neither microcysts nor macrocysts reported or observed.

Myxamoebae of *D. caveatum* sp.n. prey upon cells of other dictyostelids and prevent them from fruiting; development is arrested at any stage from preaggregation to pseudoplasmodial or sorogenic migration; affected cell masses retract any preexisting tips, round up, and 1-3 days later show in turn multiple papillae, sorogens, and sorocarps of *D. caveatum* (Fig. 12-20E). Such sorocarps are often longer than indicated above and may reach 4-5 mm without proportional increases in sorophore width or sorus diameter.

HABITAT. Bat guano in darkness at 15°C and 100% relative humidity, Blanchard Springs Cavern, Arkansas.

HOLOTYPE. Strain B4-3 of Waddell (= Wisconsin strain WS-695), isolated in September 1980 by David R. Waddell, Princeton University, Princeton, N.J., and sent to Wisconsin in the winter of 1981.

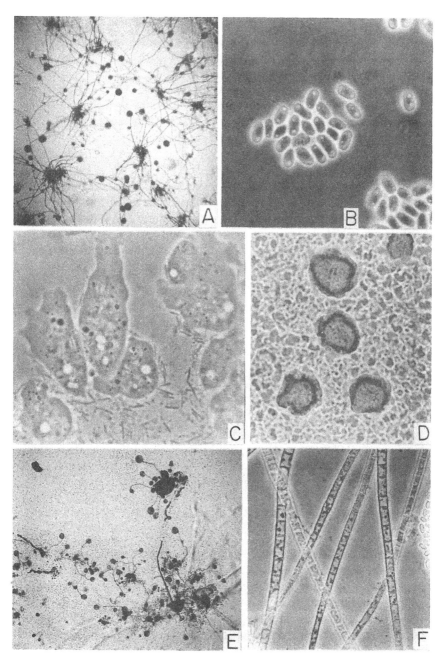

FIG. 12-20. *Dictyostelium caveatum* Waddell, Raper, and Rahn sp. nov. *A*. Clustered sorocarps produced in association with *Escherichia coli* B/r on 0.1L-P agar. × 24. *B*. Spores containing prominent granules, usually bipolar. × 900. *C*. Vegetative myxamoebae somewhat compressed by a coverglass. × 900. *D*. Young aggregations formed without inflowing streams. × 900. *E*. Sorocarps of the smaller type (larger type not illustrated) that have developed on infected segments of a fragmented sorogen of *D. rosarium*. × 10. *F*. Sorophores typical of the species with vacuolate cells in single tiers. × 385

The foregoing description of *D. caveatum* is based upon cultures grown in association with *E. coli* or *E. coli* B/r upon 0.1L-P and hay infusion agars to facilitate comparison with other dictyostelids. Additional substrates may be used to attain different objectives (Waddell, 1982), and bacteria other than those cited may be employed as nutrient by giving attention to substrate composition.

Dictyostelium caveatum sp.n. was reported recently under this name by David R. Waddell (1982) of Princeton University, but no species description was provided. To correct this omission, and in order to include the species in the current monograph, a formal description has been prepared based upon observations of the type culture in laboratories at Princeton University and at the University of Wisconsin. Since *D. caveatum* is the only dictyostelid known to prey upon other amoeboid organisms, it should be of special interest. Its discovery was serendipitous, as the following paragraph from one of Waddell's letters to KBR indicates

The species was initially isolated by mixing bat guano with a heavy suspension of *E. coli* and plating on hay infusion agar. I was originally looking for "blind" slime molds which could not migrate toward light. *D. caveatum* is fairly light insensitive although the slugs (sorogens) do migrate toward light at low cell densities if the intensity is high enough. I originally thought it was a *P. pallidum* strain that didn't form whorls. Therefore, it made sense to mix it with *P. pallidum* in different ratios to see how the light response was affected. The surprising result was that all the mixtures exhibited the weak phototactic response characteristic of *D. caveatum*. I then mixed the two species at very low ratios of *ca.* 1:10,000 (*D.c:P.p.*) and found mixed populations of slugs that either formed normal *P. pallidum* fruits or formed the characteristic *D. caveatum* fruits after the slug had migrated a certain distance. This indicated that only one *D. caveatum* cell per aggregate seemed to be sufficient to eventually cause arrest and conversion of the pseudoplasmodium.

Other dictyostelids in which inhibition of normal morphogenesis has been demonstrated include *Dictyostelium discoideum*, *D. purpureum*, *D. mucoroides*, *D. rosarium*, *D. minutum*, *D. polycephalum*, *P. violaceum*, *P. filamentosum*, and *Acytostelium leptosomum*.

Whereas *D. caveatum* is of singular interest because of its unique ability to thwart the development of other dictyostelids by consuming or disorganizing their cellular elements (Waddell, 1982), the slime mold is not without interest for other reasons. It is one of a few species that grow and fruit nicely at temperatures below 20°C, and it is one of a small number that form aggregations by a convergence of myxamoebe as individual cells rather than via inflowing streams. The former group, species that fruit at low temperatures, would include *D. septentrionalis*, *D. polycarpum*, and

P. filamentosum, none of which seem to be closely related to *D. caveatum* or to each other. Of the latter group, species whose myxamoeba aggregate as individual cells, *D. deminutivum* and *D. multi-stipes*, and to a lesser degree *D. minutum*, possess growth characteristics and produce fruiting structures that are often smaller but otherwise suggest those of *D. caveatum*. Having made this statement, we must qualify it immediately since *D. caveatum*, for reasons still quite obscure, produces sorocarps of two contrasting patterns. The more common pattern consists of small sorocarps, often about 1 mm high, that arise in fairly crowded clusters from small, primary, dome-like aggregations or on fragments of severed pseudoplasmodia or sorogens of other dictyostelids, as seen in Fig. 12-20E. Sorocarps of the second pattern, by contrast, are unlike those of any other *Dictyostelium*: sorophores are quite thin and very fragile, up to 4-5 mm in length, and usually bear minute sori or none at all. Except for slightly enlarged basal areas, these sorophores typically consist of single tiers of cells throughout, and sorophore tubes are apparently quite thin, for the cells within may vary substantially in pattern. It is an interesting and perplexing fact that these longer sorocarps are (September 1982) commonly produced on arrested aggregations or pseudoplasmodial fragments of dictyostelids upon which *D. caveatum* is predatory—not in agar plates with *E. coli* or *E. coli* B/r in the absence of another dictyostelid. It was not always so. When first received, limited numbers of long, fragile sorocarps developed at the inoculation sites when *D. caveatum* was implanted in lawns of *E. coli*, a pattern of behavior that is consistent with information provided by Waddell:

> One other observation I have made since my last communication is that the original stock of *D. caveatum* which is stored on silica gel in Princeton forms two types of colonies when it is cloned with *E. coli*. One type fruits poorly and the fruiting bodies are very stalky. The other forms dense lawns of fruiting bodies and "looks" much more normal. This strain which I am currently calling Clone 2A is the one which I am using here. (Max Planck Institut für Biochemie, Martinsried bei München, August 1982.)

Whereas fruiting in cultures grown with bacteria alone is improved dramatically by the addition of activated charcoal at the time of cell aggregation or early culmination, its presence seems not to enhance the formation of long sorocarps. Clearly, *D. caveatum* merits careful and detailed study to determine the bases of its dual nature.

Sorogen formation from single aggregations is asynchronous and may continue for periods of many hours, suggesting perhaps that the myxamoebae continue to multiply or mature physiologically after aggregation is seemingly complete. This phenomenon is not unique for *D. caveatum*, but it is particularly striking in this species.

In his original report, Waddell (1982) presented evidence of the phagocytosis of cells of other dictyostelids by the myxamoebae of *D. caveatum*. Still, it is not certain, as Waddell has indicated (1982), that something more is not involved, for there are suggestions that this species may in some way parasitize or immobilize its victims prior to actual consumption, whether by engulfment or otherwise. Furthermore, in cultures of this species with other dictyostelids, the latter may often produce seemingly normal pseudoplasmodia or sorogens only to have these cease culminating without showing any outward sign of attack by *D. caveatum*.

Where *Dictyostelium caveatum* will lead we would not presume to predict. But that it opens new fields for study is obvious already, as Waddell (1982) has amply demonstrated.

Dictyostelium fasciculatum Traub, Hohl, and Cavender, in Amer. J. Bot. *68:* 166-170, figs. 9-18 (1981).

Cultivated on low-nutrient substrates such as 0.1L-P in association with *Escherichia coli* strain B/r at 18-22°C. Sorocarps typically clustered (Fig. 12-21H), less often solitary depending upon the aggregation pattern, semi-erect or inclined, mostly 2-4 mm high but up to 1.0 cm, strongly phototropic (Fig. 12-21I). Sorophores colorless, comparatively thin, occasionally branched, with bases plane or slightly enlarged, sometimes crowded, commonly 10-18 μm near the base, gradually tapering to 3-5 μm at the sorus. Sori globose to citriform (Fig. 12-21I), milk-white to cream-colored, mostly 100-250 μm in diameter but ranging from 60 to 300 μm. Spores narrowly elliptical to somewhat oval, commonly 5-6 × 2.2-3.0 μm, less often 7.5-8.0 × 3.2-3.5 μm, with granules conspicuous (Fig. 12-21A), usually bipolar but sometimes subpolar or monopolar. Cell aggregations vary in size and pattern (Figs. 12-21C-F) some initially large and fragmenting, others relatively small and producing clustered sorogens (Fig. 12-21G). Vegetative myxamoebae not distinctive (Fig. 12-21B), commonly 15-20 × 10-14 μm, with one or two prominent contractile vacuoles. Macrocysts and microcysts not reported or observed.

HABITAT. Soils of different forest types at all altitudes in Switzerland; also in France, Germany, and Denmark, and in deciduous forests of Wisconsin and North Carolina (fide J. C. Cavender), probably cosmopolitan in distribution.

HOLOTYPE. Strain SH-3, isolated by Franz Traub in 1970 from soil of an oak forest in Osterfingen SH, Switzerland.

As seen in laboratory cultures, *Dictyostelium fasciculatum* often bears a striking resemblance to *D. mucoroides*; hence it was long overlooked as

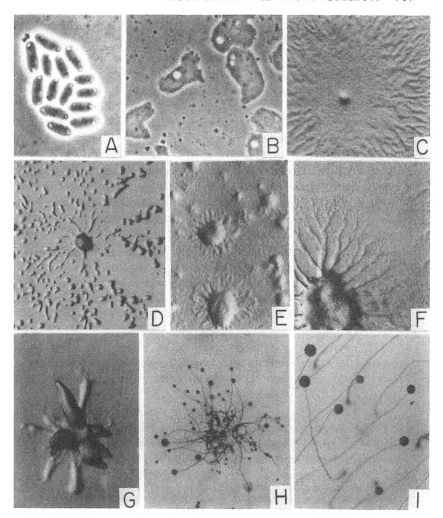

FIG. 12-21. *Dictyostelium fasciculatum* Traub, Hohl, and Cavender. *A*. Spores with prominent polar granules. × 900. *B*. Vegetative myxamoebae. × 750. *C*. Aggregation pattern I (see text). × 75. *D*. Aggregation pattern I, later and showing severed streams. × 4.5. *E*. Aggregation pattern II (see text). × 20. *F*. Aggregation pattern II with flattened center. × 10. *G*. Emerging sorogens. × 13.5. *H*. Cluster of sorocarps from single aggregation. × 9.5. *I*. Field of larger solitary sorocarps. × 10. (Rearranged from Traub, Hohl, and Cavender, 1981a)

a separate species. Despite its white to cream-colored sori and sorophores of lengths that overlap those of *D. mucoroides*, apparent similarities are largely illusory. The sorocarps of *D. fasciculatum* are more flexuous and more delicate—many of the sorophores consist of single tiers of cells (flattened, isodiametric or elongate) throughout their entire length. Cell aggregation is also different and may occur in one of three intergrading but still fairly distinct patterns. These have been described by Traub et al. (1981a) in the following way:

Aggregation pattern I (Fig. 12-20C,D) occurs when spores and bacterial suspension are spread as a band over the agar surface. A small mound first appears in the postfeeding myxamoebae (ca. 40 hr). This becomes surrounded by a broad band of inflowing myxamoebae which produce rather vigorous streams. These broad patterns are 20-30 mm in diameter. The streams soon begin to disintegrate and many smaller, secondary centers begin to form within this area. The larger primary center remains stumplike with truncated streams. Around the secondary centers, some radiate streaming may occur. . . . Not uncommonly a corona in which myxamoebae do not stream may develop around primary centers and between (these and) the outer region of secondary centers. Later secondary centers form in this region.

Aggregation pattern II (Fig. 12-20E,F) develops when spores are inoculated on a bacterial streak. The primary center continues to grow by means of short, thick or at times longer and thinner streams. The central mass is often flattened with a slight depression in the middle (Fig. 12-20F) and is of a relatively large diameter. While aggregation is continuing, the aggregate divides into many culmination papillae (Fig. 12-20G). Synchronous culmination may occur but is not the rule. Culminating bodies may migrate apart slightly before building stalks. This type of aggregation produces the bushy habitus characteristic of this species.

Aggregation pattern III is intermediate in nature between those of I and II. Vigorous streaming produces a thick center in the form of a bulge which subsequently divides into many culmination papillae. Along the periphery smaller streams dissociate and form secondary centers. The bushy form results also from this aggregation pattern.

Dictyostelium fasciculatum is of special interest, since strains now included under this name apparently played a key role in the evolution of Traub and Hohl's "new concept of the taxonomy of the family Dictyosteliaceae" (1976; see also Traub et al., 1981a, p. 170). Although *D. fasciculatum* remained undescribed for five years, the strain later cited as its type, SH-3, was listed among the first cultures considered to be "*Polysphondylium*-related strains" based upon their clustered habit, spores with polar granules, and failure of the myxamoebae to aggregate in response to cAMP. Despite such evidence of kinship, we believe Traub, Hohl, and Cavender (1981a) acted wisely by retaining their new species in *Dictyostelium*.

Judging by appearances only, one may question if the small, clustered sorocarps seen in Figure 12-21H can belong to the same species as the larger, solitary structures shown in Figure 12-21I. The answer is yes.

Furthermore, sorocarps of contrasting dimensions are encountered in some other species, to wit: strains ID-9 and ID-10 of *D. mucoroides*, and in the newly described species that we have just considered. Unfortunately, the explanation for such dual behavior is no clearer here than there.

Substrates such as 0.1L-P or 0.1L-P/2 agars are quite satisfactory for cultivating *D. fasciculatum*, and very dense fruiting can be obtained by substituting an even smaller amount of yeast extract (0.025%) for the peptone. Good growth and development occurs between 15 and 23°C, with optimum rates at 18-22°C. Cell aggregation can be accelerated and sorocarp formation markedly improved by adding activated charcoal to the cultures at the first sign of postvegetative cell alignment.

Dictyostelium tenue Cavender, Raper, and Norberg, in Amer. J. Bot. 66: 207-217, figs. 18-28 (1979).

Cultivated on low-nutrient substrates such as 0.1L-P/2, or 0.1G-P/2, or dilute hay infusion agar in association with *Escherichia coli* or *E. coli* B/r at 24-27°C. Sorocarps typically clustered (Figs. 12-22A,F), erect or semi-erect in diffuse light, phototropic in unidirectional light, unbranched, branched near the base, or forming lateral branches from groups of cells abstricted from the distal ends of rising sorogens (Fig. 12-22C). Sorophores very slender and delicate, hyaline, consisting of single tiers of cells except for slightly bulbous bases, 1-6 mm in length, slightly tapered, ranging from 8-24 μm in diameter near the base to 3.0-5.5 μm at the apex. Sori globose, milk-white (Fig. 12-22B), commonly 50-100 μm in diameter (average ca. 75-85 μm). Spores ellipsoid, relatively short, mostly 4.5-6.0 × 2.5-3.5 μm, rarely 7 μm in length, with prominent granules, usually bipolar (Fig. 12-22J). Cell aggregations may develop as streamless mounds (Fig. 12-22D) or as small radiate patterns 0.5-2.5 mm in diameter, inflowing streams generally thin, commonly anastomosed (Fig. 12-22E) and ending in raised and somewhat scalloped peripheral ridges. Myxamoebae somewhat rounded when feeding (Fig. 12-22H), 10-13 × 8-10 μm in diameter, becoming very irregular and much more elongate prior to aggregation (Fig. 12-22I). Microcysts produced by unaggregated myxamoebae in some strains (Fig. 12-22K) but not in others; macrocysts not reported or observed.

HABITAT. Soils of tropical rain forests in Mexico, Panama, Costa Rica, East Africa, and Southeast Asia; apparently common in the tropics.

HOLOTYPE. Strain PJ-6, isolated by J. C. Cavender in 1970 from soil of a tropical rain forest, Peutjang Island, Java.

Other representative cultures include PJ-2 from a similar source; MBII-4 isolated by JCC in 1966 from soil of tropical coastal forest, Mombasa, Kenya; PR-4 isolated by JCC from soil of tropical rain forest, Poza Rica,

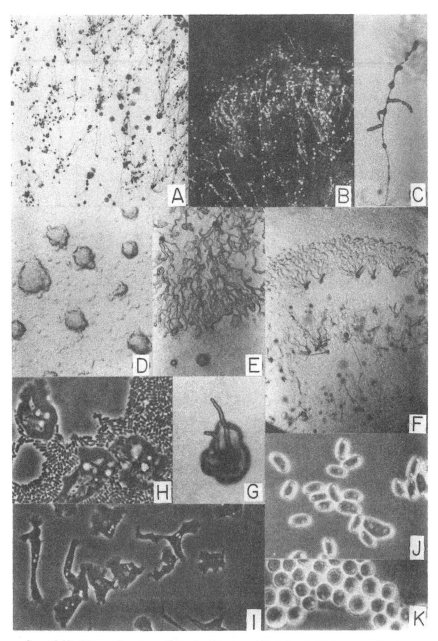

Fig. 12-22. *Dictyostelium tenue* Cavender, Raper, and Norberg. *A*. Sorocarps produced on 0.1L-P/2 agar; note clustered habit and thin sorophores. × 6.75. *B*. Sorocarps in denser culture photographed with surface light. × 15. *C*. Sorocarp with branches in process of formation. × 26. *D*. Mound-like aggregations formed without inflowing streams. × 26. *E*. Part of an aggregation with anastomosing streams. × 9. *F*. Portion of plate culture showing three successive fruiting zones: mature sorocarps, below, developing sorocarps, center, and aggregations and emerging sorogens, above. × 7.5. *G*. Sorogens arising from completed aggregation. × 25.5. *H*. Myxamoebae feeding on *Escherichia coli*. × 765. *I*. Postfeeding myxamoebae; note irregular elongate shape. × 350. *J*. Spores showing characteristic shape, irregular dimensions, and polar granules. × 560. *K*. Microcysts representing unaggregated myxamoebae that have encysted as individual cells. × 765. (From Cavender, Raper, and Norberg, 1979)

Veracruz, Mexico; and Pan-52 isolated in 1958 by KBR from soil of tropical rain forest, Barro Colorado Island, Panama.

Dictyostelium tenue as presented here embraces a continuum of forms gathered from widely separated areas, and while appreciable differences exist among these, they all possess certain characteristics in common, including delicate, colorless sorophores that are remarkable for their length relative to their diameter and the limited number of cells expended in their construction; a tendency to produce clustered sorocarps; and the inconstant pattern of cell aggregations that range from small mounds to rather elaborate networks of interlacing myxamoebal streams. In addition to being clustered, the sorophores not uncommonly bear lateral branches somewhat as in *D. aureo-stipes*, but with each branch typically formed from a single small cell mass. In some strains sorocarps of a single cluster will develop over a period of time, while in others they tend to develop simultaneously. There is no set number of sorocarps that arise from a single aggregation, and this may reach twenty or more in some strains.

While not formally described until 1979, the kind of dictyostelid now assigned to *D. tenue* was first isolated by KBR from Panama soil in 1958. It was again encountered, and frequently, by Cavender in the course of screening soils from different areas of Central America during the early 1960s and was reported and illustrated by Cavender and Raper (1968) as "a delicate variant of *D. mucoroides*(?)" found to be very abundant in the soils of some tropical forests of Mexico and Costa Rica. It was again reported from tropical soils collected at several locations in East Africa (Cavender, 1969b) and subsequently from similar environments in Southeast Asia, including Peutjang Island, Java, the source of strain PJ-6, later selected as the species type. It is believed to occur throughout the tropics.

In its habits of growth, cell aggregation, and sorocarp construction *D. tenue* bears considerable resemblance to Cavender's *D. multi-stipes* (1976). It is, however, appreciably larger than that species, particularly in the dimensions of its mature sorocarps. It also bears some resemblance to *D. aureo-stipes*, particularly in the irregular manner in which its sorophores occasionally support lateral branches, but its sorophores are usually much more delicate and they show no yellow pigmentation. It also bears some resemblance to Hagiwara's *D. monochasioides* (1973b) but fails to show the singular pattern of sorophore branching that is a principal hallmark of that species.

Microcysts are formed in some strains but not in others, at least under our conditions of culture. Of the microcyst-producing isolates, strain PR-4 is perhaps the most productive of this simple resting stage. Suboptimal culture conditions appear to accentuate microcyst production, which can be triggered by an increase in temperature from 25 to 27°C or by cultivation on suboptimal substrates such as cerophyll agar. When formed, microcysts are globose to subglobose and measure about 3-6 μm in diameter.

Cell aggregation and sorocarp formation may be enhanced by adding activated charcoal to cultures as consumption of the associated bacteria nears completion.

Dictyostelium delicatum Hagiwara, in Bull. Natl. Sci. Museum *14*: 359-361, fig. 5A-F (1971).

Cultivated on substrates such as 0.1L-P, 0.1L-P (pH 6.0), and dilute hay infusion agars in association with *Escherichia coli* or *E. coli* B/r at 22-24°C. Sorocarps erect or semi-erect, solitary, clustered or occasionally coalesced, varied in pattern and dimensions, simple or irregularly branched (Fig. 12-23A), commonly 2.5-5.0 mm in height but sometimes more, becoming interwoven and tangled, weakly phototropic. Sorophores hyaline or very faintly yellow, sinuous or occasionally spiral, gently tapered from 12-25 μm in diameter near the bases (Fig. 12-23E) to 4-8 μm at the sori, branches arise near the bases or more often subterminally (Fig. 12-23A), bases slightly enlarged, usually clavate (Fig. 12-23E). Sori translucent, white, or cream-colored, globose to subglobose, mostly 75-150 μm in diameter, occasionally more. Spores narrowly ellipsoid, plane or slightly reniform, commonly 4.5-7.0 × 2.0-3.0 μm, with prominent polar granules (Figs. 12-23B,G). Cell aggregations at first radiate in pattern usually 1.5-3.0 mm in diameter but occasionally more, converging streams comparatively thick and often ending abruptly (Fig. 12-23D), soon becoming fragmented with segments fruiting separately; sorogens often quite elongate when formed but contracting to 200-400 × 50-75 μm or less as sorocarp construction approaches midpoint. Vegetative myxamoebae inconstant in form (Fig. 12-23H), often 12-15 × 9-12 μm and showing vesicular nuclei with 2-4 peripheral nucleolar lobes, one or more contractile vacuoles, and food vacuoles with bacteria in process of digestion. Neither microcysts nor macrocysts reported or observed.

HABITATS. Leaf mould from deciduous forest and bird dung, Japan.

HOLOTYPE. Strain #67 isolated by Hagiwara from leaf mould in 1971 and sent to us as the species type in October 1974.

Of several species of *Dictyostelium* that produce sorocarps of intermediate size, *D. delicatum* would seem to be most closely allied with *D. aureo-stipes* Cavender, Raper, and Norberg (1979). It differs from that species, however, in a number of particulars: its sorocarps are generally smaller, less phototropic, more sinuous and en masse become intertwined and quite tangled. The sorophores are less profusely branched but equally patternless and range from colorless to pale yellowish-white rather than lemon yellow to golden yellow as in *D. aureo-stipes*. Branches are gen-

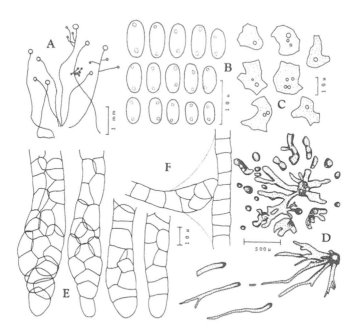

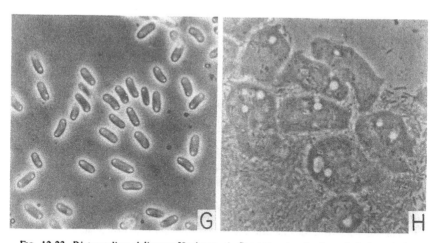

Fig. 12-23. *Dictyostelium delicatum* Hagiwara. *A.* Sorocarps showing irregularly branched and unbranched sorophores. *B.* Spores of varying forms and dimensions with polar granules. *C.* Myxamoebae containing one or more contractile vacuoles. *D.* Two aggregations, the upper partially fragmented and the lower showing discontinuous inflowing streams. *E.* Sorophore bases. *F.* Branch attachment; note lack of continuity with upright stalk. *G.* Spores as seen by phase microscopy. × 900. *H.* Vegetative myxamoebae at edge of bacterial colony. × 900. (*A-E*, camera lucida drawings that accompanied Hagiwara's description of *D. delicatum*, fig. 5, 1971; *G,H*, photographs of the type strain made in Wisconsin)

erally more delicate and bear smaller sori than in *D. aureo-stipes*, while terminal sori tend to be cream-colored rather than milk-white. Aggregations average somewhat smaller than in *D. aureo-stipes* and consist of comparatively thick converging steams that quickly break up into segments (Fig. 12-23D) which fruit separately. Under optimal conditions it is a delicate and attractive species that grows and fruits quite satisfactorily; under less favorable conditions aggregation and fruiting are delayed and developing sorocarps show still further irregularities. A much greater proportion of the cell population is used up in seemingly uncontrolled branching, largely secondary in nature but not infrequently primary as well. Additionally, late-forming sorocarps tend to be appreciably longer and to produce proportionately smaller sori.

Dictyostelium bifurcatum Cavender, in Amer. J. Bot. *63*: 66-68, figs. 36-46 (1976).

Cultivated on 0.1L-P and 0.1L-P/2 agars in association with *Escherichia coli* at 23-25°C. Sorocarps erect or semi-erect, not phototropic, arising singly or more often in groups, sometimes with lateral branch(es) near the base, main axes commonly bifurcating once or twice into more or less equal divergent branches bearing sori (Figs. 12-24G-I). Sorophores colorless or slightly yellow, comparatively short and stout, occasionally up to 1.2-1.3 mm in height, commonly 20-35 μm in diameter near the base, without a characteristic taper, branches often wider at midpoint than at site of divergence, narrowed to 8-12 μm within the sori, apices substantially enlarged (Fig. 12-24L). Sori globose to subglobose (Fig. 12-24H,J,K), white to faintly yellow, relatively large, commonly 75-200 μm in diameter and rarely up to 300 μm. Spores ellipsoid to slightly reniform, mostly 5.0-6.5 × 2.0-3.0 μm, occasionally less and rarely more, with conspicuous polar granules (Fig. 12-24A). Cell aggregations small, wheel-like, 1.0-2.0 mm in diameter or occasionally more (Fig. 12-24C) with centers raised and inflowing streams often comparatively wide, larger aggregations (Fig. 12-24D) fragmenting into smaller units that fruit separately (Fig. 12-24E). Vegetative myxamoebae characteristic of the genus but comparatively small (Fig. 12-24B), mostly 12-15 × 8-10 μm in diameter when moving on an agar surface. Macrocysts and microcysts not reported or observed.

HABITAT. Humus and leaf mould from tropical rain forest and semi-deciduous forest, Java and Bali, Indonesia.

HOLOTYPE. Strain UK-5, isolated in the summer of 1970 by J. C. Cavender from soil collected in a tropical rain forest at Udjung-Kulon, Java, and sent to us early in 1971.

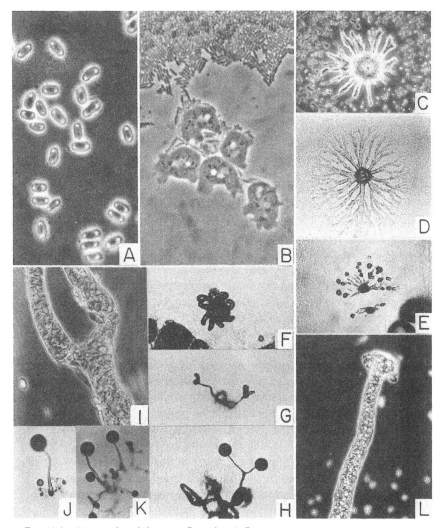

FIG. 12-24. *Dictyostelium bifurcatum* Cavender. *A*. Spores showing relatively small dimensions and prominent polar granules. × 830. *B*. Vegetative myxamoebae feeding on cells of *E. coli*. × 830. *C*. Young, small aggregation; note the truncate inflowing streams. × 50. *D*. An older and larger aggregation of typical wheel-like form. × 30. *E*. A still older aggregation after converging streams have broken up preparatory to sorocarp formation. × 13. *F*. Multiple sorogens arising out of a single aggregation. × 21. *G*. Rising sorogens shortly after bifurcation. × 18. *H*. A simple bifurcate sorocarp. x 27. *I*. Enlarged view showing detail of branching in a bifurcate structure. × 335. *J* and *K*. Unbranched sorocarps; note large sori borne on short sorophores. × 20. *L*. Enlarged terminus of sorophore branch, a characteristic feature of *D. bifurcatum*. × 215. (*D, F-K* from Cavender, 1976a)

The foregoing account is based upon Cavender's original description of *D. bifurcatum* supplemented by recent observations of strain UK-5 made in our laboratory. As emphasized by Cavender, and as indicated by the name selected for the taxon, sorogens of the species possess a singular tendency to divide dichotomously and to produce sorocarps with branched stalks. This property of aerial primary branching is shared by *Dictyostelium rosarium*, and analogous behavior is occasionally seen in other species when sorogens are in contact with the substrate, but nowhere else is it so commonplace or diagnostic as in *D. bifurcatum*, where it often occurs at two successive levels. Because of this, and especially when two or more sorocarps arise from a common site, the total picture may be that of a bush-like fructification.

The species possesses other interesting attributes. For example, the sorophores are not tapered progressively from base to sorus as in most dictyostelids. Instead they often show an enlargement some distance above the base, and the same may be seen in the sorophore branches midway between the sites of bifurcation and their terminal sori. The sorophores are withal quite short and stout when compared with other species, and in general appearance *D. bifurcatum* might suggest E. W. Olive's *Dictyostelium brevicaule* (1901) except for its branched sorocarps. The sorophores are of further interest in possessing prominent terminal enlargements from which the sori are suspended (Fig. 12-24L). Lesser swellings may be seen occasionally in several species, but those of *D. bifurcatum* may be compared only with the inverted cone-like sorophore termini of *D. mexicanum*. Still, the two structures are not strictly comparable, for those of *D. bifurcatum* seemingly consist of moribund myxamoebae and cellular debris, while in *D. mexicanum* they consist of small but fully differentiated stalk-like cells. The two species may or may not be closely related.

Sorophore bases are somewhat unusual in that they often seem not to have a fixed point of initiation. This is suggested by the common differentiation of rows of vacuolated stalk-like cells on the agar surface well in advance of an obvious cellulose tube to mold and delimit the emerging stalk. Present also are slightly enlarged bases of more conventional pattern, as in *D. mucoroides*.

Cell aggregations are generally small and seldom exceed 2.0 mm in diameter, the larger of which regularly break up into segments that fruit individually. Such behavior is not unique for *D. bifurcatum*, but the phenomenon is illustrated particularly well by this species. Smaller aggregates may give rise to sorogens without prior disruption.

Microcysts have not been observed. On richer media, however, cells that do not aggregate often burrow into the agar surface, where they become very irregularly lobed and in this state reflect light in a manner that suggests massed microcysts when viewed at low magnifications. Such distorted cells are usually most abundant along the margins of the streaks.

Myxamoebal growth may be enhanced by using substrates buffered to pH 6.0 with phosphates, and cell aggregations tend to be somewhat larger if less numerous. The number of sorocarps arising from single aggregates is also increased, but they are commonly not so well formed as on unbuffered agars containing similar amounts of lactose and peptone. Whereas Cavender (1976a) noted that *D. bifurcatum* "develops fairly well on 0.1% lactose-peptone and 0.1% lactose-0.05% yeast extract media," he apparently preferred weak hay infusion agar or nonnutrient agar to which pregrown cells of *E. coli* were added along with spores of the slime mold. In our experience concurrent cultivation of the dictyostelid and bacteria on glucose-peptone media (0.1 or 0.05% each) has proved quite satisfactory. In fact, some of the best-formed, if not the most numerous, sorocarps encountered in this study were produced on glucose-peptone agar of the lower nutrient concentration.

Dictyostelium monochasioides Hagiwara, in Bull. Natl. Sci. Museum *16*: 494-496, fig. 7A-F and pl. 1A-E (1973); also in Mem. Natl. Sci. Museum Tokyo (7): 82-83 (1974).

The description of the species that follows is abstracted in substantial part from Hagiwara's original publication.

Cultivated upon pregrown cells of *Escherichia coli* deposited on nonnutrient agar and incubated at room temperature. Sorocarps white, solitary or gregarious, erect or inclined, usually branched and often rebranched in a singular pattern where secondary and tertiary branches arise from preexisting branches to give a clustered appearance (Figs. 12-25A,D and Figs. 12-26C,D). Sorophores generally delicate, consisting of single tiers of cells except in basal areas (Fig. 12-25E), sinuous and becoming intermingled in dense cultures, mostly 0.5-3.0 mm in length and ranging from 7.5-25 μm in diameter at the bases, narrowing to 1.5-3.5 μm in the sori (Fig. 12-25C). Sorogens long and thin, vermiform, often 200-300 μm × 50 μm or less. Sori globose, hyaline to white, variable in size, mostly 50-150 μm in diameter, sometimes less or rarely more. Spores hyaline, variable in form and dimensions, ranging from elliptical, 4.4-6.4 × 2.8-3.8 μm (Fig. 12-26A), to elongate ellipsoid or recurved, 7.6-9.6 × 3.4-4.6 μm, the latter presumed to be of increased ploidy (*fide* Hagiwara), usually showing polar granules (Fig. 12-25B). Cell aggregations may or may not show definite inflowing streams, the former about 0.5-2.5 mm in diameter (Fig. 12-25F) but soon fragmenting with segments fruiting separately. Myxamoebae about 12-15 × 8-12 μm, showing vesicular nuclei with peripheral nucleoli and one or two contractile vacuoles (Fig. 12-26B). Neither macrocysts nor microcysts reported or observed, albeit myxamoebae in great numbers may burrow into the agar, become nodulated, and superficially resemble massed microcysts when viewed at low magnifications.

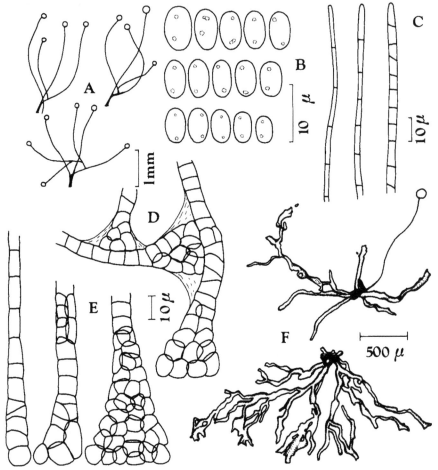

FIG. 12-25. *Dictyostelium monochasioides* Hagiwara. *A.* Three sorocarps showing singular pattern of branching. *B.* Spores with prominent polar granules. *C.* Apical portions of sorophores showing characteristic delicacy and cellular composition. *D.* Detail of branching habit. *E.* Sorophore bases. *F.* Cell aggregations, from the upper of which a sorocarp has already developed and a second sorogen is now forming. (From Hagiwara, fig. 7, 1973b)

HABITAT. Tropical soil and leaf mould, New Guinea and Papua.

TYPE. Strain TNS 364, isolated by Hagiwara from soil collected in Ambunti, New Guinea, in December 1971. The type was not seen by us, our observations having been made upon Hagiwara's strain 653, which was sent to us in October 1974 as *D. monochasioides*.

Superficially, *D. monochasioides* bears a striking resemblance to *D. tenue* Cavender, Raper, and Norberg (1979) but differs from the latter in

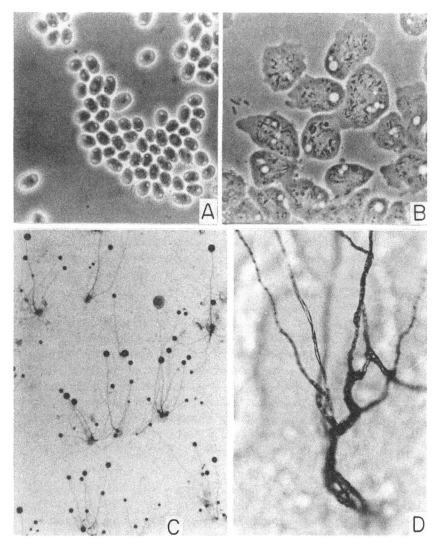

FIG. 12-26. *Dictyostelium monochasioides* Hagiwara. *A*. Spores showing characteristic form and presence of polar granules. × 900. *B*. Vegetative myxamoebae at the edge of a bacterial colony. × 900. *C*. Habit of species showing clustered sorocarps. × 16. *D*. Detail of branching habit that characterizes the species. × 200. (*A,B*, photomicrographs of strain 653, which was sent to us by Dr. Hagiwara as representative of the species; *C,D*, photographs used by Hagiwara in describing *D. monochasioides*, 1973b)

its irregular pattern of rebranching from preexisting branches to create, in some cases, an almost cyme-like effect (Fig. 12-25A). Patterns of aggregation appear much the same, and in each species, particularly on 0.1L-P agar, there is a marked tendency for larger but essentially centerless secondary aggregates to develop with richly and closely anastomosing streams that may or may not later develop additional sorogens from fractional parts. It is of interest, and possibly significant, that the type in each case stems from Southeast Asia.

Of several combinations of host bacteria and substrates investigated in our laboratory, optimum growth and fructification of strain 653 was obtained upon dilute hay infusion agar, perhaps reflecting some approximation of its natural habitat. Growth was somewhat greater on 0.1L-P (pH 6.0) than on unbuffered 0.1L-P agar, but fruiting was generally less satisfactory. On 0.1L-P/2 agar the formation of secondary, closely anastomosing aggregations was increased with a consequent reduction and/or delay in sorocarp development. As in many of the other delicate species, fruiting was substantially improved by the addition of activated charcoal at the onset of cell aggregation.

Dictyostelium monochasioides as reported in 1973 was based upon a strain (type) isolated from soil collected in New Guinea. In the following year, three additional strains showing monochasium-like branching were obtained from soil and leaf mould of evergreen forests of the Yeayama Islands, Okinawa, including strain 653 studied by us (see above). These strains, according to Hagiwara (1974), showed only spores of the smaller size-range, thus suggesting that the larger spores cited in the species description may have emerged during laboratory cultivation of the type. Possibly so, but spores of strain 653 photographed in our laboratory also showed marked differences in size, albeit all were small (Fig. 12-26A).

Dictyostelium polycephalum Raper, in Mycologia *48*: 192-196, figs. 3 & 5 (1956); also Raper, in J. Gen. Microbiol. *14*: 716-732, figs. 1-20 (1956). See also Whittingham and Raper, in Amer. J. Bot. *44*: figs. 1-3 (1957).

Cultivated on dilute hay infusion or 0.5% glucose, 0.1% peptone agar in association with a mixed flora consisting of *Klebsiella pneumoniae* and *Dematium nigrum* at 25-30°C. Sorocarps arise from migrating pseudoplasmodia, very small, often solitary, but typically clustered to form coremiform fructifications (Fig. 12-27B), variable in number, rarely up to nine or ten per cluster, commonly 350-650 μm in height (Figs. 12-27A,B). Sorophores erect, tapered, appressed throughout most of their length (Fig. 12-28D,E), diverging terminally to bear whitish or hyaline globose sori 40-90 μm in diameter (Fig. 12-28E). Spores elliptical to reniform (Fig. 12-27C), usually 6.0-7.5 × 3.0-3.5 μm granules often evident but not consistently polar. Cell aggregations typically radial in pattern (Fig. 12-27E), often 5-10 mm in diameter, producing few to many long, thin, and

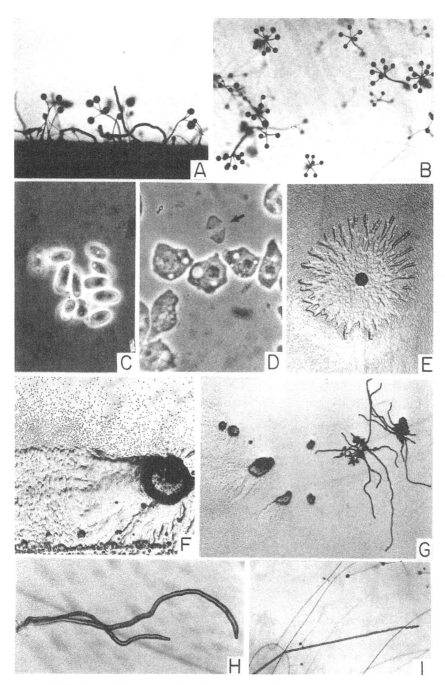

Fig. 12-27. *Dictyostelium polycephalum* Raper. *A.* Coremiform sorocarps and newly formed pseudoplasmodia on an agar block viewed from the side. Culture growing in association with *Klebsiella pneumoniae* and *Dematium nigrum* on 0.5% glucose, 0.1% peptone agar. × 25. *B.* Sorocarps formed in association with *K. pneumoniae* on agar containing galactose (0.5%) and asparagine (0.2%) in a clay-covered Petri dish (see text). × 25. *C.* Spores, showing characteristic shape and commonly a single polar granule. × 900. *D.* Myxamoebae beneath coverglass; note empty spore case (arrow) and how it was broken to release the protoplast. × 900. *E.* A developing cell aggregation showing typical radial pattern. × 9. *F.* Aggregation at edge of a bacterial streak; note ridges that reflect waves of converging myxamoebae. × 16.5. *G.* Aggregations in process of formation (left) and delicate migrating pseudoplasmodia emerging from two centers of aggregation (right). × 8. *H.* Two typical migrating pseudoplasmodia. × 16.5. *I.* Migrating pseudoplasmodium moving through a colony of *Absidia*; note its minimal support. × 8. (Rearranged in part from Raper, 1956)

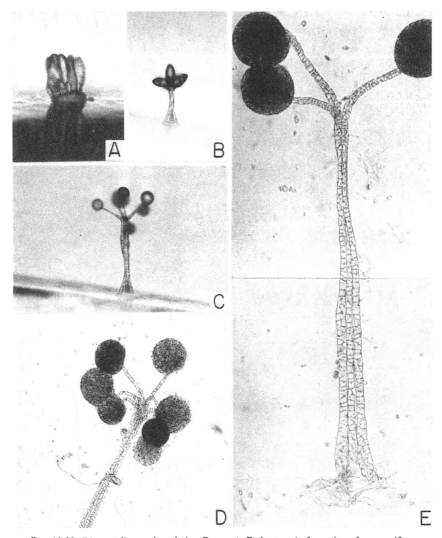

FIG. 12-28. *Dictyostelium polycephalum* Raper. *A*. Early stage in formation of a coremiform fructification. The pseudoplasmodium ceased migration, rounded up, and became papillate; a sorogen then emerged beneath each papilla and initiated sorocarp formation (side view). × 50. *B*. A more advanced stage of a different cluster. × 40. *C*. This coremiform fructification developed from the sorogens of *A* as seen 1 day later. × 50. *D*. Terminal portion of coremiform cluster showing seven sori and supporting sorophores (sori fixed and stained). × 150. *E*. Coremiform fructification showing cellular detail in its three adherent sorophores (sori fixed and stained). An apron of slime anchors the structure and holds it in an upright position. × 215. (Rearranged from Raper, 1956)

fragile migrating pseudoplasmodia (Fig. 12-27G-I) up to 1.0 cm in length, not phototropic. Myxamoebae comparatively small (Fig. 12-27D), usually 8-12 × 5.0-7.5 μm, sometimes larger or smaller. Unaggregated cells form microcysts, globose or nearly so, 4.0-6.5 μm in diameter; macrocysts reported but not observed.

Alternatively, *Dictyostelium polycephalum* may be cultivated successfully in two-membered culture with *Klebsiella pneumoniae* (or *Escherichia coli*) as the bacterial associate on a 0.5% galactose, 0.2% asparagine agar (Fig. 12-27B) if porous clay covers are substituted for glass Petri dish lids at the onset of cell aggregation. Coremiform clusters of up to 15 sorocarps have been obtained from single pseudoplasmodia. Evidence points to proper control of atmospheric moisture as especially significant for optimal fructification (Whittingham and Raper, 1956).

HABITAT. Widely distributed in soils of deciduous forests from eastern Canada to subtropical Central America; found also in soils from East Africa and Southeast Asia, possibly global in distribution.

HOLOTYPE. Strain S-4, isolated by KBR in 1951 from well-rotted leaf mould of a mixed hardwood forest near Holly Hill, South Carolina.

Represented also by strain WS-93 from deciduous forest soil, University Arboretum, Madison, Wisconsin, and by isolates from Michigan, Illinois, Texas, and several stations overseas.

Dictyostelium polycephalum is a unique cellular slime mold with several distinguishing characteristics. Particularly striking are the coremiform fructifications formed through the coordinated differentiation of sister sorogens, each forming its sorophore separately but in synchrony with the others so that they rise together and stand as a single column composed of few to many contiguous stalks—contiguous that is, until sporogenesis approaches. At this time the sorogens bend apart and build outward at a substantial angle, which serves to remove each sorus as far as possible from its sibling neighbors. For the present we can only guess what manner of controls are operative here.

Coremiform fructifications are also found in *D. polycarpum*, but the intersorogenic association is far less intimate and the contiguous condition persists at best for a very limited time before the sorogens build apart (Fig. 12-29F). Furthermore, the sorogens arise directly from the aggregation rather than from a migrating pseudoplasmodium, which in turn represents only a fractional part of the initial aggregation in *D. polycephalum*.

The second and equally singular feature of *D. polycephalum* is the seemingly primitive and yet very remarkable migrating pseudoplasmodia. Unlike *D. discoideum*, where an aggregation typically produces a single migrating body, in *D. polycephalum* few to many migrating pseudoplasmodia normally arise from one aggregation, the number being roughly dependent upon its size. Typically these are long and thin (Figs. 12-27G-I), quite fragile, and inconstant in shape. They are often much contorted and show little evidence of apical dominance. Still, they may move across the agar surface for distances of 2 or 3 cm, seemingly unaffected by light and other identifiable stimuli, before rounding up to form multiple sorocarp initials. The migrating body is enclosed by a thin slime sheath, the continuity of which may be easily severed. When this occurs the body separates into two or more fractions, each of which retains the capacity to produce one or more sorocarps in proportion to its mass.

Comparatively little attention has been given to the movement of these migrating pseudoplasmodia, but we can nonetheless marvel at their performance. Not uncommonly they move over an agar surface with only a small fraction of their body in contact with the substrate. Or they may move through a loose fungus colony touching only a few hyphae (Fig. 12-27I). With only a small posterior section touching the agar, and standing essentially erect in a Petri dish, they can reach the cover a centimeter away and subsequently migrate across the underside. All the while, by differential cell movements within the body, or in response to minimal outside pressures, the migrating structure may quickly break apart. Structurally, it appears to be far less highly organized than its counterpart in *D. discoideum*.

The spores of *D. polycephalum* merit special attention also. They germinate not by longitudinal splitting from one end, as in other species with elliptical spores, but by rupture at midpoint so that the two halves often separate to release the protoplast (Fig. 12-27D). A suggestion of such germination may be observed among spores of *D. mexicanum* (Fig. 12-37J), but not in the striking manner displayed here.

Cultivation of *D. polycephalum* requires some special attention and is deserving of much more. For reasons still obscure, it has been found convenient to grow and to describe the slime mold in association with a bacterium (e.g. *K. pneumoniae*) as a nutrient source and with a black yeast-like fungus (*Dematium nigrum*) as a second associate. Perhaps the fungus reduces atmospheric moisture, or adsorbs some gaseous compound that inhibits fruiting, or serves in some other capacity still unsuspected. Fortunately, the species can be grown satisfactorily in two-membered culture with *Escherichia coli* by using a substrate based upon galactose and asparagine, and by covering the fruiting cultures with porous clay covers (Coors) rather than glass Petri dish lids (Whittingham and Raper, 1957). Previously unreported, but also satisfactory for some strains, is the use of more conventional substrates plus the addition of a generous amount of Norite in the form of a ridge between two parallel streaks previously inoculated with a mixed suspension of nutritive bacteria and slime mold spores. Chips of activated charcoal may be used also, in which case the migrating pseudoplasmodia are attracted to the chips and at times build sorocarps on their surfaces. Sorocarps are generally more numerous per cluster and of smaller dimensions when *D. polycephalum* is cultivated in association with *Klebsiella* and *Dematium* on glucose-peptone agar than when hay infusion agar is employed.

Whereas cell growth occurs over a pH range of about 5.0-8.0, the optimum for sorocarp formation is about pH 6.5, or close to that of most dictyostelids. The optimum temperature for fruiting, 28-30°C, however, is well above that of most species and a full 10°C higher than for *D. polycarpum* and *D. septentrionalis*.

Dictyostelium polycarpum Traub, Hohl, and Cavender, in Amer. J. Bot. 68: 162-171, figs. 1-8 (1981).

Cultivated on low-nutrient substrates such as 0.1L-P (pH 5.5-6.0) and thin hay agars in association with *Escherichia coli*, strain B/r, at 18-20°C. Sorocarps typically clustered (Fig. 12-29F), commonly 8-12 per cluster, often more or less, clusters adherent near the base, mostly 2-3 mm high, often less or up to 5-6 mm. Sorophores thin and delicate, unbranched, usually consisting of single tiers of cells except for some bases, mostly 10-29 μm wide below to 1.5-2.5 μm at the sori, colorless or faintly yellow, erect and adherent to about ⅓ ultimate length, then diverging so that sori stand apart (Fig. 12-29F). Sori terminal, globose, milk-white to light gray or hyaline, mostly 80-120 μm but ranging from 40 to 160 μm in diameter. Spores narrowly elliptical (Fig. 12-29A), reniform or occasionally weakly fusiform, 8-10 × 2.5-3.0 μm, occasionally more, polar granules present but loosely arranged. Cell aggregations radial in pattern (Fig. 12-29C), variable in dimensions from 1.0 to 4 or 5 mm depending upon the strain and culture conditions; sorogens multiple, closely crowded (Figs. 12-29D,E), rising before aggregation is complete. Feeding myxamoebae commonly 15-18 × 9-12 μm (Fig. 12-29B), often with two or more excretory vacuoles and nuclei with two or three peripheral nucleolar lobes. Neither macrocysts nor microcysts reported or observed.

HABITAT. Decaying conifer needles and duff of subalpine larch (*Larix decidua*), mugo pine (*Pinus mugo*), and spruce (*Picea abies*) at 1600-1800 meters in cantons of Graubünden, Ticino, and Appenzeil, Switzerland.

HOLOTYPES. Strain GR-4, isolated in 1970 by Franz Traub from soil of a mixed conifer forest, Engadine, Graubünden; and strain VE-1b, isolated by J. C. Cavender in 1979 from soil of a larch forest, Valle, Ticino. Of

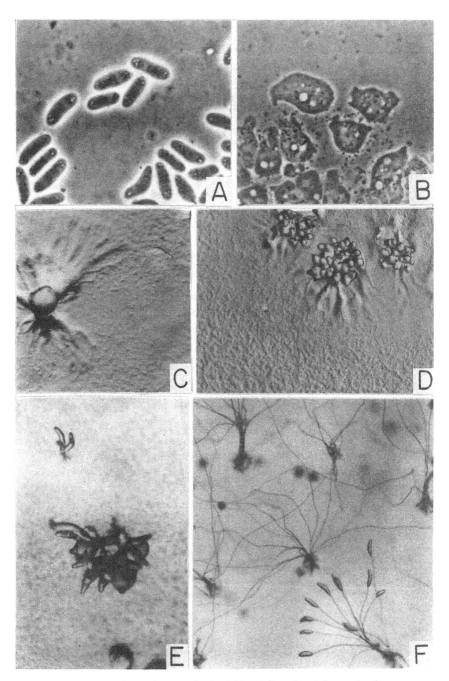

Fig. 12-29. *Dictyostelium polycarpum* Traub, Hohl, and Cavender. *A*. Spores showing narrow elliptical form and unconsolidated polar granules. × 1400. *B*. Vegetative myxamoebae on agar. × 940. *C*. Aggregations with diffuse inflowing streams. × 15. *D*. Late aggregates with many papillae. × 10. *E*. Sorogens beginning to form. × 20. *F*. Field of clustered sorocarps; note diverging sorogens at lower right. × 18. (Picture assembly from Traub, Hohl, and Cavender, 1981a)

the two cotypes, GR-4 is more delicate in form and more fastidious in culture requirements.

Dictyostelium polycarpum is a very special species. The clustered habit of its sorocarps is reminiscent of the coremiform structures produced in *D. polycephalum* that differs in many particulars. Additionally, there is no suggestion of a migrating pseudoplasmodium as found in that species. Sorocarps arise only from the sites of cell aggregation. The clustered habit is also reminiscent of certain strains of *D. rhizopodium* (e.g., strain PR-10D), but there only the crampon-like bases are adherent and the sorocarps that arise from them are separate and heavy compared with those of *D. polycarpum*.

Ecologically, *D. polycarpum* also stands apart, having been isolated only from surface soil and litter of subalpine conifer forests in Switzerland; the surface soil and its cover are quite acid and continually cool. Only two other dictyostelids are associated specifically with somewhat comparable habitats: *D. aureo-stipes* v. *helvetium*, also from subalpine forests of Switzerland (Traub et al., 1981b); and *D. septentrionalis*, from spruce-hemlock forests in Alaska (Cavender, 1978).

In the laboratory *D. polycarpum* grows demonstrably better on acidified 0.1L-P agar (pH 5.5-6.0) than upon the unbuffered medium, and at a temperature of 18-20°C rather than the more usual 22-25°C. Based upon detailed studies, Traub et al. (1981a) found growth to be fairly uniform from 14 to 23°C and observed that fruiting was "best at about 20°C." Both cell growth and the ability to produce sorocarps fall off sharply above 24°C, while virtually no growth occurs at 26°C. Initial growth from spores is comparatively slow, which may result in part from delayed germination and in part from difficulties in securing uniform spore suspensions.

The sori of *D. polycarpum* and the spores they contain are noteworthy also. When sori are removed from sorocarps by touching them with a small loop filled with water, the sori do not disperse but remain intact, floating on and in the droplet. In fact, it is difficult to obtain a uniform spore suspension, even after vigorous agitation with a Vortex stirrer. It would appear that the sori contain some type of immiscible slime that holds the spore together. This possibility is further indicated by the appearance of such sori when mounted in water and subjected to modest pressure, as by tapping the coverglass, before being examined with the compound microscope. Seen from above, such a sorus appears as a delicately fashioned rosette in which the spores are aligned with their long axes toward the center of the sorus. Such orientation may account in part for their elongate and sometimes mildly fusiform shape. The spores are of interest for other reasons as well. They are comparatively large, relatively narrow, and contain polar granules that are not concentrated in a single cluster, or even a very few conspicuous clusters; instead, many small, separate granules are scattered through the cytoplasm at either end of the spore. Using phase microscopy a subcentral clear area believed to be the nucleus is often discernible.

The addition of activated charcoal to postvegetative cultures accelerates aggregation substantially. Centers arise adjacent to the small charcoal chips and from these extend radiating streams of myxamoebae up to a few millimeters in length. Sorocarp formation is also enhanced, and, as in many other species, sorogens that arise near the chips usually build toward them until they collide. The substitution of unglazed porcelain covers (Coors) for glass culture dish lids at the time of cell aggregation improves sorocarp formation dramatically. In fact, we have seen no other case where the beneficial effect of such a shift exceeds that in *D. polycarpum*. Sorocarps are more numerous and more regular in form, and a far greater proportion of the sorogens complete the fruiting process than usually occurs in comparable cultures under glass. We can at present only speculate concerning the cause. The unglazed clay covers may permit the escape of some self-generated gas that is inimical to fruiting; or they may serve to reduce the moisture content of the air just enough to permit optimal fruiting. We do not know—but the manner in which the clustered sorogens diverge during fruiting, thus separating the sori, seems to suggest the former.

Dictyostelium aureo-stipes Cavender, Raper, and Norberg, in Amer. J. Bot. *66*: 207-217, figs. 1-8 (1979).

Cultivated on low nutrient substrates such as 0.1G-P/2, 0.1% lactose-0.05% peptone, or 0.1% glucose-0.025% yeast extract agars in association with *Escherichia coli* or *E. coli* B/r at 25-28°C. Sorocarps solitary or clustered, erect or semi-erect (Fig. 12-30A), commonly 3-7 mm in height in diffuse light or darkness, strongly phototropic and up to 1.0 cm or more in unidirectional light, branched, usually conspicuously so (Fig. 12-30B). Sorophores pale to golden yellow with pigment bleaching in bright light, tapering from 20-35 μm at the base to 6-8 μm at the top, commonly bearing 15-25 lateral branches irregularly spaced and of different dimensions. Sori white, globose, terminal sori 100-200 μm in diameter, lateral sori 30-80 μm, borne on branches mostly 100-375 μm in length. Spores elliptical to slightly reniform, mostly 4.7-7.0 × 2.5-3.5 μm, with prominent polar granules (Fig. 12-30H). Cell aggregations assume different patterns, including (1) small mounds without inflowing streams, (2) radiate patterns of small to medium proportions (Fig. 12-30C), and (3) large formations with inflowing streams up to 5-10 mm or more in length, the latter subdividing (Fig. 12-30D) to yield many sorogens. Myxamoebae commonly 8-18 × 6-10 μm as seen on an agar surface (Fig. 12-30G). Microcysts produced by some strains and on some substrates; macrocysts not reported or observed.

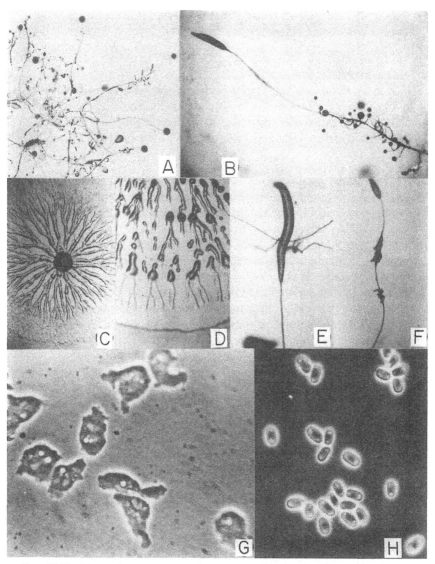

FIG. 12-30. *Dictyostelium aureo-stipes* Cavender, Raper, and Norberg. *A*. Sorocarps in different stages of development. × 5.25. *B*. Partially completed sorocarp with multiple branches near the base, each bearing a sorus. Note the sorogen within which the sorophore is being extended. × 15. *C*. Intact wheel-like aggregation. × 8.25. *D*. Aggregation in which streams have fragmented into multiple centers prior to fruiting. × 16.5. *E*. Developing sorogen before any cell masses have been abstricted. × 16.5. *F*. Sorocarp in which two irregular cell masses have been abstricted and from which branches are beginning to form. × 16.5. Postfeeding myxamoebae on an agar surface. × 765. *H*. Spores showing characteristic shape and conspicuous polar granules. × 845. (From Cavender, Raper, and Norberg, 1979)

HABITAT. Soil and leaf mould of deciduous forests in the United States, Mexico, and Central America.

HOLOTYPE. Strain I-3, isolated by KBR in 1942 from soil of a deciduous forest near Peoria, Illinois.

Of many strains studied, the following may be listed as representative: Ill-15b, from soil, Brownfield Woods, Urbana, Illinois; DC-35, from soil, deciduous forest, Squaw Creek, Missouri, isolated by David Castener; JA-1c, from soil, deciduous forest, Door County, Wisconsin, isolated by J. C. Cavender.

Dictyostelium aureo-stipes is widely distributed in soil and leaf litter of deciduous forests in temperate North America. It has been isolated frequently over a period of many years and was, in fact, referred to within our laboratory, and occasionally in lectures outside, as "the yellow-stalked *Dictyostelium*" long before it was formally described. Three characters in particular serve to distinguish the species: the pale to golden yellow pigmentation of the sorophores, the large number of irregularly spaced branches, and the milk-white sori that stand in sharp contrast to the pigmented stalks.

Sorocarps of *D. aureo-stipes* may be relatively long and complex and achieve a level of development suggesting *Polysphondylium*; however, they never possess the symmetry and regularity of form achieved by that genus. In fact, the branching seems to be haphazard, varying from one sorocarp to the next. Branches may be concentrated near the base, or in the sub-terminal region, or they may be scattered along a major part of the sorophore. Branches vary greatly in size, as do the sori they bear, but are generally longer when borne near the base. Unfortunately, the aforementioned characteristics are not always fully expressed, and isolates are occasionally found that show only limited branching. These can be identified by applying the combined criteria of yellow sorophores, white sori, patternless branching, and spores with polar granules. In a similar way, the presence of yellow sorophores alone is not sufficient warrant for species diagnosis, since this character is shared by *D. aureum* and *D. mexicanum* (Cavender, Worley, and Raper, 1981), species in which both stalks and sori are pigmented yellow. Finally, although the species name intentionally emphasizes sorophore pigmentation, we must reiterate that it varies in different isolates, is dependent upon correct incubation, and fades all too rapidly when sorocarps are exposed to bright light. The report by Huffman and Olive (1963) entitled "A Significant Morphogenetic Variant of *Dictyostelium mucoroides*" represents a case in point. The culure studied by them (CR17) and later forwarded to us is wholly representative of *D. aureo-stipes*; their text and illustrations could hardly be more accurate. Linkage with *D. mucoroides* was, we assume, based upon its milk-white sori, whereas yellow-pigmented stalks, if present, could have been overlooked or regarded as unimportant.

There is reason to believe that Hagiwara's *Dictyostelium delicatum* (1971) is also closely related to *D. aureo-stipes*, although smaller in all its proportions except for the spores. He records sorophores as "often strikingly branched" and "white to yellowish white" with sori in the same shades. Spores are ellipsoid and overlap those of *D. aureo-stipes* in both major and minor dimensions. Polar granules are not mentioned in the text but are indicated in his illustrations of spores.

Whereas some branches near the base of a sorophore may be produced by delayed streaming myxamoebae, the great majority are formed from cell masses of varying size that are abstricted from the rising sorogen and left behind on the lengthening stalk (Fig. 12-30E,F). Fractional masses of small size may form single branches. More commonly, however, the abstricted cell masses are larger, usually somewhat elongate, and may bear numerous papillae from each of which is constructed a lateral branch terminated by a sorus of proportional size. Unlike the situation in the genus *Polysphondylium*, there is no discernible pattern in the abstriction of the cell masses or in the manner in which they subdivide to form the numerous side branches.

Sorocarp formation is enhanced substantially if chips of activated charcoal are added to the culture at the time of cell aggregation. This is especially true of strains that have been cultivated in the laboratory for long periods of time.

Dictyostelium aureo-stipes var. *helvetium* Cavender, Raper, and Norberg, in Amer. J. Bot. *66*: 207-217, fig. 9-17 (1979).

Cultivated on low-nutrient substrates such as 0.1L-P/2, or 0.1G-P/2, or dilute hay infusion agar in association with *Escherichia coli* or *E. coli* B/r. Sorocarps usually solitary, erect, typically branched (Fig. 12-21A), phototropic. Sorophores golden yellow with pigmentation enhanced by dark incubation, commonly 5-10 mm in length and bearing 20-30 lateral branches (Figs. 12-31B,C), occasionally reaching 2 cm in unidirectional light, sturdy, tapering from 50-60 μm in diameter near the base to 10-20 at the apex. Sori globose, white; terminal sori 175-250 μm in diameter, lateral sori 40-80 μm and borne on branches 100-400 μm long, usually decreasing toward the apex. Spores elliptical to slightly reniform, mostly 5.5-7.5 × 2.5-3.5 μm with prominent polar granules (Fig. 12-31I). Cell aggregations small, radiate (Fig. 12-31D), evenly spaced, and generally producing single sorogens that construct upright sorocarps (Figs. 12-31E,F). Myxamoebae as in the species and with organelles easily discernible (Fig. 12-31H). Microcysts and macrocysts not reported or observed.

HABITAT. Soil of coniferous forests in Switzerland and a similar site in Alaska.

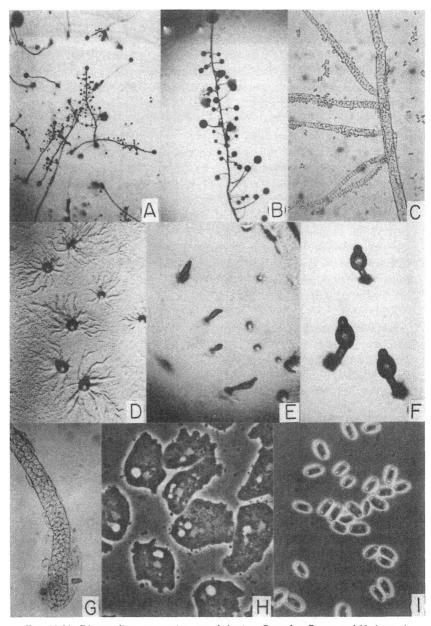

FIG. 12-31. *Dictyostelium aureo-stipes* var. *helvetium* Cavender, Raper, and Norberg. *A*. Erect sorocarps showing characteristic branching habit. × 6.75. *B*. Single sorocarp with large terminal and more than 30 irregularly spaced branches, each bearing a sorus, some of which have coalesced. × 12. *C*. Detail of a portion of sorophore showing the manner of branch attachment. × 115. *D*. Evenly spaced aggregations on 0.1L-P agar. × 13. *E*. Single sorogens arising from aggregations. × 15. *F*. Three sorocarps in early stage of development; note apical tips wherein sorophores are being extended. × 25. *G*. Base of sorophore showing simple pattern and structure. × 112. *H*. Myxamoebae on surface of 0.1L-P/2 agar. × 765. *I*. Spores showing characteristic pattern and conspicuous polar granules. × 560. (From Cavender, Raper, and Norberg, 1979)

HOLOTYPE. Strain GE-1, isolated by Franz Traub from soil collected in coniferous forest, Etang de la Gruyere, Switzerland. Representative too is strain VS-1, also isolated from soil of a coniferous forest in Switzerland.

Structurally, the variety *helvetium* differs from the species in its generally larger, robust sorocarps that commonly stand erect or nearly so, and in the more uniform disposition of its lateral branches (Figs. 12-31B,C). The yellow pigmentation of the sorophore is equally as intense as that of the species, if not more so, and is subject to the same environmental constraints relative to its formation and extinction.

Normal growth and development is limited to a narrow temperature range between 20°C and 25°C. Branching is sparse and sorocarps are poorly formed at 20°C, while at 25°C growth is slowed and the sorocarps are stunted. *D.a.* var. *helvetium* thus tends to be a low-temperature slime mold in the laboratory, as it seems to be in nature, for it has been isolated only from soils collected in cool coniferous forests of Switzerland by Franz Traub (1977) and from a similar source in Alaska by J. C. Cavender (1978).

Unlike the species where the pattern of cell aggregation varies widely, and is dependent on strain to a considerable degree, the process in var. *helvetium* is essentially similar in each of several isolates examined. Aggregations are relatively small and wheel-like, and generally produce single sorocarps (Figs. 12-31D-F).

As in the species, fructification is facilitated by the addition of activated charcoal at the onset of cell aggregation.

Dictyostelium coeruleo-stipes Raper and Fennell, in Amer. J. Bot. *54*: 515-528, figs. 27-37 (1967).

Cultivated on low-nutrient agars such as 0.1L-P, 0.1L-P/2, or 0.05% lactose 0.025% yeast extract in association with *Escherichia coli* at 23-26°C. Sorocarps erect or semi-erect in diffuse light (Fig. 12-32A,F), usually solitary but occasionally clustered, commonly 1-3 mm in height, rarely branched, phototropic and often longer in unidirectional light. Sorophores arising from expanded, usually flattened crampon bases (Fig. 12-32G), often complexly digitate, sometimes more conical with digitation limited or lacking (Fig. 12-32H), strongly tapered, 25-35 μm (rarely 50 μm) near the crampon bases to 4-7 μm at the sori, pale blue to bluish purple in color, pigmentation more pronounced in light-grown cultures. Sori globose to subglobose, white to cream-colored in the light, pale yellow in darkness, usually 100-200 μm in diameter, occasionally larger and often smaller. Spores elliptical to capsule-shaped (Fig. 12-32B), some slightly reniform, mostly 5.5-7.5 × 2.5-3.5 μm, occasionally larger, with prominent granulation, primarily polar. Cell aggregations wheel-like (Fig. 12-32D,E), yielding single sorocarps, sometimes larger with streams fragmenting to

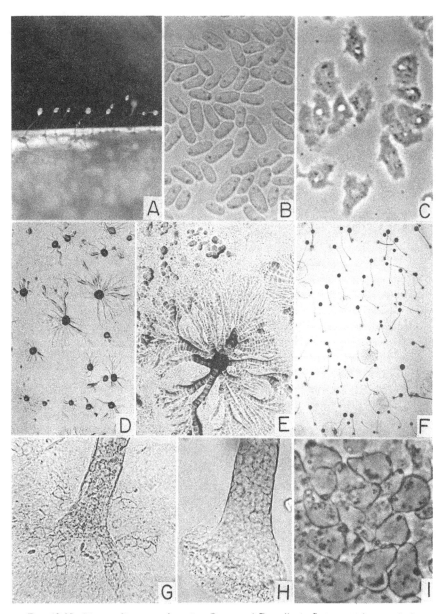

Fig. 12-32. *Dictyostelium coeruleo-stipes* Raper and Fennell. *A*. Sorocarps photographed with reflected light. × 9. *B*. Spores showing prominent granulation, mostly bipolar. × 1150. *C*. Preaggregative myxamoebae. × 525. *D*. Typical aggregations, the smaller of which will form single sorocarps. × 8. *E*. A much larger aggregation with concentric ridges, indicating probable outward relay of an unidentified chemotactic attractant. × 8. *F*. Sorocarps in a spread culture photographed from above, transmitted light. × 6. *G*. Typical crampon-type sorophore base photographed in situ. × 300. *H*. Pyramidal base of sorophore with minimal digitate extensions. × 300. *I*. Strongly vacuolate cells on agar surface representing loosely aggregated myxamoebae that have differentiated in a manner suggesting stalk or crampon cells. × 1050. (Reassembled from Raper and Fennell, 1967)

produce more and smaller fructifications. Myxamoebae resemble those of the smaller size range in *D. rhizopodium* (Fig. 12-32C). Macrocysts and microcysts not reported or observed.

HABITAT. Soil and decomposing leaf litter of forests in Florida and Mexico.

HOLOTYPE. Strain MH-2 isolated by J. C. Cavender in 1964 from subtropical hammock soil, Mahogany Hammock, Everglades National Park, Florida.

Other representative strains include RP-2a, from soil, Royal Palms Hammock, Everglades National Park, Florida; OK-3, from soil, hammock forest near Lake Okeechobee, Florida; and ST-10, from soil, seasonal evergreen forest near San Andres Tuxtla, Veracruz, Mexico. All were isolated by Cavender.

This species is noteworthy for its generally erect and relatively short sorocarps with stalks that are strongly tapered and pale blue to light bluish purple in color. In contrast, the sori are white or only lightly pigmented in cream to pale yellow or occasionally faint lavender shades. The crampons are generally well formed and often symmetrical. The sorogens are phototropic and in one-sided light often produce relatively long sorophores. Pseudoplasmodia may show a slight yellow pigmentation in dark-grown cultures.

Dictyostelium coeruleo-stipes is apparently less widely distributed than *D. rhizopodium* or the two smaller crampon-based species, *D. lavandulum* and *D. vinaceo-fuscum*. This may reflect in part the limited number of surveys reported to date. However, the strains now at hand grow vegetatively at 10°C and fruit normally between 15 and 30°C, the time required being greater at lower temperatures. Additionally, most of the strains can be grown successfully on substrates of somewhat greater nutrient content than those listed in the species description; and *Klebsiella pneumoniae* or *E. coli* B/r may be employed as a bacterial associate if preferred. Stated differently, the species does not appear to be particularly fastidious.

Neither macrocysts nor microcysts have been reported; but in some cultures, and for reasons unknown, clusters of imperfectly aggregated cells may become strongly vacuolate to resemble poorly differentiated crampon or stalk cells (Fig. 12-32I).

Although the identity of the acrasin for this species remains unknown, as it does for the other crampon-based species, the occasional presence of concentric ridges in the streams of aggregating myxamoebae (Fig. 12-32E) suggest that chemotactic stimuli are relayed outward from the aggregation centers in a manner analogous to that in *D. discoideum*.

Dictyostelium laterosorum Cavender, in J. Gen. Microbiol. *62*: 113-122, pls. 3 & 4 (1970).

Cultivated on 0.1L-P, 1L-P (pH 6), or hay infusion agars in association with *Escherichia coli* at 22-25°C. Sorocarps typically erect or inclined (Fig. 12-33F), sometimes prostrate, rarely branched, variable in size and proportions. Sorophores lightly pigmented in pale lavender and very light brown shades, usually 5-10 mm in length but ranging from 2 to 20 mm or more, commonly 15-35 μm in diameter near the base to 8-15 μm below the sorus; sorophore bases somewhat enlarged and rounded, flattened or weakly digitate. Sori of two types, terminal and lateral; terminal sori globose to subglobose, mostly 100-200 μm in diameter, bluish gray to lavender with brownish cast; lateral sori smaller, 50-150 μm, sessile, subglobose, numbering from one to ten when present and distributed fairly evenly on the sorophore (Fig. 12-33F), often lacking. Spores elongate, elliptical to reniform or recurved, mostly 8-10 × 3-4 μm but ranging from 7.0 to 12.5 × 2.5 to 4.5 μm, with conspicuous polar granules (Figs. 12-33A,C). Cell aggregations arise from clearly defined centers toward which streams of myxamoebae converge, or, alternatively, as cell clumps or mounds that may either fruit directly or merge to produce larger, radiate aggregations (Fig. 12-33D), the latter usually breaking up prior to culmination. Myxamoebae generally rounded or ovoid when feeding, ca. 9-12 μm in diameter, becoming irregularly elongate or somewhat triangular in outline when actively moving on an agar surface, ca. 15-20 × 10-15 μm in linear dimensions (Fig. 12-33B). Neither macrocysts nor microcysts observed or reported.

HABITAT. Leaf mould and surface soil of moist tropical forests, Trinidad and Tobago (West Indies) and Colombia, South America.

HOLOTYPE. Strain TBII-1, isolated by J. C. Cavender from soil collected in Tobago during the summer of 1968 and sent to us later that year. Strain AE-4, also from Cavender, was isolated in 1968 from soil of a seasonal evergreen forest in Trinidad. Additional strains have been isolated from forest soils of Colombia, South America (see Cavender, 1970).

The lateral sori arise by periodic abstriction of small masses of myxamoebae from the posterior end of the rising sorogen (Fig. 12-33E) as in *D. rosarium*, and one might surmise from this behavior that *D. laterosorum* is closely allied with that species. However, such a view is not supported by other characteristics. The sori and to a lesser degree the sorophores are pigmented, albeit lightly, rather than colorless or milk-white; furthermore, the spores are ellipsoid to reniform rather than globose and have conspicuous polar granules, suggesting that the acrasin is something other than cAMP (see Traub and Hohl, 1976). From the standpoint of pigmentation,

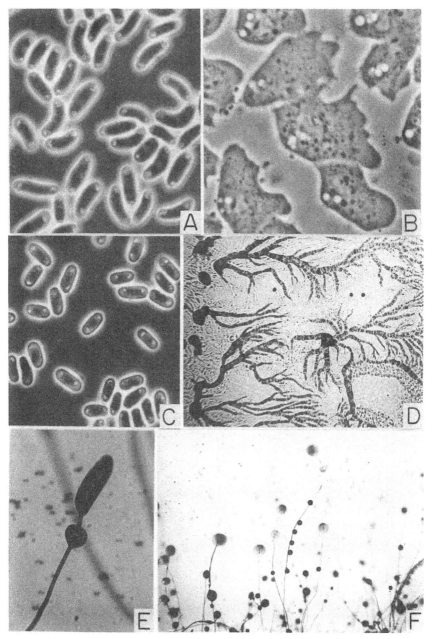

Fig. 12-33. *Dictyostelium laterosorum* Cavender. *A*. Spores of the type strain TBII-1. Note their inconsistent form and the presence of polar granules. × 960. *B*. Vegetative myxamoebae actively moving on an agar surface. × 1000. *C*. Spores of strain AE-4. Note their smaller size, more constant dimensions, and the presence of conspicuous polar granules. × 960. *D*. A later stage in aggregation. Such patterns may arise from primary centers or they may develop by coalescence of smaller preexisting aggregations as some centers achieve dominance. × 7.5. *E*. A developing sorogen that has abstricted a small mass of myxamoebae that will differentiate into a sessile sorus. × 40. *F*. Sorocarps of different dimensions showing disposition of terminal and lateral (sessile) sori. In most cases the terminal sori have touched the agar and collapsed. × 18. (*D,E,F*, photomicrographs from J. C. Cavender)

one is led to consider a possible relationship with *D. lavandulum* and other crampon-based species (Raper and Fennell, 1967). Such kinship seems logical from Cavender's original report (1970) that the sorophores of *D. laterosorum* had digitate bases, amounting in some cases to well-developed crampons. In the present study, and working with Cavender's strains, we have found no sorophore bases that might be described as constituting crampons. But strains do change during continued laboratory cultivation, and it is possible that the property has been lost or suppressed during more than a decade since the species was first described.

Of the two cultures examined in this study, strain AE-4 now conforms to the original species description somewhat better than strain TBII-1, although the latter is the type. For this reason we suggest that AE-4 be regarded as a cotype of *D. laterosorum*. Whereas developing sorocarps of both strains show positive phototropic responses, that of AE-4 is substantially stronger; as a consequence, the sorophores are usually longer, average somewhat thinner, and bear sori that are smaller than their counterparts in strain TBII-1. In supplemental notes accompanying the species description Cavender (1970) noted that the spores of *AE-4* were larger than those of TBII-1. During eight months of laboratory cultivation, however, spores of the latter came to equal in size and form those of AE-4, presumably through diploidization. In the present study we have found the opposite to be true; that is, the spores of AE-4 are smaller, as our photomicrographs demonstrate (Fig. 12-33A,C). The significance of this observation is not known, but it does emphasize that spore dimensions are subject to change and alone do not constitute a reliable basis for species recognition.

The number of sessile sori/sorophore in *D. laterosorum* is far less than in *D. rosarium*. Furthermore, Cavender (1970) reported that such sori were occasionally borne on short stalks, a pattern never encountered in *D. rosarium*. Could it be that in the abstriction of small masses of rising myxamoebae *D. laterosorum* constitutes a sort of bridge between genera *Dictyostelium* and *Polysphondylium*?

Among several bacteria investigated as possible nutrient sources, Cavender (1970) found *Escherichia coli* to be best, an opinion in which we concur. Of the difference substrates employed, 0.1L-P agar buffered to pH 6.0 with phosphates supports somewhat greater and more rapid growth than unbuffered 1L-P, whereas fruiting seems to be neither enhanced nor reduced.

Dictyostelium rhizopodium Raper and Fennell, in Amer. J. Bot. *54*: 515-528, figs. 1-26 (1967).

Cultivated on low-nutrient substrates such as 0.1L-P, 0.1L-P (pH 6.0), and hay infusion agars in association with *Escherichia coli* or *Klebsiella pneumoniae* at 23-25°C. Sorocarps erect or inclined, unbranched or sometimes bearing one or more lateral branches, solitary (Fig. 12-34A) or in clusters (Fig. 12-34H,I), variable in size and proportions, commonly 3-5 mm in height but sometimes less and often up to 1 cm or more, phototropic. Sorophores arising from well-developed and often strongly dissected crampon-like bases (Fig. 12-34G), yellowish to light brownish purple, mostly 25-50 μm in diameter near the base to 5-15 μm at the sorus. Sori globose to citriform, commonly 100-250 μm in diameter, white to cream-colored when formed but shading to yellowish olive or pale brownish olive in age. Spores variable, in some strains ellipsoid or capsule-shaped, rarely bent, mostly 6.0-8.0 × 3.0-3.5 μm (Fig. 12-34B), in other strains appreciably larger, narrowly elongate, often reniform or recurved, commonly 10-13 × 3-4 μm, all spores with prominent polar granules. Cell aggregations typically radiate in pattern (Figs. 12-34D,E), commonly 3-5 mm in diameter but may be larger or smaller; streams tend to dissociate in subcentral areas (Fig. 12-34E) as sorogens form. Myxamoebae irregular to roughly triangular (Fig. 12-34C), variable in size, about 15-20 × 10-13 μm in strains with smaller spores and 18-25 × 12-15 μm in strains with largest spores. Microcysts abundant in some strains and under some conditions; macrocysts not reported or observed.

HABITAT. Surface soil and leaf litter from forests of Panama, Mexico, Costa Rica, Florida, and several sites in Southeast Asia (Cavender, 1976b).

HOLOTYPE. Strain Pan-33, isolated in 1958 by KBR from forest soil collected on Barro Colorado Island, Panama, by Dietrich Kessler.

Other isolates include strain Pan-15, also from Panama soil; PR-10D and PR-10b, isolated by J. C. Cavender from soil of seasonal evergreen forest, Poza Rica, Mexico; and MH-3, also isolated by Cavender, from soil, Mahogany Hammock Forest, Everglades National Park, Florida. Of these isolates, Pan-33, Pan-15, and PR-10D produce spores of the larger dimensions and variable forms mentioned above, whereas HM-3 and PR-10b produce spores of the smaller range and form as illustrated.

Dictyostelium rhizopodium appears to be fairly widely distributed in the tropics and subtropics. It has been isolated from forest soils collected in southern Florida, in various parts of Central America, and in several areas of Southeast Asia, including Indonesia, Singapore, Malaysia, and Thailand. It has not been isolated insofar as we know, from more temperate areas either in North America or elsewhere. Perhaps reflecting this distribution, virtually all strains can grow and fruit at 30°C, which is above the favorable range for *D. discoideum* and many other cellular slime molds.

As recovered from nature, strains vary appreciably in habits of growth, in pigmentation, and in the dimensions of their spores. For example, the sorocarps of most isolates tend to develop singly, whereas those of strain

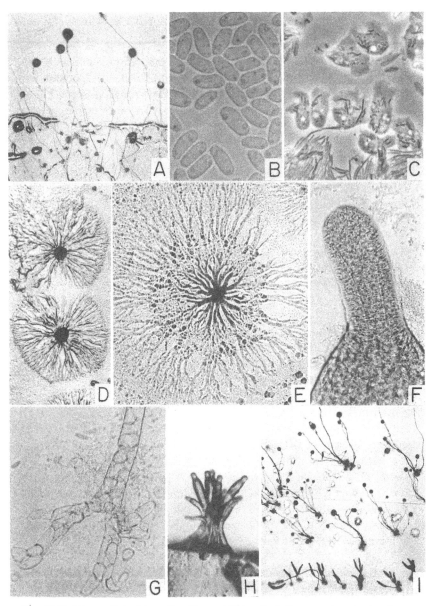

Fig. 12-34. *Dictyostelium rhizopodium* Raper and Fennell. *A*. Sorocarps in a streak culture of *E. coli*. Sori on erect stalks appear gray; those that have collapsed on agar appear dark and larger. × 5. *B*. Spores containing obvious polar granules (transmitted light). × 1150. *C*. Vegetative myxamoebae feeding at colony edge. × 525. *D*. Two characteristic aggregations. × 8. *E*. Larger aggregation; note that inflowing streams have broken up in subcentral area. Fruiting by such dissociated myxamoebae must await emergence of new centers. × 8. *F*. Apical area of developing sorocarp showing terminally expanded sorophore sheath and characteristic orientation of myxamoebae that surround it. × 300. *G*. Crampon base and sorophore showing similar differentiation of constituent cells. × 300. *H*. Clustered sorogens arising from a single aggregation. × 30. *I*. Mature sorocarps of the same strain and young sorogens along colony edge (below). × 8. (Reassembled from Raper and Fennell, 1967; see also Fig. 8-18)

PR-10D are usually clustered and at times develop almost as coremiform fructifications, the sorophores being adherent for some distance above their intermeshed bases (Fig. 12-34H). Pigmentation of sori and sorophores alike varies among different strains and to a lesser degree within single strains depending upon age, the substrate, and in some measure the associated bacteria. Particularly striking are differences in the form and dimensions of spores as cited above. Whether these represent differences in ploidy, as we are led to suspect, remains an open question—chromosome numbers have not been determined for any of the crampon-based forms insofar as we know.

Whereas the species description as originally written (and as presented above) is based upon low-nutrient substrates for comparative purposes, *D. rhizopodium* can be grown successfully on somewhat richer substrates, including some of those used for cultivating *D. discoideum, D. mucoroides*, and *D. purpureum*. For example, Bonner's agar medium diluted to one-fourth strength, together with *Klebsiella pneumoniae* as bacterial associate, was employed in our earlier studies of crampon formation in this slime mold. By giving special attention to the substrate, bacteria other than *E. coli* and *K. pneumoniae* may be used as food sources. For example, large and well-formed sorocarps may be produced in association with *Pseudomonas fluorescens* growing on hay infusion agar, the substrate on which *D. rhizopodium* was first cultivated. Although optimal at pH 5.8-6.2, the pH of the culture medium seems not to be especially critical, since this and the other crampon-based species can grow and fruit satisfactorily over a range from about pH 5.4 to 7.0. But differences in the substrate can alter culture patterns.

No mention was made of microcysts in the original description of *D. rhizopodium*, but in the present study we have found them to be produced in great abundance on phosphate buffered substrates by the type strain, Pan-33. No explanation is offered. Microcysts might have been overlooked in previous investigations, or some unrecognized difference(s) in culture conditions may be responsible. It is of some interest that phosphates often limit or preclude macrocyst formation while promoting the formation of microcysts (Toama and Raper, 1967a).

As in many other species, sorocarp formation is markedly enhanced by substituting clay for glass Petri dish covers once cell aggregation is in progress. The addition of activated charcoal also has a positive if somewhat lesser effect.

In addition to being fascinating structures in their own right, the crampons of *D. rhizopodium* and related species are believed to approximate the type of basal support described by van Tieghem in 1884 for his then new genus, *Coenonia*. This, unfortunately, has not been rediscovered, although it is still searched for with unabated hope and curiosity after a full century.

Dictyostelium lavandulum Raper and Fennell, in Amer. J. Bot. *54*: 515-528, figs. 38-49 (1967).

Cultivated on low-nutrient agars such as 0.1L-P, 0.1L-P/2, or 0.05% lactose, 0.025% yeast extract (or peptone) agars in association with *Escherichia coli* at 20-24°C. Sorocarps comparatively delicate (Fig. 12-35A), usually single but loose clusters may develop from single aggregations, erect or prostrate, the former developing early and up to 4-5 mm high, the latter strongly phototropic, up to 2 cm in length and often stolon-like along the agar surface (Fig. 12-35G). Sorophores pale lavender-purple in color, sinuous, becoming tangled, often one cell thick except in basal portions, usually arising from well-formed but narrow crampon bases (Fig. 12-35F). Sori globose to subglobose, pale grayish lavender, mostly 100-200 μm in diameter, occasionally 250 μm. Spores ellipsoid or capsule-shaped, occasionally reniform, variable in size, mostly 6.0-7.5 × 3.0-3.5 μm but ranging from 5.0 × 3.0 to 11-12 × 4.5 μm, with prominent polar granules (Fig. 12-35C). Cell aggregations usually radial in pattern and 2-4 mm in diameter, mound-like in richer cultures, becoming lemon yellow to pale gold as cell convergence progresses and sorogens are formed. Myxamoebae not distinctive, commonly 15-18 × 9-12 μm as seen on the agar surface (Fig. 12-35B). Unaggregated myxamoebae commonly form microcysts (Fig. 12-35H), globose to subglobose, thin-walled 4.0-7.0 μm in diameter; macrocysts not observed or reported.

HABITAT. Soil and leaf mould of wet forests in Costa Rica, Central America, and various collection sites in Southeast Asia and Africa.

HOLOTYPE. Strain Tu-5a, isolated by J. C. Cavender, 1962, from soil of subtropical wet forest, near Turrialba, Cartago, Costa Rica.

Other cultures examined in our study include FO-2, isolated by Cavender from soil of tropical rain forest near Fortuna, Limon, Costa Rica, and VQ-9a, also by Cavender, from wet forest soil near Villa Quesada, Alajuela, Costa Rica.

Cavender has subsequently obtained additional strains of *D. lavandulum* from soils collected in Indonesia, Malaysia, and Thailand (1976b), and from semi-evergreen forests at Ituri and Budongo in East Africa (1969b), Some, but not all, of the latter have been examined in our laboratory.

Two characteristics in particular serve to separate *D. lavandulum* from the other crampon-based Dictyostelia: the overall lavender and somewhat floccose appearance that results from its pigmented sori and from the abundant development of long, tangled, lightly pigmented sorophores; and the development of a yellow to pale golden pigmentation during cell aggregation that persists in the young sorogens. Such pigmentation is en-

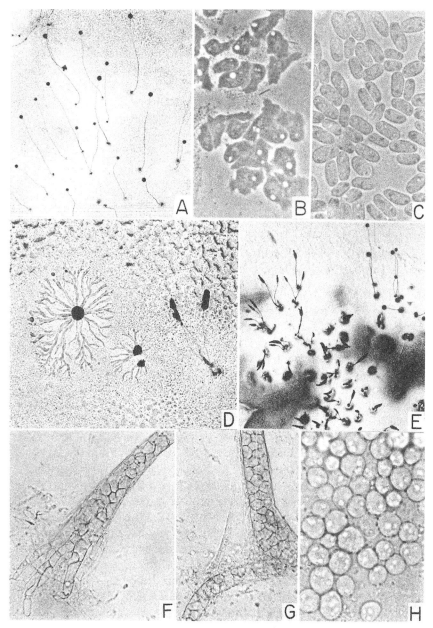

Fig. 12-35. *Dictyostelium lavandulum* Raper and Fennell. *A.* Sorocarps in a spread culture with *E. coli.* × 8. *B.* Preaggregative myxamoebae. × 525. *C.* Spores showing internal granulation, mostly polar (transmitted light). × 1150. *D.* Field showing radial aggregations (one split) and two sorogens that have fallen to the agar surface. × 8. *E.* A richer culture showing mound-like aggregations and fructifications in different stages of development, young sorogens (lower center) to mature sorocarps (upper right). × 8. *F.* Crampon base consisting of three "fingers." × 300. *G.* Stolon-like fruiting structure touched agar and sorogen started anew to build an upright stalk; note wider base. × 300. *H.* Microcysts, thin-walled resting stage. × 1050. (Rearranged from Raper and Fennell, 1967)

hanced by incubation in darkness and tends to fade when cultures are exposed to light. The pigment may be the same or comparable to that present in the sorophores of *D. aureo-stipes* (Cavender et al., 1979), for it too is enhanced by incubation in darkness and fades when exposed to light. In neither case has the pigment been identified.

The very long sorophores of *D. lavandulum* are of some special interest because they develop stolon-like as the sorogens build toward a light source, touching the agar, then rising again only to fall back, and at each point of contact forming a new but nondigitate base (Fig. 12-25G).

Sorocarp formation is improved substantially if activated charcoal is added at the time of cell aggregation, or if clay dish covers are substituted for glass.

Dictyostelium vinaceo-fuscum Raper and Fennell, in Amer. J. Bot. *54*: 515-528, figs. 50-64 (1967).

Cultivated on low-nutrient substrates such as 0.1G-P/2 or 0.05% lactose, 0.025% yeast extract (or peptone) in association with *Escherichia coli* at 23-26°C. Sorocarps delicate, variable in size, upright when formed (Fig. 12-36A), later becoming tangled, not clustered but often formed in groups from large aggregations, typically unbranched but may bear a few irregularly spaced branches, phototrophic in unidirectional light. Sorophores arise from small but typically digitate crampon-like bases (Fig. 12-36F) or from expanded masses of vacuolate cells, flexuous, sinuous or spiral, variable in length, mostly 1-4 mm, less commonly 6-7 mm or when stolon-like up to 1.5 cm in unidirectional light, mostly one cell wide and showing a slight taper from 8-12 μm below to 1.5-2.5 μm at sori, hyaline to faintly purple with age. Sori globose (Fig. 12-36A), at first pale pinkish lavender, then mauve, and finally almost purple-black, mostly 75-150 μm in diameter, infrequently 200 μm. Spores variable in size and shape, mostly ellipsoid to reniform, commonly 6.0-7.0 × 3.0-3.5 μm (Fig. 12-36B) but up to 10.0 μm or more in some strains, granular inclusions lacking or inconspicuous. Cell aggregations of different size and pattern, small with radiate converging streams (Fig. 12-36D,E), and large complex patterns where small wheel-like aggregations are surrounded by large coronas, or "halos," outward from which extend scores or hundreds of nearly parallel inflowing streams (Fig. 36E). Myxamoebae not distinctive, about 15-20 × 8-12 μm as seen on an agar surface (Fig. 12-36C). Microcysts commonly produced under adverse conditions, thin-walled, globose to subglobose, 4.5-7.0 μm in diameter (Fig. 12-36H); macrocysts not reported or observed.

HABITAT. Soils and leaf litter from forests of Texas, Florida, Mexico Costa Rica, Thailand.

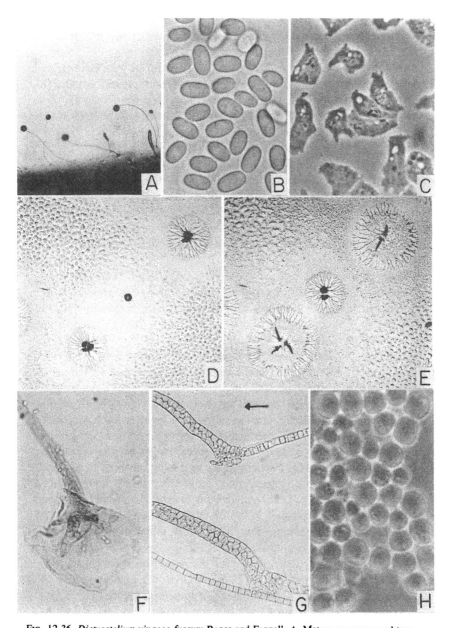

FIG. 12-36. *Dictyostelium vinaceo-fuscum* Raper and Fennell. *A*. Mature sorocarps and two rising sorogens. × 10.5. *B*. Spores typical of the species. Note absence of polar granules. × 1150. *C*. Preaggregative myxamoebae. × 525. *D,E*. The same aggregations at 2:45 and 5:30, respectively. Note presence of sorogens in upper and lower aggregations in *E*, and areas of disassembled streams surrounding them. × 7. *F*. Crampon base of sorophore showing digitate pattern (stained). × 300. *G*. Two sorophores with suggestions of crampons where sorogens touched the agar and reinitiated efforts to build erect stalks. Note how the stalk diameters increased at points of ascension. × 240. *H*. Microcysts photographed on the agar surface. × 1056. (Rearranged from Raper and Fennell, 1967)

HOLOTYPE. Strain Li-2, isolated by J. C. Cavender in 1962 from soil of a tropical dry forest, Liberia, Guanacaste, Mexico.

Of many strains examined, three additional ones may be cited as representative: CC-4, from soil of mesquite scrub forest, Alice, Texas; MR-2, from soil of tropical deciduous forest, Matias Romero, Veracruz, Mexico; and Br-8, from soil of tropical dry forest, Barranca, Puntarenas, Costa Rica. All were isolated by Cavender.

Among the crampon-based Dictyostelia, *D. vinaceo-fuscum* produces the most delicate sorocarps. It appears also to be the most widely distributed in nature with the possible exception of the much larger species *D. rhizopodium*. Like that species, and the other crampon formers as well, it seems to be confined to tropical and subtropical regions; but unlike the others it tends to occupy relatively dry rather than very moist to wet habitats.

Dictyostelium vinaceo-fuscum may be aligned with *D. coeruleo-stipes, D. rhizopodium*, and *D. lavandulum*, since it possesses a crampon-like base, albeit this is not always fully expressed. It also shares with the latter two species a measure of pigmentation that ranges through lavender and mauve shades to the deep vinaceous smoke colors that characterize its mature sori. It seemingly differs, however, in the character of its spores, for they show little if any evidence of polar granulation—a striking character in the other three species.

As in other delicate species, cell aggregation in *D. vinaceo-fuscum* may take different forms, in the one case producing small radial aggregations (Fig. 12-36D,E) that subdivide and give rise to a very few sorogens; in the other case the aggregates become more expansive, the converging streams break up, and the fractional parts fruit independently. In still other cases, and at an appreciable distance from the primary aggregation, a circular ridge, or "halo," of myxamoebae will appear, outward from which scores of essentially parallel inflowing streams extend into the lawn of previously unaligned myxamoebae. Meanwhile, within the matte-like subcentral region a few small radial aggregations may appear, whilst many fruiting centers emerge along and near the peripheral ridge (Fig. 7-9A). While such complex patterns can be spectacular, the smaller primary aggregations are generally more effective in sorocarp production.

As in related species, some sorocarps may become very long and stolon-like when cultures are incubated in unidirectional light. The sorogens build toward the light but because of their weight fall to the surface, pause, and then reinitiate stalk formation in a semi-vertical direction, again fall to the surface, pause, etc. At each point of contact with the agar the suggestion of a crampon is formed and the stalk becomes substantially thicker as formation in a more vertical direction is resumed (Fig. 12-36G).

Sorocarp formation in *D. vinaceo-fuscum* is enhanced substantially when activated charcoal is introduced at the time cell aggregation is in progress, while equal or greater improvement may be obtained by replacing glass with clay dish covers at a similar stage of development.

Dictyostelium mexicanum Cavender, Worley, and Raper, in Amer. J. Bot. *68*: 378-381, figs. 13-23 (1981).

Cultivated on low-nutrient substrates such as 0.1L-P, 0.1L-P (pH 6.0), or 0.1% G, 0.05% Yxt agars in association with *Escherichia coli* at 22-28°C. Sorocarps conspicuously yellow throughout or with sori pale yellow in some strains, erect or semi-erect, usually limited in number, solitary or in clusters consisting of one or more primary structures around which smaller sorocarps later develop; primary sorocarps simple or branched, rarely 2.5-3.0 mm high (Fig. 12-37A,B), secondary sorocarps 1.0-1.5 mm. Sorophores strongly tapered, the larger ones 30-60 μm at the base and narrowed to 12-20 μm at the sorus, sometimes anchored in rudimentary to well-formed basal disks (Fig. 12-37D,E), main axes and branches expanded terminally and often flattened above (Figs. 12-37F,G). Sori globose to citriform (Figs. 12-37A,B), deep yellow in some strains, less pigmented in others, up to 250-300 μm in larger structures, 100-150 μm in smaller sorocarps. Spores elliptical to weakly reniform, mostly 5.5-8.0 × 2.5-4.0 μm, rarely more, with conspicuous polar granules (Fig. 12-37I), often germinating by lateral rupture of spore cases (Fig. 12-37J). Cell aggregations small, radiate in pattern (Fig. 12-37C), mostly 2-4 mm in diameter; pseudoplasmodia and sorogens yellow in color. Myxamoebae not distinctive, 15-20 × 9-12 when moving on an agar surface (Fig. 12-37H). Cells not entering aggregations form microcysts, globose to subglobose (Fig. 12-37K), mostly 5-7 μm in diameter; macrocysts are produced when appropriate mating types are paired, small, yellowish, and ca. 15 μm in diameter.

HABITAT. Soils of relatively dry to dry forests in different areas of Mexico.

HOLOTYPE. Strain PR-7a, isolated by J. C. Cavender in January 1962 from soil of a deciduous forest near Poza Rica, Vera Cruz, Mexico.

Additional strains obtained from Mexican soils by JCC include PR-5, isolated concurrently from a site similar to that of Pr-7a; OB-3, from desert soil, Obregón, Sonora; AL-3b, from thorn forest near Alamos, Sonora; TC-10, from tropical deciduous forest near Tehuantepec, Oaxaca; and Cu-3a, from thorn forest near Culiacán, Sinaloa.

The specific name, which reflects the geographic area of species origin, was first applied by Franz Traub while working with these strains at the University of Wisconsin in the summer of 1974. It was later used, but without a species description, in his doctoral dissertation at the University

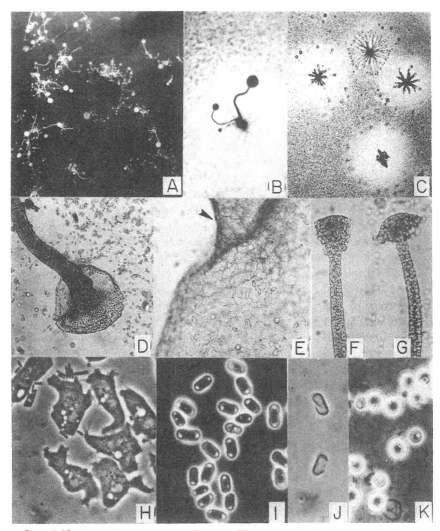

Fig. 12-37. *Dictyostelium mexicanum* Cavender, Worley, and Raper. *A*. Culture showing primary sorocarps with large sori surrounded by smaller secondary sorocarps. × 6. *B*. Single sorocarp with lateral branch. × 20. *C*. Cell aggregations forming in dense field of myxamoebae. × 6. *D*. Basal area of primary sorocarp showing well-formed disk. × 70. *E*. Portion of same disk enlarged to show similar cellular structure of sorophore (arrow) and disk. × 280. *F,G*. Expanded apices of two sorophores characteristic of this species. × 175. *H*. Vegetative myxamoebae beyond margin of bacterial colony. × 740. *I*. Spores showing characteristic form and prominent polar granules. × 740. *J*. Two empty spore cases where germination occurred by lateral rather than longitudinal rupture. × 740. *K*. Globose microcysts formed by unaggregated myxamoebae; note empty cyst wall (lower center) with triangular rift made during germination. × 740. (From Cavender, Worley, and Raper, 1981)

of Zurich (1977). Use of the epithet was formalized by Cavender, Worley, and Raper in 1979.

Dictyostelium mexicanum possesses a number of distinguishing characteristics that separate it from *D. aureum*, including yellow pigmentation of pseudoplasmodia and sorogens; short and often strongly tapered sorophores; occasional presence of well-formed basal disks; enlarged terminals of sorophore axes and branches; and formation of macrocysts by compatible mating types. These merit some attention.

In contrast to the developing fructifications of *Dictyostelium aureum*, where only the sorophores and sori are yellow in color, in *D. mexicanum* such pigmentation is apparent during the aggregative process, becomes intensified as the cells consolidate to form the sorogens, and typically persists in the completed sorophores and sori alike. Differences in pigmentation occur among strains as these are isolated from nature, and in some strains the pigment tends to fade as the sorocarps mature (e.g. AL-3a); in other strains sorocarps remain deeply pigmented. For reasons unknown pigmentation is intensified on acidic substrates. Yellow pigmentation in advanced aggregations and sorogens is seen also in the crampon-based *D. lavandulum*, but in that species it fades during sorocarp formation, particularly in light, and is replaced by a pale vinaceous to grayish lavender coloration in stalks and sori.

The sorophores of *D. mexicanum* are quite distinctive, being yellow in color, relatively short, and often sharply tapered. The apices of central axes and branches are typically expanded into the form of inverted cones from which the sori are suspended. Whereas limited terminal swelling may be seen in some other species, including some strains of *D. aureum*, in no other dictyostelid does one find structures comparable to these. Additionally, the larger sorophores of *D. mexicanum* commonly arise out of masses of stalk-like cells that in some cases constitute clearly defined basal disks. Such structures are reminiscent of the basal disks in *D. discoideum*, while the occasional presence of yellow sori in the latter species might further suggest a measure of close relationship. This, however, is not supported by other considerations. There is no hint of a migrating pseudoplasmodium in *D. mexicanum*, aggregating myxamoebae show no response to cAMP, spores contain conspicuous polar granules, and microcysts are freely formed. Such similarities as exist probably result from parallel but phylogenetically unrelated evolution.

In limited studies by Franz Traub (1977), small yellowish macrocysts, measuring about 15 μm in diameter, were produced when strain OB-3 was paired with strains TC-10 or AL-3b. In contrast, no macrocysts were formed in pairings with strains PR-5 or Cu-3a. Upon this evidence, and since no macrocysts have been reported for single isolates, it is presumed that *D. mexicanum* is basically heterothallic.

Microcyst production may be especially heavy in this species when cultivated in the usual manner. In fact, aggregations may be relatively few in number and develop as isolated islands in an otherwise dense field of nascent microcysts. The addition of small chips of activated charcoal increases markedly the process of cell aggregation and subsequent sorocarp formation, while a lesser positive effect may be realized by pouring plates several days prior to inoculation.

Dictyostelium rosarium Raper and Cavender, in J. Elisha Mitchell Sci. Soc. *84*: 32-47, figs. 1-8 (1968). *See also* Benson and Mahoney, Amer. J. Bot. *64*: 496-503 (1977).

Cultivated on low-nutrient agars such as phosphate buffered 0.1L-P or 0.1G-P (pH 5.8-6.0) in association with *Escherichia coli* or *Klebsiella pneumoniae* at 23-25°C. Sorocarps solitary or clustered (Fig. 12-38A), typically erect or inclined, sometimes prostrate, unbranched or at times showing primary dichotomous branching (Fig. 12-39C,D), variable in size and proportions, consisting of sorophores bearing terminal sori and numerous, usually smaller sessile sori along the sorophore axes, hence beaded in appearance (Fig. 12-38A). Sorophores variable in diameter, mostly 12-25 μm near the base but often less and sometimes up to 35 μm, tapering gradually from base to terminal sorus, commonly 3-6 mm in length but ranging from 2 mm to 1 cm or more, bases rounded or wedge-shaped but not enlarged. Sori globose, milk-white; terminal sori mostly 100-250 μm in diameter, sessile sori usually smaller, mostly 50-150 μm (Fig. 12-39A,B) and numbering few to 25 or more, uneven in size on the same sorophore. Spores globose to subglobose, 4.5-6.0 μm in diameter (Fig. 12-38B), with content finely granular and walls comparatively thin. Cell aggregations often appearing first as broad sheet-like bands of converging myxamoebae (Fig. 12-38D), then developing conspicuous streams (Fig. 12-38E) ranging from a few millimeters to a centimeter or more in length, and finally becoming nodular (fig. 12-38F) and breaking apart to form rounded masses of cells that give rise to one or more sorocarps. Myxamoebae generally rounded when feeding and measuring about 12-15 × 10-12 μm in diameter (Fig. 12-38C), appearing larger when moving. Macrocysts formed when compatible mating types are paired (Fig. 12-38I); microcysts have not been observed or reported. The acrasin is cyclic AMP.

HABITATS. Dry prairie soil in Wisconsin and soils of semi-arid regions in Texas, California, and Washington State; also dungs of horse and deer in Mexico and Washington State and bat dung from caves in Arkansas and Missouri.

HOLOTYPE. Strain CC-7, isolated in 1962 by J. C. Cavender from soil collected in a mesquite forest near Alice, Texas.

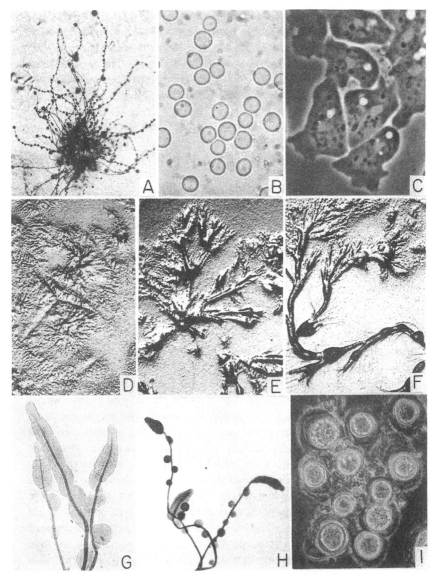

FIG. 12-38. *Dictyostelium rosarium* Raper and Cavender. *A.* Clustered sorocarps showing characteristic beaded appearance. × 7.2. *B.* Mature spores showing typical globose form. × 950. *C.* Vegetative myxamoebae photographed when moving. Note vesicular nuclei, contractile vacuoles, granular endoplasm, and clear ectoplasm at advancing fronts. × 950. *D-F.* Progressive stages in cell aggregation. × 8. *G.* Sorocarps in process of formation. Note apical sorogens and masses of cells being abstricted to form sessile sori. Stained with chloroiodide of zinc to reveal origin and cellulosic character of sorophores. × 38. *H.* Sorocarps in process of formation with some sessile sori and sorogens wherein the stalks are being extended. × 25. *I.* Macrocysts produced by pairing two compatible mating types, strains DC-31 × WS-519. × 300. (Rearranged in part from Raper and Cavender, 1968)

Additional cultures investigated include strain CS-4 (Fig. 12-39E), isolated by Cavender in 1962 from horse dung collected at Ciudad Serdan, Pueblo, Mexico; strain FH-8, isolated from Carmen Stoianovitch in 1965 from deer dung collected in the state of Washington; strain WS-587, isolated by John Sutherland in 1973 from dry prairie soil collected near Cassville, Wisconsin; strain WS-689, isolated by David Waddell from bat dung collected in Blanchard Spring Caverns, Arkansas; and numerous strains from Daniel Mahoney representing isolates obtained during an extensive survey of the dictyostelid populations of soils from many locations in southern California (Benson and Mahoney, 1977).

Dictyostelium rosarium is distinguished by two characteristics in particular (1) its globose spores, a pattern that is found elsewhere only in the very delicate *D. lacteum* van Tieghem (1880) and in two or three species of the genus *Acytostelium*; and (2) its sessile sori borne at intervals along the upright sorophores, a character that is shared with, but less well expressed by, Cavender's *D. laterosorum* (1970). The manner in which the sessile sori arise is shown in Fig. 12-38G and H. Since this has been discussed elsewhere (see Chapter 8) we need not elaborate here.

Whereas the description of *Dictyostelium rosarium* is based in the main upon strain CC-7, it applies quite well to other strains examined during this study, allowing for the variation one may expect among the isolates of a cosmopolitan species.

Species of bacteria other than those cited can be substituted as a food source but offer no advantage. The pigment of *Serratia marcescens* is digested by the feeding myxamoebae so that the resulting pseudoplasmodia and sorocarps remain uncolored, as do those of *D. giganteum* and most isolates of *D. mucoroides*. Culture media of increased nutrient content may be used successfully when buffered adequately, thus providing increased growth and more luxuriant fructification. Myxamoebae can grow over a pH range of 4.0 to higher than 7.0, but for fruiting the useful range is appreciably less, being approximately 5.0-6.2 with an optimum at pH 5.8-6.0. Fruiting in most strains is accelerated and enhanced in the presence of activated charcoal, a response that is especially striking in the type strain, CC-7. Significant improvement also occurs when glass Petri dish lids are replaced by porous porcelain covers (Coors), or when similar cultures are inverted over sterile mineral oil. The fruiting process is also accelerated by exposure to light, and this too is particularly striking in strain CC-7. Fortunately, many strains are less fastidious and fruit satisfactorily when handled without special manipulation.

Dictyostelium rosarium was thought to be rare when the species was described in 1968, for we then had in hand only the first three strains listed above. Since that time, several investigators in this laboratory and elsewhere have isolated the slime mold, and we now know it to be widely distributed and in some regions quite common. The studies of Benson and

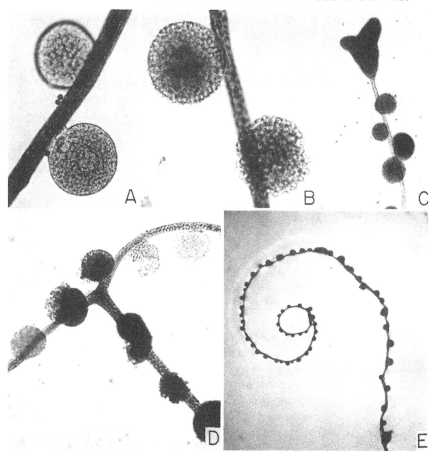

Fɪɢ. 12-39. *Dictyostelium rosarium* Raper and Cavender. *A*. Detail of two young sessile sori. Note manner of attachment and that the nascent sori are bounded by continuous slime envelopes. × 240. *B*. A more advanced stage; the slime covering has disappeared coincident with spore maturation. × 240. *C*. A sorogen in proces of dichotomous but unequal primary branching. × 70. *D*. An example of dichotomous branching where the two branches are equal. × 135. *E*. A prostrate sorocarp of strain CS-4 where the sessile sori lie on the agar surface. × 12. (Rearranged from Raper and Cavender, 1968)

Mahoney (1977) represent a case in point, for they found *D. rosarium* to be present in 22 of 26 plant communities in southern California that yielded dictyostelids. It was frequently isolated from soils collected in mixed chaparral and Joshua tree deserts, thus confirming its tolerance for drier soils in contrast to *D. mucoroides.*

Although not yet studied in depth, macrocysts have been obtained in *D. rosarium* by pairing compatible mating types as reported by Chang and Raper (1981).

When compared with other larger species, *Dictyostelium rosarium* is a very poor competitor. In studies, previously unreported, R. F. Pescatore

(1972) found that its spores failed to germinate when implanted with those of *D. discoideum* or *D. giganteum*; it grew to only a fraction of its solo potential when intermixed with *D. discoideum* in liquid shaken cultures; and it failed to aggregate when equal numbers of pregrown myxamoebae were intermixed with those of *D. discoideum* on agar plates.

As in *D. discoideum*, cyclic AMP is the acrasin effecting cell aggregation, albeit the interval between pulses is substantially longer, this being about 20 rather than 5 minutes (Konijn, 1972).

Dictyostelium lacteum van Tieghem, in Bull. Soc. Bot. Fr. *27*: 317 (1880). Also E. W. Olive, in Proc. Amer. Acad. Arts Sci. *37*: 339-340 (1901), and in "Monograph of the Acrasieae," Proc. Boston Soc. Natur. Hist. *30*: 451-513 (1902).

Van Tieghem's brief description follows in translation: "The spore mass forms a milk-white drop at the summit of a stalk that I have always seen formed of a single file of cells. The spores are colorless, spherical, and very small, measuring 2 to 3 μm in diameter. The plant has been encountered several times on agarics undergoing deterioration."

A description based upon current observations follows:

Cultivated on low-nutrient agars such as 0.1L-P/2 or 0.1G-P/4 in association with *Escherichia coli* or *E. coli* B/r at 22-25°C. Sorocarps very small and delicate (Fig. 12-40A), solitary or more often clustered in groups of 5-10 or up to 15-16, commonly 0.5-1.0 mm long, but ranging in length from 0.35 to 1.5 mm and occasionally 2.0 mm, rarely branched. Sorophores very thin, composed of a single tier of cells, usually including the base (Fig. 12-40G,H), tapering from about 5-10 μm below to 2.5 μm or less at the apex, bearing very small pale to milk-white sori. Sori globose, commonly 25-60 μm but up to 120 μm in diameter. Spores globose to subglobose (Fig. 12-40J), appearing slightly granular, typically 3.5-5.0 μm but ranging from 3.0 to 5.5 μm in diameter. Cell aggregations typically developing as small rounded mounds (Fig. 12-40C), without stream formation except in very thin cultures, giving rise to multiple sorogens (Fig. 12-40D-F). Myxamoebae not distinctive, appearing much as in other species of the genus, about 12-16 × 8-12 μm when feeding on bacteria (Fig. 12-40B). Microcysts have been reported, macrocysts have not.

HABITATS. Decomposing agarics, France (*fide* van Tieghem) and humus-rich soils from deciduous and boreal forests, United States, Canada, Spain.

NEOTYPE. UI-14, isolated by KBR in 1950 from soil of a virginal oak forest known as Brownfield Woods, Urbana, Illinois; it is believed to conform with van Tieghem's description quite well despite its slightly greater spore measurements. Other representative strains have come from forest soils collected in Illinois, Indiana, Virginia, and Wisconsin. Indi-

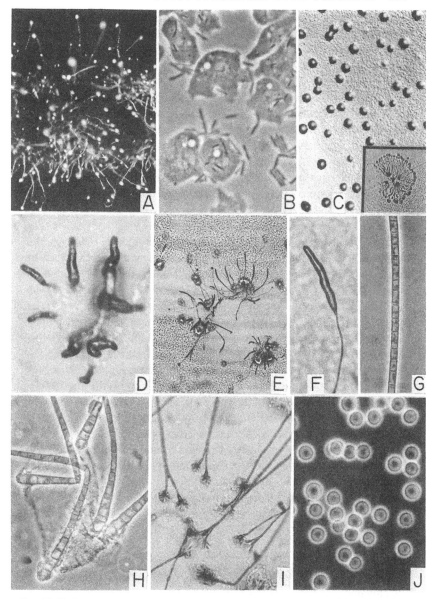

FIG. 12-40. *Dictyostelium lacteum* van Tieghem. *A.* Sorocarps photographed by surface light on low-nutrient agar with *Escherichia coli* B/r. × 14. *B.* Vegetative myxamoebae feeding on *E. coli*; note nuclei, contractile vacuoles, and food vacuoles containing bacteria. × 900. *C.* Small, mound-like aggregations formed without stream formation on weak hay infusion agar. × 14. Insert, aggregation via streams on agar with *very low* nutrient (0.1L-P/6). × 8. *D.* Multiple sorogens arising from a single small aggregation. × 70. *E.* Three aggregation areas from which mutliple, thin sorogens are developing. × 8. *F.* A single sorogen in process of fruiting. × 70. *G.* Typical sorophore composed of a single tier of vacuolate cells. × 240. *H.* Bases of clustered sorocarps showing cellular structure. × 240. *I.* Bases of sorocarps photographed to show expanded aprons of slime that anchor structures to substrate. × 70. *J.* Globose spores characteristic of this species. × 750

cations are that the species is widely distributed in the forest soils of North America. The fact that the species was overlooked for so many years can be attributed to the study methods previously in vogue and the small dimensions of its sorocarps.

Dictyostelium lacteum is distinguished particularly by its globose spores, a character shared with *D. rosarium* and *Acytostelium leptosomum* but without evidence of close kinship. Based upon patterns of growth, cell aggregation, and sorogen formation, its closest relative may be *D. deminutivum*. However, the dimensions of its sorocarps, small as they are, would still average somewhat larger in *D. lacteum*. It is interesting to note that in each species cell aggregation occurs without evident stream formation except in the thinnest cultures on very low-nutrient substrates (Fig. 12-40c).

Macrocysts have not been observed in any strain to date. However, this should not be taken to suggest that other isolates might not produce the sexual stage, or even strains now in hand under altered conditions of culture. We can say that in pairings of six strains on 0.1L-P agar and with incubation in the dark, presumed to be favorable, no macrocysts were produced. Microcysts have been reported but not observed in our cultures.

As the sorogens of *D. lacteum* arise from recently segmented, dome-like aggregations, they become quite elongate and remain so until the last of the convergent myxamoebae are free of the substrate. After this they assume a more conventional cartridge-like form (Fig. 12-40F) and as development advances form definite nipple-like apices wherein the sorophores are lengthened. We presume the process to be basically similar to that in *D. purpureum* or *D. mucoroides* but on a far more delicate scale. As in other species with plain or slightly enlarged bases, these are anchored to the substrate by expanded aprons of slime left in place as the sorogens lift off the substrate (Fig. 12-40I).

The acrasin effecting cell aggregation in *D. lacteum* has been identified as a pterin derivative by Haastert et al. in Leiden (1982). See Chapter 7.

Dictyostelium lacteum van Tieghem var. *papilloideum* Cavender, in Amer. J. Bot. *63*: 68-69, figs. 45-52 (1976).

Cultivated on low-nutrient substrates such as 0.1L-P/2, dilute hay infusion, or 0.1% lactose, 0.05% yeast extract agars in association with *Escherichia coli* at 22-25°C. Sorocarps erect or inclined, delicate, unbranched, 0.6-1.4 mm in height (Fig. 12-41A); Sorophores colorless, very thin, and flexuous, consisting of a single tier of cells, 7-10 μm in diameter at the base and tapering to 2.5-3.0 μm above. Sori white or translucent, globose, mostly 40-100 μm in diameter. Spores globose, small, typically 2.7-4.0 μm in diameter (Fig. 12-41B). Cell aggregations mound-like,

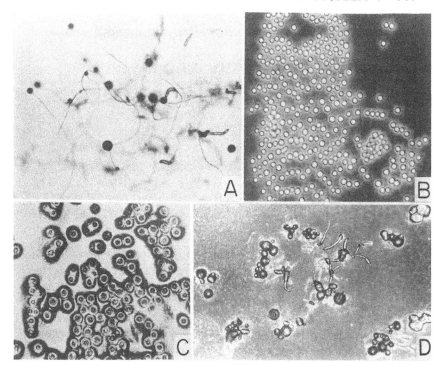

FIG. 12-41. *Dictyostelium lacteum* V. Tiegh. var. *papilloideum* Cavender. *A.* Sorocarps and sorogens shown in silhouette. Note the delicate, flexuous nature of the sorophores. × 37. *B.* Massed spores showing their small dimensions and characteristic spherical shape. × 585. *C.* Clustered, mound-like aggregations preparatory to fruiting. × 25. *D.* Less crowded aggregations showing presence of papillae and the emergence from these of elongate sorogens. × 25. (From Cavender, 1976a)

formed without inflowing streams (Fig. 12-41D) or with very weak streams, producing numerous papillae, each of which gives rise to a single sorogen and sorocarp. Neither macrocysts nor microcysts reported.

HABITAT. Humus and leaf mould of tropical rain forest, Philippines.

HOLOTYPE. Strain LB-3, isolated in 1970 by J. C. Cavender from forest humus and leaf mould collected in Lambuan, Panay, Philippines, and sent to us in January 1971.

While possessing most of the characteristics of van Tieghem's *Dictyostelium lacteum*, the variety is distinguished especially by the densely clustered papillae that develop on the surface of mound-like aggregations and from which single, long, thin sorogens develop that subsequently give rise to delicate sorocarps (Fig. 12-41C).

CHAPTER 13

Dictyosteliaceae: Polysphondyliun

TECHNICAL DESCRIPTION

Polysphondylium Brefeld, in Unters. Gesammtgeb. Mykol. *6*: 1-34, Taf. I & II, figs. 1-35 (1884). Also Olive (E. W.), in Proc. Amer. Acad. Arts Sci. *37*: 341-342 (1901) and Proc. Boston Soc. Natur. Hist. *30*: 451-513, pls. 6-8 (1902); Harper, in Bull. Torrey Bot. Club *56*: 227-258, figs. 1-44 (1929), and ibid. *59*: 49-84 (1932).

Vegetative stage consisting of free-living, independent myxamoebae which feed upon bacterial cells; fruiting stage characterized by the aggregation of these myxamoebae into pseudoplasmodia and the subsequent development from them of branched sorocarps with cellular stalks that, in the main, bear globose sori terminally and at the tips of lateral branches, the branches typically but not uniformly arranged in regularly spaced whorls. Dimensions and patterns of sorocarps vary substantially in different species and within single strains.

GENERAL DESCRIPTION AND COMMENTS

The genus *Polysphondylium* has been known and admired for almost a century since Brefeld described *P. violaceum* in 1884—admired not only for the delicacy and symmetry of its form but more especially for the singularity of its developmental cycle, which, with the addition of whorled branching, follows the same general course as *Dictyostelium mucoroides* described by Brefeld in 1869. The 1884 paper, in which both slime molds were discussed, is noteworthy for several reasons: it corrects errors in the earlier report and sets forth in great detail the complete asexual cycle of these slime molds; it contains beautifully delineated illustrations of *P. violaceum* (Figs. 1-2 and 1-3) that are remarkably accurate if, in some cases, somewhat idealized; it clearly describes the construction of symmetrically branched fructifications by the divergent differentiation of large populations of previously independent myxamoebae; and it records that this study first enabled Brefeld to attain "complete mastery of the artificial cultivation of fungi." Whereas van Tieghem had in 1880 reported fructification to occur without cell fusions in his *Acrasis* and in species of *Dictyostelium* as well, it remained for Brefeld to introduce the term *schein-plasmodium* (pseudoplasmodium) to emphasize properly the uniqueness of this multicellular structure, and to describe how its constituent cells became transformed into the stalk cells and spores of the emergent fructification.

In addition to *P. violaceum*, characterized by its generally robust sorocarps and sori of vinaceous to pale violet color, six other species of *Polysphondylium* have been described, all with white to hyaline sori and uncolored sorophores. Of these, *P. pallidum* Olive (1901), marked by its more delicate and typically symmetrical sorocarps, is, like *P. violaceum*, quite cosmopolitan in distribution. *P. album* Olive (1901), based upon a single isolate, is thought not to differ from *P. pallidum* sufficiently for species recognition. The others are known from restricted geographic areas but may have wider distribution than is presently known. Three have been described by Hagiwara from forest soils and leaf mould in Japan: *P. candidum* (1973a), *P. pseudo-candidum* (1979), and *P. tenuissimum* (1979). The other, *P. filamentosum*, described by Traub, Hohl, and Cavender (1981a), was obtained from numerous forest soils in Switzerland.

The close relationship between *Dictyostelium* and *Polysphondylium* is clearly apparent, and as early as 1884, in writing about *Coenonia*, van Tieghem questioned the wisdom of recognizing two genera "when in one and the same species the fructifications may be simple or branched according to their dimensions." Potts (1902) expressed similar doubts, and Olive (1902) noted the presence of *Dictyostelium*-like fruiting structures in cultures of *Polysphondylium*, especially when grown in van Tieghem cell cultures. Having cited this as evidence of close relationship, he then added, however, that "the almost constant character seen in *Polysphondylium* involving the bearing of whorls of branches all of about equal length, indicates that the two organisms (*Polysphondylium* and *Dictyostelium*) possess important physiological as well as structural differences." The significance of branching was again raised by Rai and Tewari in 1961, and in 1976 Traub and Hohl published their "A New concept for the taxonomy of the Dictyosteliaceae," wherein they showed that many species now classified as *Dictyostelium* possess a number of characteristics in common with *Polysphondylium*, particularly *P. pallidum*. They did not propose the demise of *Polysphondylium*; instead they suggested the need for a taxon intermediate between the existing genera. In the following year Traub (1977) introduced but did not formally describe *Heterosphondylium* gen. nov. to accomodate forms lacking whorls of branches but possessing a number of "*Polysphondylium*-like characteristics," including, among others, spores with clusters of polar granules, a lack of aggregative response to cAMP, and the frequent formation of microcysts. Although such criteria may in fact provide evidence of interspecific kinships, we do not at this time find the correlations sufficiently firm to warrant the recognition of a new genus. We are continuing, therefore, to recognize as *Dictyostelium* isolates with unbranched or irregularly branched sorophores and as *Po-*

lysphondylium those strains with sorophores bearing lateral branches in whorls, recognizing that with sparse growth or suboptimal conditions a particular strain may not attain its full fruiting potential. In other words, whereas culture conditions may determine whether or not whorls of branches are actually produced in a *Polysphondylium*, the potential for doing so is an inherited character possessed and retained by the cell population, myxamoebae and spores, a character that only awaits the return of optimal conditions for full expression.

Finally we should note that although the aggregative process in *P. violaceum* is seemingly much like that in the larger species of *Dictyostelium* such as *D. discoideum*, the acrasin of *P. violaceum* is not cAMP. Much attention has been directed toward the isolation and characterization of a *P. violaceum* chemotactic substance in Bonner's laboratory (see Chapter 7), and the work has recently come to fruition with the isolation and identification of an unusual dipeptide, designated glorin, which is the acrasin for *P. violaceum* and which is chemotactively active for *P. pallidum* as well.

Cognizant of the diversity among natural isolates and of their responses to different cultural environments, the reader should consider as somewhat elastic the dimensions cited for the sorocarps of the several species recognized and described herein. Upon this word of caution, a key to the genus follows.

KEY TO SPECIES OF *Polysphondylium*

A. Sorocarps generally robust; sori in vinaceous to pale
   violet shades, darkening in age; sorophores lightly
   pigmented in similar shades; strongly phototropic .. *P. violaceum* Brefeld
B. Sorocarps comparatively delicate; sori white to hyaline; sorophores uncolored, weakly phototropic as a
   rule
  1. Terminal segments of sorophores not elongate; terminal sori comparatively large
    a. Whorls of lateral branches usually 3-8 in number,
       sometimes more; sorocarps erect or semi-erect,
       typically symmetrical .... *P. pallidum* Olive
    b. Whorls of lateral branches often 10-15 or more in
       number; branches short and sori generally small .. *P. tenuissimum* Hagiwara
  2. Terminal segments of sorophores often quite long
     with terminal sori usually minute or lacking
    a. Branches typically rebranched, terminal segments
       long, filament-like .... *P. filamentosum* Traub, Hohl,
                                                        and Cavender
    b. Sorocarps fairly compact, symmetrical; branches
       seldom rebranched, usually 2-4 whorls, 3-8
       branches per whorl, occasionally more; terminal

segments short to long (if long, sori small or lack-
ing); spores 8-10 × 3.5-5.0 μm. . . . . *P. candidum* Hagiwara
c. Sorocarps less compact, delicate; sorophores thin,
with terminal segments occasionally quite long
and terminal sori minute or lacking; spores 6.2-
7.9 × 3.0-3.7 μm . . . . *P. pseudo-candidum*
Hagiwara

SPECIES DESCRIPTIONS

Polysphondylium violaceum Brefeld, in Unters. Gesammtgeb. Mykol. *6*:
1-34, Taf. I & II, figs. 1-35 (1884). Also Olive, in Proc. Amer. Acad.
Arts Sci. *37*: 341 (1901), and Proc. Boston Soc. Natur. Hist. *30*: 451-
513, figs. 81, 94, 112-118 (1902).

Cultivated on media such as 0.1L-P(pH 6.0), 0.5G-0.1P, and hay infusion agars in association with *Escherichia coli* at 22-25°C. Sorocarps
usually solitary, sometimes clustered, variable in length, up to 1.0 cm or
more when incubated under overhead light, strongly phototropic and often
much longer in unidirectional light, typically bearing branches in whorls
(Fig. 8-19), the number and dimensions of these varying with the strain
and the culture environment. Sorophores weakly pigmented vinaceous or
pale violet, commonly ranging from 15 to 40 μm in diameter at the base
and gradually narrowing to 8-10 μm near the apex, bases slightly clavate,
unexpanded or occasionally narrowed; whorls per sorocarp often 3-8, ranging from 1 to 10, occasionally more, evenly spaced or interrupted, sometimes ill-formed; branches per whorl commonly 2-5, rarely more, variable
in length. Sori globose, vinaceous to pale violet, darkening in age and of
two classes, terminal and lateral (on branches), the former commonly 100-
200 μm in diameter, sometimes larger, the latter about 35-75 μm as a
rule. Spores narrowly elliptical, mostly 5-7 × 2.5 - 3.0 μm (Fig. 13-1A)
but occasionally 8.0 μm in length, with clusters of polar granules usually
prominent. Cell aggregations radiate in pattern (Fig. 13-1C), ranging from
2-3 mm in diameter to quite large with converging streams up to a centimeter in length, the larger aggregations subdividing into several parts
with each producing one or more sorocarps. Vegetative myxamoebae (Fig.
13-1B) irregular in form and variable in dimensions, commonly 15-20 ×
10-15 μm, showing filose pseudopods when actively moving. Macrocysts
relatively rare, usually produced by pairs of compatible mating types, less
commonly formed by self-compatible strains, commonly 25-60 μm in
diameter (Fig. 13-1D); microcysts not reported or observed. The acrasin
is a newly identified dipeptide named glorin (Shimomura, Suthers, and
Bonner, 1982).

HABITATS. Cosmopolitan in distribution, common on dungs of many

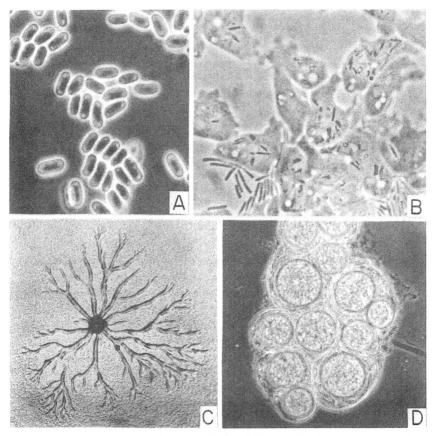

FIG. 13-1. *Polysphondylium violaceum* Brefeld. *A*. Spores showing characteristic form and prominent polar granules. × 900. *B*. Vegetative myxamoebae at the edge of a bacterial colony. × 900. *C*. Aggregation typical of the species. × 13.5. *D*. Cluster of macrocysts in the self-compatible strain ID-10. × 260. (See also Fig. 8-19)

animals (the classic source) and in leaf mould, humus, and forest soils— especially abundant in deciduous forests of the Northern Hemisphere.

No type culture is extant, and strains as isolated from natural sources and cultivated in the laboratory vary so much as to render any designation of a neotype doubtfully useful. Nonetheless, by mentally reconstructing Brefeld's species from his text and, more especially, from his scaled drawings (Figs. 1-2 and 1-3), we believe the following strains conform reasonably well with his original specifications: strain V-6, isolated in October 1937 by KBR from surface soil and leaf litter of a deciduous forest near Vienna, Virginia; strain P-6, isolated in March 1931 by KBR from leaf mould collected in a pine forest at La Plata, Maryland; and strain

ID-10, isolated in the summer of 1970 by J. C. Cavender (1976a) from forest soil collected at Ipo Dam, Philippines.

Of these strains, P-6 was included by Raper and Thom (1941) in their investigation of interspecific mixtures among the Dictyosteliaceae and provided early proof that the chemoattractants for cell aggregation in *P. violaceum* and *D. discoideum* were different (Fig. 7-10). Derivatives of strain V-6 were employed by Shaffer (1961b) for his pioneering studies on the "Founder Cells," shown to be responsible for initiating cell aggregation in *P. violaceum* (Fig. 7-7). Cells of a similar type have since been found to be functional in *P. pallidum* and several other species as well. Strain ID-10 was studied by Nickerson and Raper (1973a,b) in their investigations of macrocyst formation and germination among the Dictyosteliaceae. Additionally, the study of cyst ultrastructure that yielded the first cytological evidence suggesting the sexual character of the macrocysts was centered upon strain ID-10 (Erdos, Nickerson, and Raper, 1972). It is self-compatible and produces macrocysts abundantly on a wide variety of substrates.

The description of *P. violaceum* given above is based upon its growth and culmination on substrates of comparatively low nutrient content. This is done deliberately to facilitate comparison of this and other species under reasonably comparable conditions. *P. violaceum* is, however, one of the dictyostelids that can be cultivated quite successfully on richer substrates with a consequent increase in cell growth, which is in turn reflected in somewhat larger and more numerous sorocarps. Since the use of such media involves the concurrent growth of the slime mold and a bacterial associate, steps may be necessary to offset or remove any excess toxic products of the latter (see Chapter 6). Although *P. violaceum* can grow and fruit reasonably well over a wide range of pH from 5.0 to 7.0, the optimum pH is about 6.0-6.5, and this should be maintained if possible.

Variation among strains makes it difficult to cite specific dimensions for sorocarps in any dictyostelid, and this is particularly true for *P. violaceum*. Not only do strains vary as they are isolated from soil, or other natural habitats, but the same strain will often present quite different pictures on substrates of unlike composition. Other problems may arise as well. When cultures are maintained over long periods by infrequent transfer they often tend to become progressively atypical—in fact, the best proportioned sorocarps are often seen in primary isolation plates, where the slime mold not only feeds upon indigenous bacteria but also competes with a diverse microflora.

Light profoundly affects the pattern and dimensions of the sorocarps, for virtually all strains of *P. violaceum* are strongly phototropic. In unidirectional light, even of very low intensity, the sorophores become very long, and in building these the sorogens commonly fall to the substrate and move for some distance on its surface before reassuming a near vertical orientation. As a consequence long gaps may occur in which no branches are formed, either in whorls or otherwise. As in other dictyostelids, wherever the sorogen touches the substrate, pauses, and then rises again, the sorophore will show some thickening—thus the prior history of the advancing mass can be traced on the substrate surface. An extreme case of elongate sorophores is represented by strain WS-690, received early in 1981 from David Waddell of Princeton as an isolate ("A-10") from bat dung deep in a cave in Arkansas. Few lateral branches are produced, and these only near the distal ends of sorophores several centimeters in length.

Considering their abundance in nature, it seems strange that only a single species, *P. violaceum*, has been described from among the pigmented polysphondylia. Clearly, it is not because they are all alike. On the contrary, additional species may not have been proposed because of their extreme variability and the paucity of breakpoints upon which to hinge new species. Even so, we know of two exceptions, both isolated from soil by J. C. Cavender. The first, from the southern Appalachian Mountains, was cited and illustrated as a very small lavender–pigmented form said to be deserving of species recognition but not formally described (Cavender, 1980). The other, from Puerto Rico, was first reported verbally and subsequently submitted for examination some years ago. It is of near normal pattern and dimensions for *P. violaceum* but characterized by sori of a light tan to greenish pigmentation (fide JCC) rather than vinaceous purple, and spores that are comparatively large and devoid of polar granules.

In addition to providing cytological evidence that macrocysts could be the sites of true sexuality in the dictyostelids (Erdos et al., 1972), *P. violaceum* was one of the first species in which macrocyst formation was shown to result from the pairing of compatible mating types (Clark, Francis, and Eisenberg, 1973). Two syngens or breeding groups were identified, with no cross-reaction between members of different groups. The isolation of compatible mating types from a single gram of soil suggested to Francis that breeding occurred also under natural conditions. Further studies on mating types and macrocyst formation in the species were conducted by Clark and Speight (1973) and by Clark (1974) as reported in Chapter 9. Whereas mating types capable of producing macrocysts in *P. violaceum* appear to be fairly common in nature, reports of only two self-compatible strains have come to our attention: strain ID-10, cited and discussed above, and WS-577, isolated in 1972 from decaying oak leaves by John Sutherland, University of Wisconsin.

As noted above, strain ID-10 produces macrocysts abundantly upon a wide variety of substrates. It is of further interest that they develop abundantly, if slightly later, in illuminated cultures and on substrates containing levels of PO_4 that inhibit macrocyst formation in the much studied strain DM-7 of *D. mucoroides* (Nickerson and Raper, 1973a).

Polysphondylium pallidum Olive, in Proc. Amer. Acad. Arts and Sci. *37*: 341-342 (1901), and in Proc. Boston Soc. Natur. Hist. *30*: 451-513, figs. 61-65, 86, 119-120 (1902). *Synonym: Polysphondylium album* Olive, in same reports.

Cultivated upon substrates such as 0.1L-P, 0.1G-P, 0.1L-P(pH 6.0)/2, and weak hay infusion agars in association with *Escherichia coli* at 22-27°C. Sorocarps solitary or in small clusters, white, erect or semi-erect (Figs. 13-2A and 13-3D), richly branched, typically symmetrical, commonly 3-8 mm high but sometimes more, weakly phototropic; sorophores uncolored, often 2-4 cells thick near the base and 12-30 μm in diameter, tapering upward to a single tier of cells 5-8 μm wide at the terminal sorus, bearing 3-8 or more fairly evenly spaced whorls of lateral branches (Fig. 13-3D), mostly 3-6 branches per whorl (Fig. 13-3D,E); branches often 300-400 μm in length, occasionally rebranched with each branch or branchlet bearing a small sorus at its tip. Sori white to hyaline, of two classes, terminal sori globose or slightly citriform, mostly 75-150 μm in diameter; lateral sori globose and smaller, about 40-65 μm in diameter. Spores elliptical to slightly reniform (Fig. 13-2B), mostly 5.0-7.0 × 2.5-3.0 μm, occasionally up to 8 × 4 μm, polar granules unconsolidated but clearly evident. Cell aggregations radiate in pattern, generally small, fairly compact, with streams often ending abruptly (Fig. 13-2E), commonly 1-2 mm in diameter and producing one or more sorocarps; pre-aggregative myxamoebae 13-18 × 9-14 μm, often with two or more contractile vacuoles that may or may not coalesce before discharging (Fig. 13-2C); aggregating cells elongate and oriented uniformly toward aggregation centers (Fig. 13-2G-I). Macrocysts produced by paired strains of compatible mating types, less commonly by self-compatible strains (Fig. 13-3F); unaggregated myxamoebae commonly form globose to subglobose microcysts 4.0-5.5 μm in diameter (Fig. 13-3E). Cells aggregate in response to glorin, the acrasin of *P. violaceum*.

Parallel with his description of *Polysphondylium pallidum* E. W. Olive (1901, 1902) described a second species, *P. album*, which was said to differ in possessing more numerous and somewhat larger sori, slightly smaller spores, and stalks that are "rather constantly weak at the base, so that the fructifications lie close to the substratum in a characteristic fashion." In our experience, newly isolated strains of unpigmented polysphondylia present such a continuous spectrum of sorocarp size and pattern that we are unable to draw lines of separation upon the bases cited by Olive. Furthermore, sorocarp patterns and dimensions of parts are strongly influenced by culture conditions, and we know not how to evaluate those employed by Olive. For these reasons, we have considered *Polysphondylium album*, based upon a single isolation, to be a synonym of the much more broadly based *P. pallidum*.

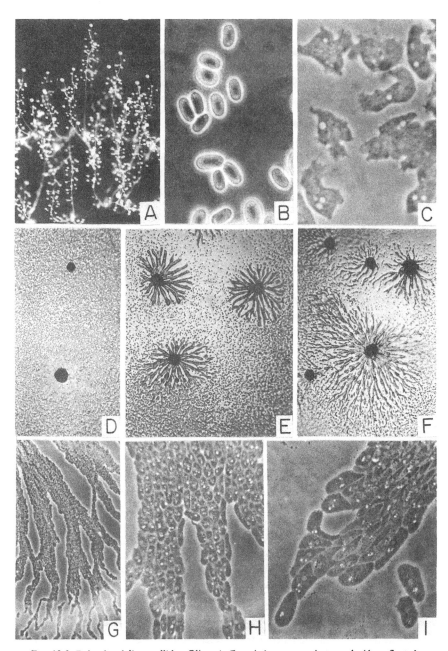

FIG. 13-2. *Polysphondylium pallidum* Olive. *A*. Crowded sorocarps photographed by reflected light. × 6. *B*. Spores showing characteristic form and the presence of polar granules. × 900. *C*. Pre-aggregative myxamoebae; note filose pseudopodia. × 900. *D*. Early stage in cell aggregation. × 12. *E*. Somewhat later stage; note abrupt termination of inflowing streams. × 12. *F*. More expansive aggregation with longer, less compact streams. × 12. *G*. Portion of an aggregation such as the preceding, appreciably enlarged. × 85. *H*. Portion of the same further enlarged to show form and orientation of converging myxamoebae. × 330. *I*. Detail of developing stream. Myxamoeba at lower left has just joined the stream and the two at lower right are about to do so. × 575. Strain WS-320

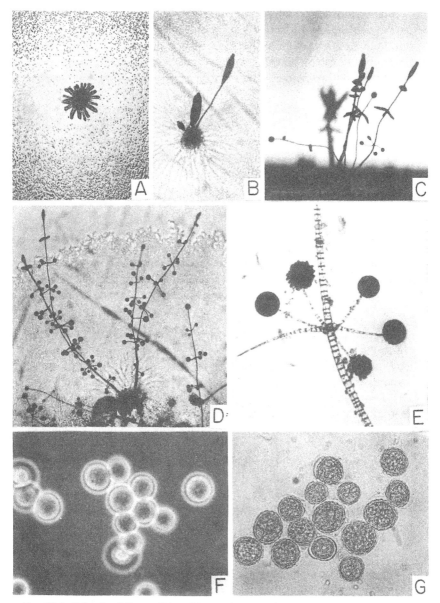

FIG. 13-3. *Polysphondylium pallidum* Olive. A. Late stage in cell aggregation just prior to emergence of sorogen(s). × 12. B. Two large and one small sorogen emerging from site of cell aggregation. × 12. C. Cluster of developing sorocarps of differing age; note progressive stages in whorl formation on sorocarp at right. × 15. D. Sorocarps in terminal stages of development, photographed with transmitted light. × 10. E. A single whorl showing sorophore axis and attachment of six lateral branches, five with sori attached. × 85. F. Microcysts. × 800. G. Macrocysts. × 225, strain EB-4.1 × EB-6.1, from D. Francis

HABITATS. Animal dungs (the classic habitat), and very common in leaf mould, humus, and surface soil of forested areas, deciduous and coniferous, from the Arctic to the tropics in North America and from India, Japan, China, East Africa, Europe, Indonesia, and Southeast Asia. Apparently worldwide in distribution.

No type culture exists, and strains as recovered from nature vary sufficiently to render designation of a neotype difficult if not meaningless. Still, we believe it proper that we cite certain strains as representing our concept of the species based upon the examination of many strains during the past half-century. Three of these are listed herewith:

Strain WS-320, received in September 1955 from Dr. V. Agnihothrudu (1956) of Madras, India, as an isolate from the rhizosphere of a growing plant.

Strain V-1, isolated by KBR in October 1937 from leaf mold and surface soil collected in a hardwood forest near Vienna, Virginia.

Strain WS-543, isolated by John Sutherland in September 1969 from soil and humus of a deciduous forest, University Arborteum, Madison, Wisconsin.

In contrast to the above, one encounters rather frequently strains that are strongly phototropic and, especially in uneven light, produce quite long, often prostrate sorocarps with irregularly spaced whorls of lateral branches, which in turn often number fewer per whorl than in V-1 or WS-320. Representative of such cultures is strain Pan-17, isolated in 1958 by Dietrich Kessler from forest soil collected on Barrow Colorado Island, Panama. It is possible that Olive's *P. album* was based upon a culture of this general type.

Many strains of *Polysphondylium pallidum* grow and fruit well at 30°C; some do well at 34°C also, and a single strain, designated "Salvador," grows and fruits quite well at 37°C. It was isolated in 1955 by Dorothy Fennell from soil collected in San Salvador. Considering its unusual temperature tolerance, it should find useful applications in genetic studies of the unpigmented polysphondylia.

Polysphondylium pallidum has been investigated more intensively than any other member of the genus, and in such studies strains WS-320 and Pan-17 have played important roles, both with regard to axenic cultivation and the formation and structure of microcysts. Both strains were employed by Hohl and Raper (1963a,b,c) in their successful efforts to develop substrates for axenic cultivation of the myxamoebae, while derivatives of the former strain were used by Sussman (1963) in pursuing the same objective. Hohl et al. (1970) again used strain WS-320 for their study of microcyst ultrastructure, as did Hohl (1976) in his investigation of the comparative structure of microcysts and spores. He concluded that the latter differed primarily in possessing walls consisting of a double rather than single layer of cellulose, and in the deposition of an inner mucopolysaccharide layer. Strain Pan-17 proved to be especially useful in determining factors that induce and/or enhance microcyst formation in *P. pallidum*. Of many possibilities investigated by Toama and Raper (1967a), increased osmolarity of the culture substrate, liquid or solid, was found to be especially important; and of the different methods tested for achieving this end, the addition of KC1 at 0.12 M proved to be most productive (ibid, 1967b). Other investigators have proposed different methods for inducing microcyst formation, including exposure of myxamoebae to volatile compounds such as ammonia (Lonski, 1976).

In a comparative study of many unpigmented polysphondylia, Joyce Becker (1959) reaffirmed the need for substrates of comparatively low nutrient content, and demonstrated quite dramatically the deleterious influence of gaseous compounds liberated by bacteria such as *Escherichia coli* when implanted on nutrient-rich media adjacent to but separate from the developing slime mold. The noxious gas(es) was not identified, but ammonia seemed to be a reasonable candidate, since the pH of the substrate rose sharply. From her work and other studies it is known that the unpigmented polysphondylia grow and culminate best under mildly acid conditions and that fruiting can often be improved by the addition of Norite or the substitution of porous clay covers for glass lids on culture dishes. Of the media cited in the description 0.1L-P(6.0)/2 is frequently best.

For nearly two decades David Francis has explored various facets of the biology *P. pallidum*, demonstrating in 1965 that the myxamoebae respond to its acrasin in much the same manner as do the cells of *Dictyostelium discoideum* to cAMP. Additionally, he deduced that cell aggregation was initiated by a single cell, presumably of the "founder" type, as Shaffer had reported for *P. violaceum* four years earlier. Cell differentiation in *P. pallidum* ultimately follows one of three divergent pathways, either to stalk cells, spores, or microcysts, and Francis has demonstrated (1979) that these may show identical sequences of events up to a branch point, after which the sequences are quite different. Lastly, he has explored, and to a very considerable degree elucidated, the sexual cycle of *P. pallidum* by analyzing the progeny of germinated macrocysts obtained by pairing mutants of compatible mating types that carry different markers (Francis 1975b, 1980). This work is continuing and offers great promise, since means are at hand for securing higher percentages of macrocyst germination in *P. pallidum* than are currently realized for *D. discoideum*.

Note should be taken of some recent studies by Spiegel and Cox (1980). Working with haploid and diploid strains of *P. pallidum*, they observed that whereas the intervals between whorls (internodes) were of equal length and the diameters of the stalks were essentially the same, the number of cells per unit distance in the diploid was roughly half that of the haploid. In other words, spacing was determined by some mechanism other than cell counting and was seemingly related to cell volume vis-à-vis the volume of the sorogen at the site of sorophore formation.

Although imperfectly understood, whorl formation in *Polysphondylium* represents a very special type of morphogenesis (Fig. 13-3C,E). As in *Dictyostelium* lateral branches arise from abstricted masses of myxamoebae left in position by the advancing sorogen, but with this important difference. In *Dictyostelium* branches arise singly and are independent of adjacent ones, if such are present. In contrast, the abstricted, doughnut-like cell mass in *Polysphondylium* first cleaves vertically into a number of essentially equal parts, and then within each part the process of culmination is rein-itiated separately but in synchrony with the other segments of the nascent whorl. As a result, the emergent branches are of nearly similar size and pattern, each consisting of a delicate, tapering stalk terminated by a small, globose sorus. Through successive repetition of this process the sorocarp becomes a beautifully symmetrical structure, and in no other species is this exemplified better than in the erect fructification of *P. pallidum*.

Polysphondylium tenuissimum Hagiwara, in Bull. Natur. Sci. Museum Ser. B (Bot.) 5: 69-72, figs. 2 and 3D-I (1979).

Hagiwara's technical description follows, preceded by information con-cerning his conditions of culture. Figure references have been inserted.

Cultivated in association with pregrown cells of *Escherichia coli* de-posited on nonnutrient agar (2%) and incubated at 20°C.:

Sorocarps white, arising singly or in groups, erect or inclined, some-times prostrate (Fig. 13-4A). Sorophores sinuous, arising from clavate or slightly digitate bases, gradually tapering from 10-40 μm in diameter near the base to 2-5.5 μm in diameter near the tip (Fig. 13-4C), with subulate tips, 3.6-9.3 mm in length, 3-21-noded (Figs. 13-4A and 13-5A); nodes bearing 3-10 branches; internodes mostly 250-440 μm in length. Branches straight or somewhat curved, arising from clavate bases (Fig. 13-4D), gradually tapering from 6-15 μm in diameter near the base to 1.5-5 μm in diameter near the tip, with subulate tips, mostly 120-190 μm in length. Terminal sori globose, mostly 30-80 μm in diameter. Lateral sori globose, mostly 20-50 μm in diameter. Spores hyaline, ellipsoid, mostly 1.7-2.0 times longer than broad, smooth, mostly 5.0-6.2 × 2.7-3.5 μm, having polar granules (Figs. 13-4B and 13-5B).

Habitat: In surface soil and leaf mould.

Type specimen: M 50008 (TNS), ex Hagiwara 120.

Strains examined: Hagiwara 120 and 132, pine coppice, Higashiyama,

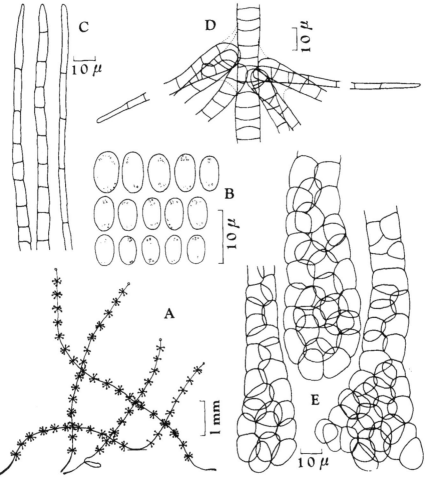

FIG. 13-4. *Polysphondylium tenuissimum* Hagiwara. *A*. Sorocarps showing very large numbers of whorls and the short branches that comprise them. *B*. Spores of varying dimensions, all with unconsolidated polar granules. *C*. Tips of sorophores. *D*. Detail of branch attachment in whorl. *E*. Bases of sorophores. (Courtesy of H. Hagiwara, 1979)

Kyoto Pref., July 4, 1971; Hagiwara 297, *Rhododendron kiusianum* community, about 1500 m. alt., Mt. Kuju, Oita Pref., May 22, 1972.

Hagiwara's further observations follows:

This new species is characterized by sorophores which bear many whorls of tiny branches and small lateral sori. Therefore sorocarps look very thin and branchy. Spores are smaller than those of *Polysphondylium pallidum* but mostly similar in shape. It is very noteworthy that the sorophore bases are often conically expanded (Fig. 13-4E) or somewhat digitate.

The pattern of aggregation is not distinguishable from that of *Polysphondylium pallidum*.

In studies completed after the species description was published, Dr. Hagiwara succeeded in demonstrating the formation of macrocysts by pairing compatible mating types. Two of these mating types were sent to us for examination and study late in 1981, and it is from them that the photographs in Fig. 13-5B-E were obtained. Figure 13-5A is a composite of two photographs previously used by Hagiwara in his species description (1979). As illustrated, the central sorocarp bears 14 whorls of side branches. The maximum number we have observed for a single sorocarp is 35—19 on a nearly vertical stalk and 16 after the sorogen touched the lid of the Petri dish and altered its direction.

Polysphondylium tenuissimum is clearly distinguished by its long thin sorophores, the large number and relatively close spacing of the whorls of side branches, the limited dimensions of these branches, and the comparatively small size of the sori they bear and of the spores the latter contain (Fig. 13-5B).

Over the years we have occasionally collected, but designated as *P. pallidum*, strains that replicate reasonably well Hagiwara's description and illustrations of *P. tenuissimum*. One of these, WS-406 was isolated from forest soil and leaf litter collected near Holly Hill, South Carolina, in November 1957. When regrown recently upon weak hay infusion agar or 0.1L-P/2 in association with *Escherichia coli* at 22-24°C it was not unusual to observe sorophores bearing as many as 15-20 closely spaced whorls, each consisting of 2-5 quite short lateral branches bearing sori 50 μm in diameter or less while the terminal sori commonly reached 80-100 μm. The spores of WS-406 seemed to conform reasonably well with Hagiwara's description and illustrations. They were broadly elliptical, usually 5.0-6.5 × 3.0-4.5 μm, or occasionally more, and contained relatively inconspicuous polar granules. Sorophore bases in WS-406 were generally clavate to slightly bulbous, as depicted in Hagiwara's drawings (Fig. 13-4E). We observed no digitate bases as Hagiwara (1979) described and photographed for the type strain.

No mention is made of microcysts in the original description, and we assume that they were not observed. In contrast, strain WS-406 produces microcysts abundantly within 2 weeks upon all substrates tested, especially on 0.1L-P agar. Microcysts are globose or nearly so, measure about 4.5-7.0 μm, and commonly occur in raft-like clusters. All things considered, we cannot claim species identity with *P. tenuissimum* but regard WD-406 as probably closely related to it.

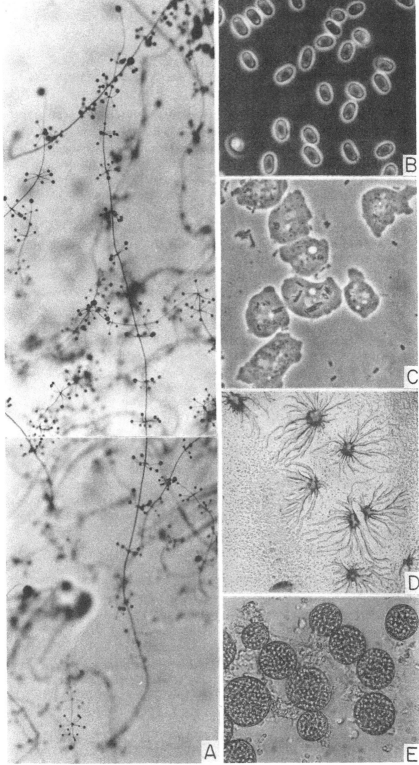

735. *C*. Vegetative myxamoebae outside margin of bacterial colony. × 735. *D*. Cell aggregations developing from mixed planting of bacteria and spores. × 18.5. *E*. Macrocysts
developed in culture of paired mating types, Hagiwara's strains C-95 × C-97. × 300. (Fig.
A consists of two photographs provided by Dr. Hagiwara that were previously used in his
description of *P. tenuissimum*, 1979; *B-E* represent new pictures from our laboratory

Polysphondylium filamentosum Traub, Hohl, and Cavender, in Amer. J. Bot. 68: 169-171, figs. 19-27 (1981).

Cultivated upon substrates such as 0.1L-P, 0.1L-P (pH 6.0), 0.5G-0.1P, and hay infusion agars in association with *Escherichia coli* B/r at 20-23°C. Sorocarps white to hyaline, erect or semi-erect, solitary or clustered, relatively large, with regular whorls of primary and secondary branches, the main and lateral axes of the primary and secondary whorls becoming greatly elongated and filamentous (Fig. 13-6E), up to 10-12 mm or more in length. Sorophores colorless, commonly 20-35 μm in diameter near the base and narrowed to about 3 μm at the summit, the main axis commonly bearing 2 or 3 whorls of primary lateral branches, each of which bears one or occasionally two whorls of 2-4 secondary branches (branchlets), each branch or branchlet bearing a globose sorus (Fig. 13-6E,F); terminal segments long and very thin, sometimes bearing minute sori but often not; primary branches variable in length, comparatively heavy and bulbous at point of origin, 18 to 25 μm in diameter and equal to or wider than the central axis, tapering rapidly; secondary branches often 150-350 μm in length and also tapered. Sori borne primarily on secondary branches, uncolored, globose, commonly ranging from 50 to 125 μm in diameter, occasionally more, remaining intact when flooded with water or 70% ethanol. Spores variable in pattern, narrowly elliptical and 8-10 × 3.0-3.5 μm (Fig. 13-6A) to broadly elliptical and 7.0-8.5 × 4.5-5.0 μm, frequently somewhat reniform, with polar granules evident and unconsolidated. Cell aggregations radiate in pattern (Fig. 13-6C), commonly 1-3 mm in diameter but up to 5 mm or more, the smallest forming single sorocarps and the largest forming either clusters (Fig. 13-6D) or fragmenting with parts fruiting separately. Myxamoebae 15-18 × 10-15 μm with characteristic nuclei, food vacuoles, and often two or more contractile vacuoles that empty independently (Fig. 13-6B). Neither macrocysts nor microcysts reported or observed.

HABITAT. Reported to be widespread in mor and mull humus of both coniferous and diciduous forests of Switzerland. Reported also from forest soil in Ohio, and from bird droppings in New Jersey, U.S.A.

HOLOTYPE. Strain SH-1, isolated in 1970 by Franz Traub from leaf mould of a beech and pine forest, Wilchingen, Switzerland, and brought to our laboratory in 1974.

Polysphondylium filamentosum is withal a very striking slime mold. Nowhere else among the Dictyostelidae do we find a comparable pattern of branching; and in no other species do we find the terminal portions of sorophores, largely branches in this case, so consistently elongate and so

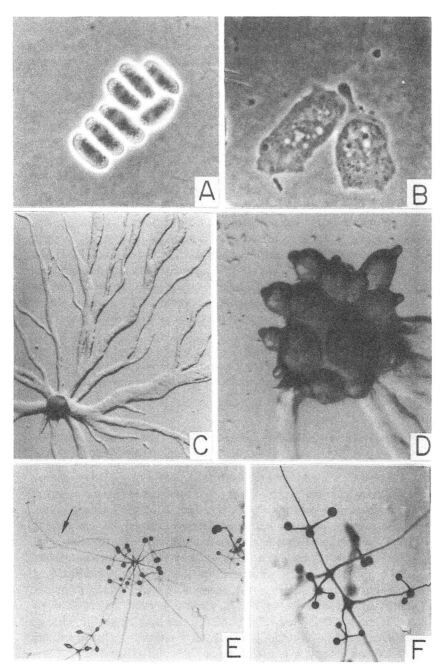

Fig. 13-6. *Polysphondylium filamentosum* Traub, Hohl, and Cavender. *A*. Spores with unconsolidated polar granules. × 1600. *B*. Vegetative myxamoebae on an agar surface. ×
1250. *C*. Large aggregation showing characteristic radiate pattern. × 6.5. *D*. Cluster of
sorogens arising from a single aggregation. × 16.5. *E*. Branched sorocarps with characteristic
long terminal filaments (arrow). × 7.5. *F*. Portion of sorocarp at somewhat larger magni-
fication showing rebranched nature. × 14. (*A-F* from Hohl as used in Traub, Hohl, and
Cavender, 1981)

frequently devoid of sori. There is evidence of kinship with *P. candidum* and probably with *P. pseudo-candidum* as well. In each of these species the terminal segment of the sorophore is often quite long and, as in *P. filamentosum*, may bear a very small sorus or none at all. Culture conditions seem to determine whether the terminal segment in *P. candidum* is extended or remains short and bears a sorus somewhat larger than those on the lateral branches, as is generally true of *P. pallidum*. Such is not the case in *P. filamentosum*, however, for in our experience the mature sorocarps assume the same multi-branched pattern irrespective of the substrate upon which they develop.

In the original description of *P. filamentosum* Traub et al. (1981) reported 0.1L-P agar to be the most satisfactory substrate for the conjoint growth of the slime mold and its bacterial associate, and in our experience it is at least the equal of any other tested. As noted by Traub et al. (1981), the preferred temperature is about 18-20°C, perhaps reflecting the cool temperature of the species's natural habitat. As reported by Traub et al. (1981), satisfactory growth and development can occur at 24°C but not at 26°C or above. Strain SH-1, the type and that studied by us, appears to be relatively insensitive to light.

Polysphondylium filamentosum is a very handsome species, and for the developmental biologist interested in tracing the metabolic changes that accompany predictable alterations in pattern it should offer some challenging opportunities.

Polysphondylium candidum Hagiwara, in Rept. Tottori Mycol. Inst. (Japan) *10*: 591-595, figs. 1-6 (1973).

Cultivated upon substrates such as 0.1L-P, 0.1L-P/2, 0.1G-P, and weak hay infusion agars in association with *Escherichia coli* B/r or on nonnutrient agar streaked with pregrown bacterial cells at 20-25°C. Sorocarps white, usually solitary, sometimes clustered, erect or inclined, not phototropic, mostly 3-6 mm in height exclusive of terminal sorophore segments (Fig. 13-7A). Sorophores uncolored, 15-35 μm near the base and narrowing to 3.0-5.5 μm at the summit, usually bearing 2-4 whorls of lateral branches, commonly 3-8 branches per whorl, occasionally more, typically borne at near 90° to the main axis (Fig. 13-7E); terminal segments may be relatively short with sori 70-150 μm (Fig. 13-8C) or very long, up to 4-8 mm, and sori only 20-25 μm or lacking (Fig. 13-8D); lateral branches commonly 100-200 μm long, conspicuously tapered (Fig. 13-7D) and bearing sori 35-75 μm in diameter. Spores elliptical to slightly reniform, mostly 8-10 × 3.5-5.0 μm, occasionally less or rarely more, with polar granules present but often indistinct (Figs. 13-7B and 13-8A). Cell aggregations radiate in pattern (Fig. 13-7F), commonly about 1.0 mm in diameter and giving rise to single sorocarps, but sometimes larger, 1.5-3.0 mm in diameter, and

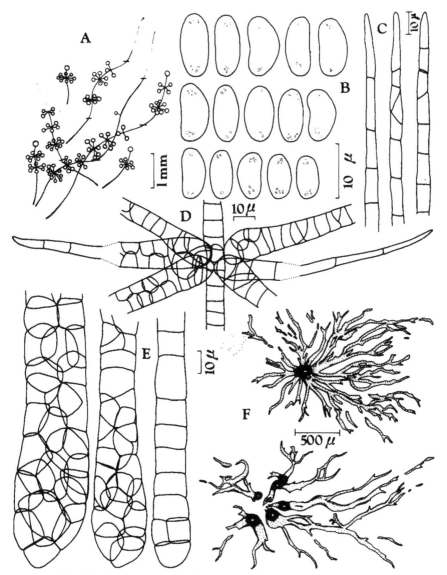

Fig. 13-7. *Polysphondylium candidum* Hagiwara. *A*. Habit of sorocarps showing some with very long terminal sorophore segments and some with short segments bearing larger sori. *B*. Spores showing forms and dimensions. *C*. Sorophore tips. *D*. Detail of lateral branch attachment in a whorl. *E*. Sorophore bases. *F*. Aggregations, one still with single center, the other following partition to yield six centers for sorocarp formation. (Courtesy of H. Hagiwara, 1973)

subdividing with fractions fruiting separately. Myxamoebae about 16-22 × 12-15 μm, with characteristic nuclei and multiple vacuoles (Fig. 13-8B). Neither macrocysts nor microcysts reported or observed.

HABITAT. Leaf mould and surface soils of mountain forests in Tottori and Fukuoka Prefectures, Japan.

HOLOTYPE. Strain 168 was received from Hiromitsu Hagiwara as the type of the species in October 1974; it had been isolated by him in October 1971 from leaf mould and soil collected in a yew forest near the summit of Mt. Daisen, Tottori Prefecture, Japan. A second strain was obtained from the same source and a third was isolated the following year from Mt. Kosho in Fukuoka Pref.

The foregoing description is based in substantial part upon our observations of the type strain cultivated on substrates comparable to those used in cultivating *P. pallidum*. For this reason it differs from Hagiwara's description in certain details, but these are for the most part relatively unimportant. Of the substrates employed in our laboratory, 0.1L-P/2 was usually most satisfactory.

Polysphondylium candidum is singularly attractive when its sorocarps develop under optimal culture conditions. Although the number of whorls seldom exceeds four, the branches in these are evenly spaced and possess a regularity of form that is seldom equaled in other species. The characteristic of very long terminal segments (Fig. 14-8D), once thought to be unique, is shared with two other species, *P. pseudo-candidum* and *P. filamentosum*. These terminal segments are nonetheless interesting, however, and raise puzzling questions as to how and why they develop. We know that their formation is strongly influenced by the culture environment, as Hagiwara has noted (1973a), for under conditions otherwise similar virtually all the sorocarps on one substrate (e.g., 0.1L-P/2) may show terminal sorophore segments up to 4-5 mm in length with sori either minute or lacking, while on a different agar the terminal segments in many sorocarps are little longer than the branches of the uppermost whorl but bear sori of larger size (Fig. 13-8C). Interesting questions are raised about the possible demarcation of prestalk versus prespore cells in the rising sorogens. However produced, the formation of such long terminal segments obviously follows a pattern other than that advanced by Raper and Fennell (1952) for stalk formation in *Dictyostelium discoideum*.

Whereas the species *P. candidum* is based upon strains isolated from forest soils in Japan, it is significant that a culture of slightly larger dimensions, but of the same basic pattern (strain FR-47), was isolated in 1957 by KBR from forest soil in Gif-sur-Yvette, France. For several years

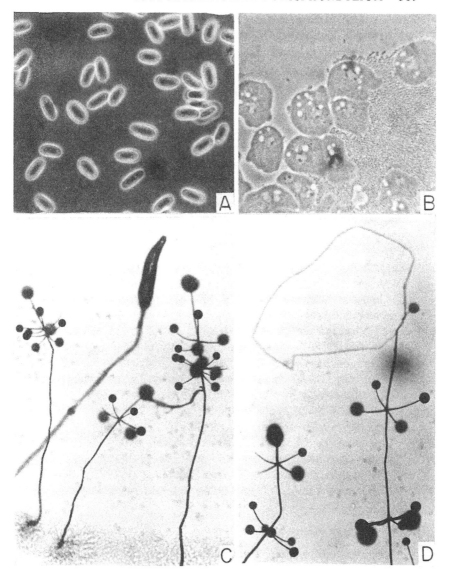

FIG. 13-8. *Polysphondylium candidum* Hagiwara. *A*. Spores showing characteristic form and unconsolidated polar granules. × 900. *B*. Myxamoebae feeding on bacteria at colony margins; cells are slightly compressed by coverglass. × 900. *C*. Three mature compact sorocarps and a rising sorogen. × 40. *D*. Terminal area of compact sorocarp (left) and a sorocarp with very elongate terminal sorophore segment typical of the species. × 40. (*C,D* courtesy of H. Hagiwara)

it was our intent to describe it as a new species until this need was obviated by Hagiwara's description of *P. candidum* in 1973.

Polysphondylium pseudo-candidum Hagiwara, in Bull. Natur. Sci. Museum Ser. B(Bot.) *5*: 67-69, figs. 1 & 3 A-C (1979).

Hagiwara's technical description, together with information concerning the type of substrate employed and the temperature of incubation, follows. Figure references have been inserted.

Cultivated in association with pregrown cells of *Escherichia coli* deposited on nonnutrient agar (2%) and incubated at 20°C.

Sorocarps white, arising singly or in groups, erect or inclined, sometimes prostrate (Fig. 13-9A). Sorophores sinuous, arising from clavate bases (Fig. 13-9E), gradually tapering from 10.5-32.5 μm in diameter near the base to 2.5-5.5 μm in diameter near the tip, with subulate tips, occasionally with conspicuously lengthened terminal segments (Figs. 13-9A and 13-10A), 1.3-7.3 mm in length, 1-5 noded; nodes bearing 1-13 branches; internodes mostly 460-940 μm in length. Branches straight or somewhat curved, arising from clavate bases, gradually tapering from 6.5-17.5 μm in diameter near the base to 2.5-5.5 μm in diameter near the tips (Fig. 13D), with subulate tips, mostly 140-230 μm in length. Terminal sori globose, mostly 30-100 μm in diameter. Lateral sori globose, mostly 30-80 μm in diameter. Spores hyaline, ellipsoid, mostly 1.9-2.2 times longer than broad (Fig. 13-9B), smooth, mostly 6.2-7.9 × 3.0-3.7 μm, having polar granules.

Habitat: In surface soil and leaf mould.

Type specimen: M 50007 (TNS), ex *Hagiwara 660*.

Strains examined: *Hagiwara 660, 663,* and *664,* evergreen forest, Mt. Banna-dake, Ishigaki Is., Okinawa Pref., June 11, 1973; *Hagiwara 744,* evergreen forest, Mt. Omoto-dake, Ishigaki Is., Okinawa Pref., May 31, 1973.

Hagiwara's further observations include:

Sorocarps are macroscopically similar to those of *Polysphondylium candidum* (Hagiwara, 1973) in the apical lengthening of sorophores at the expense of some or most cells which should become spores composing terminal sori (Fig. 13-10A). Moreover, in microscopical observations, spores resemble those of *P. candidum* in shape, i.e. these two species produce more elongate spores than *P. pallidum*. However, this new species differs from *P. candidum* in the following respects. Sorocarps are rather densely produced, sorophores are more delicate, branches are not only slenderer but far shorter though internodes are almost the same in length, and both sori and spores are smaller.

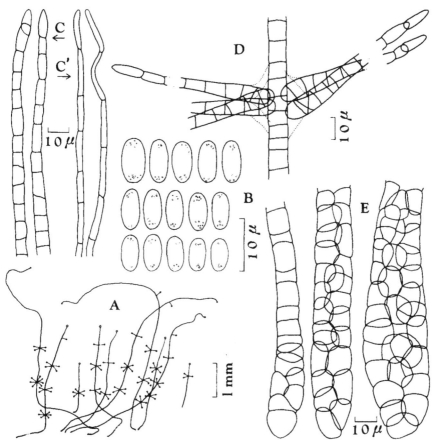

Fig. 13-9. *Polysphondylium pseudo-candidum* Hagiwara. *A*. Habit showing delicate nature of sorocarps, some with long terminal segments and others lacking them. *B*. Spores showing form, dimensions, and presence of polar granules. *C*. Sorophore tips. *C′*. Sorophore tips from long terminal segments. *D*. Detail of lateral branch attachment in a whorl. *E*. Sorophore bases. (Courtesy of H. Hagiwara, 1979)

Cell aggregation proceeds in the same way as that of *Polysphondylium pallidum* (Fig. 13-10B).

Neither macrocysts nor microcysts were reported by Hagiwara or observed in our recultivation of his type strain on 0.1L-P agar in association with *E. coli* at 22-23°C.

While tentatively recognizing *P. pseudo-candidum* as separate from *P. candidum*, it should be noted that the two species are in many ways quite similar, and arguments could be advanced for including *P. pseudo-candidum* as a variant of *P. candidum* with somewhat more delicate structures

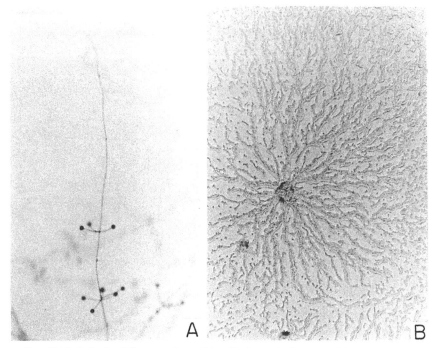

FIG. 13-10. *Polysphondylium pseudo-candidum* Hagiwara. *A*. Sorocarp with very long terminal sorophore segment that is devoid of a sorus. × 23. *B*. Developing pseudoplasmodium at about mid-stage of cell convergence. × 23. (Courtesy of H. Hagiwara)

and smaller dimensions. In other words, it is doubtful if they differ from each other more than many of the isolates that we now routinely assign to *P. pallidum*. Even so, we hesitate to combine the two under the name *P. candidum* or to recognize the former as a taxonomic variety of the latter. We shall for the present retain both and hope that future collections will resolve such questions as now exist.

CHAPTER 14

Acytosteliaceae: *Acytostelium*

TECHNICAL DESCRIPTION

Acytostelium Raper, in Mycologia *48*: 179, figs. 4 & 6 (1956).

Vegetative stage consisting of free-living, independent myxamoebae which feed upon bacterial cells; fruiting stage characterized by the aggregation of these cells to form radiate pseudoplasmodia of varying dimensions and the subsequent development from these of sorocarps composed of very thin acellular sorophores bearing masses of spores as terminal sori.

GENERAL DESCRIPTION AND COMMENTS

Acytostelium is the most recently described genus of the Dictyostelidae and differs from the older genera in several ways, particularly in the acellular nature of its sorophores. It is at the same time relatively rare in nature and is characterized by sorocarps of remarkable delicacy, factors which together probably account for the limited attention the genus has received. First discovered in the summer of 1952, and described briefly by Raper in 1956, the type of a new genus and family, *Acytostelium leptosomum*, was redescribed and reported in appreciable detail two years later by Raper and Quinlan (1958). Since that time three additional species have been described: *Acytostelium ellipticum*, by Cavender in 1970; *A. irregularosporum*, by Hagiwara in 1971; and *A. subglobosum*, by Cavender in 1976. Of these more recent discoveries, *A. ellipticum* is the most striking by far, being marked by elliptical rather than rounded spores and sorogens with elongate and conspicuously narrowed apical parts. *A. irregularosporum* and *A. subglobosum* more nearly resemble *A. leptosomum* and are in fact separated from it primarily upon the basis of their smaller sorocarps and spores that are less commonly globose.

All presently known members of the genus have been isolated from leaf mould and surface soils of deciduous and semi-deciduous forests of temperate and tropical regions. Although the genus is not common, current information would suggest that it is probably global in distribution, since it has been reported from Mexico, Trinidad, the Guyanas, and Colombia in South America, Spain, East Africa, Southeast Asia, Japan, Canada, and many localities in the United States.

Species of *Acytostelium* are easily overlooked in primary isolation plates because of their diminutive size, their lack of distinctive color, and their resemblance to many delicate filamentous fungi. Fortunately their sorocarps often appear as rather compact clones so that one first sees not the individual sorocarps but the whitish colony. Once detected, isolation into two-membered culture with an appropriate bacterial associate becomes an exercise in patience and ingenuity—patience because they tend to develop more slowly than the larger cellular slime molds, and ingenuity because of their meager size and the special culture conditions they require for optimal growth and fructification.

Substrates of very low nutrient content must be used, and agars based upon dilute hay infusions are most generally satisfactory. Very low levels of lactose (or glucose) and peptone (or yeast extract) may be used as well, or, as Cavender (1970) and Hagiwara (1971) have reported, nonnutrient agar (2%) upon which is streaked a dense suspension of pregrown bacteria, such as *Escherichia coli*, to nourish the myxamoebae. The addition of activated charcoal, powdered or as small granules, to cultures following vegetative growth commonly accelerates cell aggregation and subsequently enhances sorocarp formation. In some cases the substitution of porous clay covers for glass culture dish lids has a singularly favorable effect when added at a similar stage.

The developmental cycles of *Acytostelium* species are basically similar to those of *Dictyostelium* and *Polysphondylium*, but with some very important differences. One cannot visually distinguish between elements of the vegetative stage of an *Acytostelium* and those of a *Dictyostelium* or *Polysphondylium*, for the myxamoebae fall within the same general size ranges and appear to be very similar in structure. In each genus the cells when freely moving are characterized by filose pseudopodia, nuclei appear as clear vesicular areas with peripherally lobed nucleoli usually evident, contractile vacuoles are commonly multiple in origin and may but need not fuse before discharge, food vacuoles contain bacteria in various stages of digestion, and demarcation of the finely granular endoplasm and the enveloping hyaloplasm may or may not be evident.

Cell aggregation also follows the patterns seen in *Dictyostelium*, but optimally on a much smaller scale. Aggregations that lead to maximum fruiting in some species, such as *A. leptosomum*, are radiate in pattern, consist of well-defined streams of converging myxamoebae, and measure about 0.7-1.5 mm in diameter; in other species, such as *A. ellipticum*, they appear as small raised mounds and exhibit little or no evidence of inflowing streams, a pattern that is duplicated in *Dictyostelium deminutivum* (Anderson et al., 1968) and certain other delicate species of that genus. Whereas single sorocarps may arise from the smallest mounds, the usual pattern is one of multiple fructifications that develop progressively and sequentially.

Insofar as we are aware, no studies have been directed toward the chemical nature or isolation of the acrasin(s) of *Acytostelium*. The reader may recall, however, Hostack's demonstration (1960) that cell aggregation can be induced or in some manner stimulated by the presence of added steroids or alkaloids.

It is in sorocarp development, or culmination, that the genus differs substantially from the other dictyostelids, for it constructs acellular sorophores and thus expends no cells in sterile supportive tissue. As the myxamoebae accumulate at established centers, elongate sorogens emerge from the enlarging aggregations and in the central axis of each a lengthening tube-like stalk is laid down by the surrounding cells. Since the possible manner of its formation has been discussed elsewhere (see Chapter 8), it need not be repeated here. It is of interest, however, that the rising sorogen may assume different forms while construction of the sorocarps is in progress, these being generally clavate to fusiform in *A. leptosomum* with the apical area gradually narrowed, whereas in *A. ellipticum* the apical narrowing is more abrupt and may at first include nearly half the length of the entire body. In *A. leptosomum*, the most-studied species, the tube-like stalk is rich in cellulose, as are the spore walls (Hohl et al., 1968), and the entire structure is held in an upright position by an expanded apron of slime left on the substrate as the sorogen begins its ascent. Although published information is still lacking, our observations indicate that the same may be said for the other species as well.

Against this background we can now consider the four species that currently comprise the genus *Acytostelium*. Since *A. leptosomum* was the first to be discovered, is the most widely distributed in nature, and is the type species of the genus (and family) as well, it will be considered first and in some measure used as the basis for comparing the species more recently isolated. A key to the species follows.

KEY TO SPECIES OF *Acytostelium*

A. Spores globose, subglobose, or irregularly rounded
  1. Sorocarps gregarious, often numbering 25 or more
     per small aggregation; generally erect, commonly
     1.0 mm or more in height; spores globose, 5.0-7.0
     μm in diameter.............................*A. leptosomum* Raper
  2. Sorocarps solitary or in loose clusters of up to 10-
     12, very delicate, often ephemeral, mostly 300-500
     μm but ranging from 100 to 625 μm in height;
     spores globose or subglobose, rarely oval, com-
     monly 5.0-7.0 μm but ranging from 4.8 to 8.0 μm *A. subglobosum* Cavender
  3. Sorocarps usually solitary, sometimes gregarious,
     250-1175 μm in height, sorophores very thin and
     fragile, commonly sinuous; spores irregular in
     shape, rarely globose, 4.8-8.0 × 3.8-5.8 μm.....*A. irregularosporum*
                                                        Hagiwara

B. Spores elliptical
   1. Sorocarps solitary or gregarious, mostly 350-650
      μm but ranging from 200 to 1000 μm in height;
      sorogens with long, narrow apices; spores ellipti-
      cal, mostly 5.0-6.0 × 2.5-3.0 μm . . . . *A. ellipticum* Cavender

SPECIES DESCRIPTIONS

Acytostelium leptosomum Raper, in Mycologia *48*: 181, 196-198, figs. 4
& 6 (1956); also Raper and Quinlan, in J. Gen. Microbiol. *18*: 16-32,
figs. 1-20 (1958).

Cultivated on low-nutrient substrates such as weak hay infusion or 0.05%
lactose, 0.025% yeast extract agars in association with *Klebsiella pneumoniae* at 23-26°C. Sorocarps very delicate, gregarious (Fig. 14-1E,F),
uncolored, erect or nearly so, consisting of unbranched, acellular sorophores (Figs. 14-1F,G and 14-2B,C) commonly 750-1500 μm in length
and ranging from 1.0-3.0 μm in diameter at the base to 0.5-1.0 μm at the
summit. Sori terminal, globose (Fig. 14-1F), whitish to hyaline, mostly
30-55 μm in diameter, sometimes less, rarely more. Spores globose, hyaline, commonly 5.0-7.0 μm in diameter (Fig. 14-1A), thin-walled. Cell
aggregations typically radiate, variable in size, commonly 1.0-1.5 mm in
diameter and fruiting normally (Fig. 14-1C,E), sometimes much larger
(Fig. 14-1D), up to 1.0 cm or more and usually nonfruiting; sorogens
clavate to fusiform, commonly 125-225 μm in length (Figs. 14-1F and
14-2A). Myxamoebae variable in form and dimensions (Fig. 14-1B), uninucleate, with filose pseudopodia when moving freely, commonly 12-18
× 7.5-10 μm. Unaggregated myxamoebae may form globose microcysts,
5.0 to 8.5 μm in diameter; macrocysts not seen or reported.

HABITAT. Leaf mould and surface soils of deciduous and semi-deciduous
forests in temperate to tropical areas: United States, Canada, Mexico,
Spain, and East Africa.

HOLOTYPE. Strain FG-12A, isolated in June 1952 by KBR from leaf
mould and surface soil collected in a virginal hardwood forest, Funk's
Grove, Illinois. Additional representative strains include WS-57 and WS-
28 from surface soils of deciduous forests, University Arboretum, Wisconsin, and Gull Lake, Michigan, respectively.

In addition to the substrates cited above, *Acytostelium leptosomum* may
be cultivated successfully on agars containing very low levels of glucose
and peptone, and on nonnutrient agars streaked or spread with suspensions
of pregrown bacteria. Whereas *Klebsiella pneumoniae* seems to be the
preferred bacterial associate, other Gram-negative bacteria such as *Esch-*

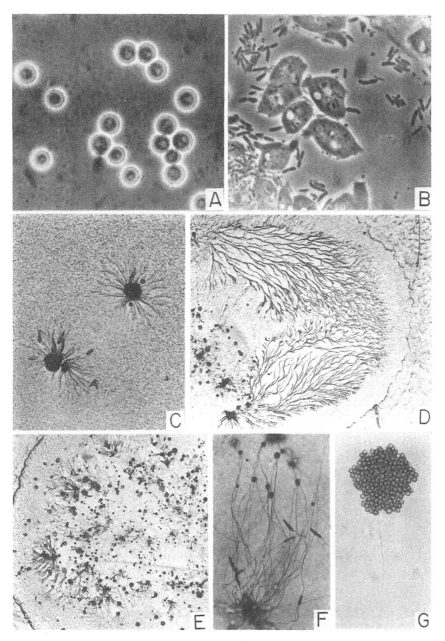

FIG. 14-1. *Acytostelium leptosomum* Raper. *A*. Globose spores. × 900. *B*. Vegetative myx-amoebae. × 875. *C*. Small aggregations with developing sorogens. × 25. *D*. Large aggregations of a type that fails to produce sorocarps and, unlike these, often shows no identifiable centers. × 9.5. *E*. Expanding colony in lawn of *Klebsiella pneumoniae* on 0.1% lactose, 0.1% yeast extract agar; note delicate sorocarps at center and aggregations near the colony margin. × 9.5. *F*. Multiple sorocarps, mature and immature, formed from a single aggregation. × 16.5. *G*. A single sorus and its very thin, unbranched sorophore; the sorus is stained and flattened by a coverglass. × 230. (From Raper and Quinlan, 1958, in substantial part)

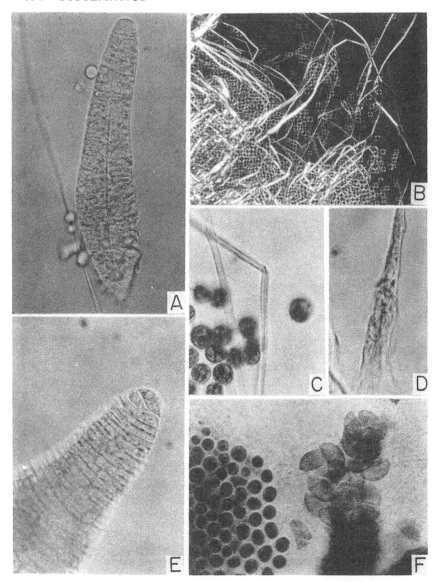

Fig. 14-2. *Acytostelium leptosomum* Raper. *A*. Developing sorocarp showing near horizontal orientation of the myxamoebae around the sorophore, which extends into the apical area. × 550. *B*. Sorophores and spores photographed with polarized light to accentuate their cellulose content. × 175. *C*. Spores stained with erythrosin together with two sorophores. × 1050. *D*. Base of a sorophore with attendant slime. × 1050. *E*. Terminus of a sorogen killed, stained, and photographed without distortion; note the orientation of the cells. × 1050. *F*. Terminal area of sorogen stained and crushed to show the flattened, "pie-shaped" character of constituent cells. × 575. (Rearranged from Raper and Quinlan, 1958)

erichia coli may be used when proper attention is given to substrate composition. If cultivated with *Serratia marcescens* on dilute hay infusion agar the tiny sori appear pinkkish, suggesting but never duplicating the coloration seen in *Dictyostelium discoideum* and *Polysphondylium violaceum*.

Cell aggregations commonly occur in two contrasting and often sequential patterns. The first to emerge are small and wheel-like in pattern and near the sites of inoculation. Such aggregations (Fig. 14-1C) normally fruit without delay, sorogens begin to develop quite early, and some sorocarps are usually fully formed before stream convergence is complete. It is not uncommon for such aggregations, about a millimeter in diameter, to produce 25 or more sorocarps (Fig. 14-1F). Other and similar aggregations often form in the surrounding area (Fig. 14-1E). Outward from these, and somewhat later, very large aggregations may arise with very long inflowing streams (Fig. 14-1D). For such aggregates clearly defined centers may or may not be apparent, streams may persist for days, and sorocarps are rarely produced. In still other cases aggregative activity fails to occur; instead, the myxamoebae become nodulated and burrow into the agar surface, where under low magnifications they resemble massed microcysts. The myxamoebae remain unwalled, however, and if properly stimulated by the addition of powdered charcoal (Norite) they may still form small aggregates and produce normal sorocarps. Equally effective at this stage, and in this species, is the substitution of porous clay covers for the glass lids of the culture dishes. In either case the fruiting response can be quite dramatic.

Not infrequently unaggregated cells enter a resting stage as individuals, forming microcysts by the deposition of thin cellulose walls. The microcysts of most cellular slime molds differ markedly in appearance from true spores, but such is not the case in *A. leptosomum*; both microcysts and spores are spherical. Whereas dimensions of the two cell types overlap, the microcysts are less uniform in size and shape, appear less refractile with phase microscopy, and on the whole average a little larger than the spores. The microcysts appear to germinate in the same manner as true spores, and the released protoplasts leave behind thin hyaline cases from which, apparently, portions of the walls have been dissolved. For reasons still obscure, myxamoebae on weak hay infusion agars seem to form microcysts somewhat earlier than those on other substrates.

Since nutrient levels must be consistently low, the pH of the substrates is not especially critical; cell aggregation occurs between pH 4.5 and 7.6, and abundant, well-proportioned sorocarps are produced over a range from 4.8 to 6.8. The optimum pH is 5.5-6.0, which is close to that of many dictyostelia.

With regard to temperature, *A. leptosomum* grows very slowly at 15°C and may or may not grow at 30°C. At 20°C cell growth and fruiting are quite good but proceed at a slower rate than at 25°C, which appears to be most favorable.

The single special character that sets the genus *Acytostelium* apart from all other Dictyostelidae is the formation of sorocarps without the expenditure of any cells in their stalks, thus conserving all the myxamoebae for producing spores. In no other species can this phenomenon be studied as effectively as in *A. leptosomum*, for, delicate though it is, it is still the largest and most vigorous member of the genus, and it fruits most readily in laboratory cultures.

It has been said that *Dictyostelium discoideum* may represent Nature's earliest example of altruism, since about 25% of the myxamoebae sacrifice themselves in stalk formation so the remainder may form spores. If this be so, *Acytostelium leptosomum* may represent one of the early examples of "intelligent" behavior, for in it no cells are lost during culmination and all become spores.

Acytostelium subglobosum Cavender, in Amer. J. Bot. *63*: 61-63, figs. 1-11 (1976).

Cultivated on low-nutrient substrates such as weak hay infusion, 0.05% lactose-0.025% yeast extract, or 0.05% glucose-0.05% peptone agars in association with *Escherichia coli* or *E. coli* B/r at 22-25°C. Sorocarps very small and delicate, erect or semi-erect, unbranched, single or in loose clusters, mostly 300-500 μm in height but ranging from 150 to 650 μm, often ephemeral. Sorophores acellular, colorless, 1.5-2.5 μm in diameter at the base to 1.0 μm or less at the summit. Sori terminal, globose or nearly so, colorless or translucent, commonly 20-40 μm in diameter, sometimes less. Spores globose to subglobose (Fig. 14-3A) or occasionally oval, commonly 5.0-7.0 μm in diameter, but ranging from 4.5 to 8.0 μm. Cell aggregations may be radiate and small, 400-850 μm in diameter, and produce up to 10 or 12 sorocarps (Fig. 14-3C), or quite large, up to 1.5 cm and often unfocused, and produce few if any sorocarps (Fig. 14-3D); rising sorogens vermiform, mostly 80-200 μm in length when formed, contracting by half or more as they ascend. Vegetative myxamoebae unusually vacuolate, 12-17 × 9-14 μm when moving (Fig. 14-3B). Macrocysts and microcysts not reported or observed.

HABITAT. Humus and leaf mould of tropical rain and semi-deciduous forests of Philippines, Indonesia, and Malaysia.

HOLOTYPE. Strain LB-1, isolated by J. C. Cavender from humus and leaf mould collected in the summer of 1970 at Lambuan, Panay, Philippines. Also represented by strains from Udjung-Kulon, Java; Panang Island, Malaysia; and Ipo Dam, Luzon, Philippines.

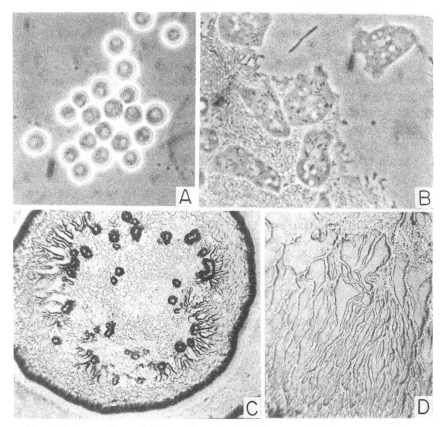

FIG. 14-3. *Acytostelium subglobosum* Cavender. *A*. Globose to subglobose spores. × 875. *B*. Vegetative myxamoebae at edge of bacterial colony. × 875. *C*. Clonal colony containing aggregations of the type that produces sorogens and sorocarps. × 30. *E*. Long inflowing streams without a definitive center. Such aggregations rarely produce fructifications. × 30

Acytostelium subglobosum, as represented by strain LB-1, bears a rather striking resemblance to *A. leptosomum* in the character of its growth and in the pattern and dimensions of its spores. It is, however, a far more delicate slime mold and should be retained as a separate species, at least for the present. Sorophores are thinner and substantially shorter than those of *A. leptosomum*, and its sori average measurably less. The sorocarps as a whole are more fragile and tend to collapse within 2-3 days, even when undisturbed. The spores are less consistently globose than those of *A. leptosomum*, perhaps due to prior interspore pressures within the relatively small sori.

Cell aggregations are of two basic types as in *A. leptosomum*. The first to appear are generally small, radiate organizations a few hundred microns in diameter; sorogens are produced within 2-3 days and sorocarps soon thereafter. As in *A. leptosomum*, myxamoebae continue to converge for some time after sorocarp formation begins. Fruiting aggregations also take the form of very small mounds with little or no streams present. Such pseudoplasmodia produce sorocarps in very limited numbers. Most spectacular are the very large aggregations, or, stated more accurately, areas of cell aggregation (Fig. 14-3D), for they often lack clearly defined centers and consist primarily of parallel and often anastomosing streams oriented in some general direction. Once such streams are formed, the myxamoebae cease to be conspicuously elongate and gradually become moribund; contractile vacuoles are still present but discharge very infrequently. If the streams fragment and fruiting is attempted the sorocarps are generally atypical, often abortive, and commonly produce no sori.

All the substrates cited in the species description have been used successfully but none triumphantly. Of the three, 0.05% lactose, 0.025% yeast extract has been the most useful, but generally with some modification or supplementation. In describing the species Cavender (1976a) recommended the use of a firm agar gel (2%) to lessen myxamoebal penetration, and in our experience we have found 3% agar to be even better. The addition of activated charcoal to preaggregative cultures provides a further advantage. The insertion of activated charcoal as small chips (6-14 mesh) is helpful in promoting aggregation and fructification in adjacent areas, but greater responses may be had by placing mounds or ridges of powdered charcoal (Norite) between or near the areas or streaks of slime mold growth. Somewhat to our surprise, we have not found clay dish covers to give a response comparable to that in *A. leptosomum* or, to a lesser degree, in *A. irregularosporum*.

We have not observed microcysts in any cultures of *A. subglobosum* even after 4 weeks, albeit rounded and seemingly near dormant myxamoebae are still present.

Acytostelium irregularosporum Hagiwara, in Bull. Natl. Sci. Museum (Tokyo) *14*: 364-365, fig. 8 (1971).

Hagiwara's original description follows, preceded by a statement of the manner in which his culture was grown.

Cultivated on nonnutrient agar plates with appropriate amounts of pregrown cells of *Escherichia coli* at room temperature (15-25°C).

Sorocarps solitary, occasionally gregarious, erect or semierect, unbranched; sorophores white, extremely delicate, 250-1175 μm in length, 1-3.5 μm in diam at the base, gradually tapering to 0.5-1.0 μm in diam near the tip; sori white, opaque, globose, 7-24 μm in diam; spores colorless, transparent, commonly subglobose or ovoid, somewhat irregularly shaped, occasionally globose, 4.8-8.0 × 3.8-5.8 μm, smooth.

Hab. From leaf mould.

Coll. strain No. 28, E-4.

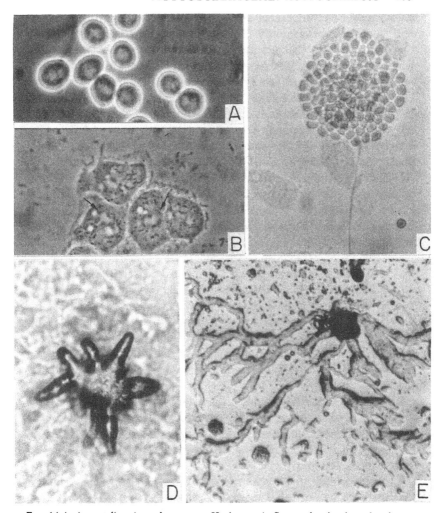

Fig. 14-4. *Acytostelium irregularosporum* Hagiwara. *A*. Spores showing irregular shapes characteristic of the species. × 900. *B*. Vegetative myxamoebae. Note the vesicular nuclei (arrows), characteristic of the dictyostelids. × 900. *C*. A sorus compressed on the agar surface to show the relative dimensions of the spores and sorophore. × 460. (Judged by their nuclei, the two amoebae in the background belong to some genus other than *Acytostelium*.) *D*. An aggregation from which multiple sorogens are emerging. × 150. *E*. A portion of a developing pseudoplasmodium. × 60 (*C-E* provided by H. Hagiwara)

Hagiwara's culture #28, the type of *Acytostelium irregularosporum*, was received in October 1974, preserved by lyophilization two months later, and reactivated for comparative examination in August 1981. It appears to be typical of the species as described. Our observations follow.

When cultivated on weak hay infusion agar and on 0.05% glucose, 0.025% peptone agar in association with *Escherichia coli* B/r at 22-26°C, or on nonnutrient agar with pregrown B/r cells, sorocarps generally as described by Hagiwara except that very few reach a height of 1.0 mm or more and many are variously bent and twisted, hence become tangled. Sorophores are very thin and fragile and seldom measure more than 2.0 μm, even at the base. Sori average somewhat larger than reported, commonly range between 20 and 35 μm, rarely more, and none as small as 7 μm (the diameter of a spore). Spores as reported (fig. 14-4A), irregular in form, and larger than in *A. leptosomum* or *A. subglobosum*. Cell aggregations vary in form and dimensions: (1) mound-like producing few sorogens (Fig. 14-4D); (2) small and radiate, 1-2 mm in diameter, fruiting directly or subdividing (Fig. 14-4E) with fractions fruiting separately; or (3) quite large with long, prominent, unfocused streams that may persist for days and never fruit. This occurs also in *A. leptosomum* and *A. subglobosum*. As pointed out by Cavender (1976a) for the latter species, optimal fruiting of *A. irregularosporum* also occurs in the first small clones to clear the bacterial lawn.

Cell aggregation is facilitated by the addition of activated charcoal. When this consists of small chips, streams radiate outward from them and sorocarps sometimes form on their sides. The addition of powdered charcoal (Norite) as small mounds or ridges is even more effective, and the substitution of clay dish covers for glass or plastic is clearly beneficial. In all these cases, unfortunately, many, and sometimes most, of the nascent sorogens never complete the fruiting process, and so relatively few well-formed sorocarps project above fields of incomplete or abortive fruiting structures.

It is of interest that Hagiwara's culture was isolated from leaf mould of a deciduous forest in Japan, since the other three species come from similar sources in different parts of the world. Currently, there is no evidence that dung is a natural habitat for this genus, as it often is for many species of *Dictyostelium* and *Polysphondylium*.

Considering their growth characteristics, their aggregative behavior, and the patterns of their spores, it appears that *A. leptosomum*, *A. subglobosum*, and *A. irregularosporum* may be closely related, and one might question whether the latter two represent more than varieties of the first. How significant are spore shapes that could result from mutual compression within the very small sori that sometimes characterize these slime molds? Cavender has recovered as few as 50 spores in some sori of *A. subglobosum*, and we have counted as few as 22 and 25 in minute sori of *A. irregularosporum*. We know that the spores have very thin walls and can be easily deformed beneath a coverglass; furthermore, irregularities in shape can often be observed within the intact sori of the latter species and may result from a paucity of enveloping slime. If suspended in the ample slime of larger sori, might not all the spores be globose? Having raised the question of spore shape and its possible bearing on species validity, we shall not attempt to settle the matter here but leave its resolution to time and additional collections.

Acytostelium ellipticum Cavender, in J. Gen. Microbiol. *62*: 119-123, figs. 31-40 (1970).

Cultivated on low-nutrient substrates such as very weak hay infusion or 0.05% glucose, 0.025% peptone agars in association with *Escherichia coli* or *Klebsiella pneumoniae* at 20-25%°C. Sorocarps extremely delicate, solitary or gregarious (Fig. 14-5E,F), unbranched, erect or semi-erect, mostly 350-650 but ranging from 200 to 1000 µm in height. Sorophores acellular, unpigmented, 1.5-2.5 µm in diameter at the base, tapering to less than 1.0 µm at the summit. Sori globose or nearly so (Fig. 14-5H), hyaline, mostly 25-45 µm in diameter, rarely more. Spores elliptical, mostly 5-6 × 2.5-3.0 µm (Fig. 14-5A), sometimes less, and rarely 7.5 × 3.5; usually but not consistently showing prominent monopolar granules. Cell aggregations differ in pattern, usually small, scattered or closely spaced and mound-like with little or no cell streaming (Fig. 14-5C,D), sometimes radiate in pattern, up to 1.0-2.0 mm in diameter, later subdividing; sorogens characterized by narrow, elongate apices, later contracting (Fig. 14-5F,G). Myxamoebae very irregular in form, mostly 10-15 × 8-12 µm (Fig. 14-5B). Neither microcysts nor macrocysts reported or observed.

HABITAT. Leaf mould and surface soil of tropical moist forests, Trinidad, West Indies; Guyana and Colombia, South America.

HOLOTYPE. Strain AE-2, isolated in the summer of 1968 by J. C. Cavender from soil collected in the Trinidad Government Forest Reserve, Arena, Trinidad. Isolates similar to AE-2 were obtained from soils of Guyana and Colombia.

Acytostelium ellipticum differs from other members of the genus in a number of ways, particularly in the form of its spores and the pattern of its developing sorogens. The spores are narrowly elliptical, or capsule-shaped (Fig. 14-5A), rather than globose or irregularly rounded, hence resemble those of *Polysphondylium* and most species of *Dictyostelium* more than those of other acytostelia presently known. The sorogens are quite distinctive also, not only within the genus but among the Dictyostelidae as a whole. They first appear as very narrow columns that project outward from the surface of the mound-like aggregations, and as they lengthen, the apical portions retain the same general diameter while the posterior portions become appreciably swollen (Fig. 14-5F), the overall pattern being ventricose-rostrate, to use Cavender's descriptive terminology. At no time do they assume the clavate or fusiform shapes that characterize *A. lepto-*

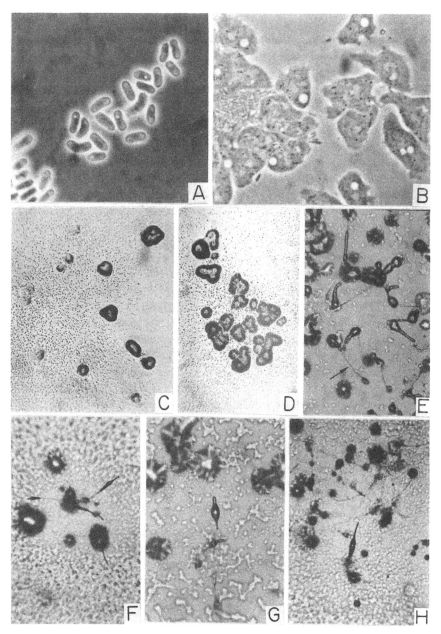

Fig. 14-5. *Acytostelium ellipticum* Cavender. *A*. Typical elliptical spores, commonly showing one, sometimes two prominent clusters of granules. × 900. *B*. Vegetative myxamoebae at edge of bacterial colony. × 875. *C*. Scattered, mound-like aggregations formed without stream formation. × 30. *D*. Clustered aggregates, perhaps derived in part by fractionation. × 30. *E*. Portion of a culture in early fruiting stage; note the young sorogens with long, narrow apices; an older one rests on its sorophore (arrow). × 85. *F*. Small aggregation from which three sorogens have emerged. × 85. *G*. Two sorogens more advanced, one in poor focus. × 85. *H*. Clustered fruiting structures; some sori have collapsed on the agar, while a young sorogen projects above it. × 85

somum. As sorocarp construction proceeds and the stalk lengthens, the narrow apical area becomes progressively shorter and in time comes to resemble the terminal papilla seen in the later fruiting stages of many dictyostelia (Fig. 14-5G).

Aggregations may be small and mound-like and give rise to one or a very few sorocarps, or they may be radial in pattern and up to 1.5-2.0 mm in diameter with a small crateriform center. In the latter case the rim of the crater may subsequently expand, then become segmented with each segment yielding one or more sorogens (Fig. 14-5D). Alternatively, developing aggregations may assume a rosette pattern with multiple lobes, each of which gives rise to one or more sorogens—the whole presenting a clustered apppearance while, in the main, the individual sorocarps tend to stand apart.

In describing *A. ellipticum*, Cavender (1970) recommended the use of weak hay infusion and 0.05% glucose, 0.025% peptone agars, the first for conjoint growth of the slime mold and associated bacteria, and the second as a base upon which to streak or spread a mixed suspension of slime mold spores and pregrown bacterial cells, either *Escherichia coli* or *Klebsiella pneumoniae*. In our reexamination of the species we have employed both procedures, and of the two we have found conjoint growth on weak hay infusion agar to be most productive. Cell growth is adequate and fruiting is much more consistent on this medium. We have also employed 0.05% lactose, 0.025% yeast extract agar for purposes of comparison, since it is used quite commonly in cultivating *A. leptosomum*, about which we have far more information. With each of the dual nutrient substrates fruiting of *A. ellipticum* has, in our experience, been limited to small isolated islands and generally delayed. To our regret aggregation and fruiting in this species are not enhanced appreciably by the addition of charcoal or the substitution of clay for glass covers on culture dishes.

It is of some interest that this singular slime mold has been isolated only from Trinidad, in the Caribbean, and from countries along the northern edge of South America. Perhaps time will reveal whether *A. ellipticum* is endemic to that general region, or if current information reflects only the limited nature of past surveys.

CHAPTER 15

Coenonia

Coenonia denticulata was described by Ph. van Tieghem in 1884, and from his written account it must have produced more highly specialized fruiting structures than any cellular slime mold presently known. As was true of *Acrasis granulata*, which he described four years earlier, *Coenonia* was not illustrated; and like that species also, it has not been rediscovered. Thus we are left with van Tieghem's brief report as the sole record of a very remarkable slime mold—one that is thought to represent a dictyostelid. What follows, therefore, represents a condensation of his original statement, for we cannot do better than to pharaphrase his description as it has been translated freely from the original French.

Coenonia was first discovered on beans (*Vicia*) which had been forgotten for some time in a glass of water. Evaporation of the liquid had partially denuded many of the uppermost seeds, and it was on the surface of the exposed integuments that the myxamoebae had gathered and fruited after multiplication in the liquid that had become alkaline.

Each fruit (sorocarp) is composed of an uncolored stalk (sorophore) 2 to 3 millimeters high, attached at the base by a ramifying foot in the form of a crampon or hold-fast and dilated at the summit into a sort of cupule or chalice with a finely dentate rim; this chalice supports a spherical, yellow and gelatinous globule (sorus) formed by the mass of spores. The basal crampon and the terminal chalice are sufficient to distinguish this organism from all the other Acrasieae.

When inoculated in a drop of nutrient broth, or in a drop of fresh or slightly ammoniacal urine, the spores germinate and, in this way, it becomes possible to follow on a microscope slide the complete development of the plant up to maturity of new fructifications.

In germinating, the spherical spore 6 to 8 μm in diameter, breaks its yellow membrane and liberates a colorless myxamoeba containing a nucleus; this (myxamoeba) later divides into two. Growth continues and such bipartition is repeated until the nutrients are exhausted. Only then, usually five to six days after inoculation, do the numerous myxamoebae begin to gather here and there at the rim of the drop or even on the glass slide a short distance from the edge. In each mass (aggregate), and as new myxamoebae continually flow in to enlarge it, some of the myxamoebae at the bottom of the heap, and in contact with the slide, become immobile, swell by absorption of water, and in pressing one against another become polyhedral, and finally secréte for themselves cellulose walls ("membranes"). This group of myxamoebae constitutes a small central plaque from which irregular branches, bifurcating here and there, radiate in all directions; thus the crampon is formed. From the central part of the crampon the pseudoplasmodium, consisting of mobile myxamoebae, grows upward and forms an obtuse cone or sorogen. At the same time, certain myxamoebae are deposited one above the other in the axis of this cone and above the central part of the crampon; they enlarge by absorption of water, crowd together, secrete cellulose walls, and finally form a more or less thick column (stalk or sorophore) depending on the dimensions of the developing fruit. As the mass of mobile myxamoebae creeps up all around it, this column grows by the accretion at its summit of more immobilized cells. When the column has attained its final height the whole mobile aggregation of cells collects at the top of the sorophore into a terminal, somewhat spherical mass. The summit then becomes dilated by the progressive immobilization at its periphery of all the myxamoebae which constitute the lower portion of the sphere. This results in the formation of a solid cupule, thicker in its center and thinner around the rim, that supports the rest of the spherical globule (sorus). Afterward, all the myxamoebae of the sorus, without absorbing water or growing, become round and wrap themselves with a membrane, the internal layer of which is cellulosic and yellowish, whereas the external layer is gelatinous and colorless. During this process each myxamoeba becomes a spore.

As the myxamoebae creep up along the stalk, and as the latter is exposed from base to summit, one can see that each peripheral cell has, midway on its external face, a small tooth-like protuberance, the acute point of which is curved upward. The crampon cells have no such teeth or, at most, they are indicated here and there by a dark granule. On the other hand, the cupule is amply provided with these; they occur on the lower cells of the cupule as on the stalk cells. Each of the marginal cells of the cupule is prolonged into a tooth, as mentioned above, while the upper cells of the cupule also extend their membranes in between the spores like so many fine, more or less long and regular teeth. The stalk teeth make easier the ascent of the myxamoebae; the teeth situated on the rim and on the upper part of the cupule prevent the spores from falling down and play a role similar to that of the teeth of the columella in *Mucor spinosus*. Because of this character, the species is named *Coenonia denticulata*. The genus name, *Coenonia* is chosen to emphasize that its fruit (sorocarp) is formed by the simple association of the myxamoebae.

In slide cultures, all fructifications appeared simple; but on bean seeds, many of the larger sorocarps were branched. At about mid-height, a stalk supported by a very large crampon had a verticil consisting of three short, equidistant branches; each of these was of the same nature as the stalk and likewise bore a small terminal cupule supporting a globule of spores. The fact that fructifications in a single and unique species can be branched or not, according to their dimensions, decreases the importance of the branching and nullifies its value as a generic character.

What seems to me a feature of general scientific interest is the striking differentiation observed in its fructifications since the fruit is formed by the simple aggregation of identical, originally free cells. Interest also stems from the fact that the nature of the differentiation of a given myxamoeba depends only upon the relative position this myxamoeba occupies in the whole aggregate. The former fact is evident; the latter can be easily demonstrated.

At the beginning of aggregation, and after the formation of the crampon, if one removes most of the pseudoplasmodium with a needle without removing the crampon and places this pseudoplasmodium in a nutrient drop, an aggregate is again formed, first developing a new crampon then a complete fruit smaller than the previous one. It is obvious that a certain number of myxamoebae, which was destined to form either the cells of the stalk or of the cupule are now transformed into crampon cells.

Later, after the column (stalk) has more or less reached its definitive dimensions, if one removes the terminal, spherical, still uncolored globule (presorus) and replaces it in the nutrient drop, this globule soon produces a new crampon, a new stalk, and finally a new complete but much smaller fruit (sorocarp). In this case, a certain number of myxamoebae which were supposed to become spores were clearly diverted in differentiation and instead produced crampon and stalk cells.

This independence and adaptability ("indifference") of the constitutive cells, which as one can see, do not preclude critical shifts in differentiation, confer to the family Acrasieae a great biological interest which is certain to increase in proportion as one learns more about these singular organisms.

In the absence of either a culture or illustrations of *Coenonia denticulata*, what follows must of necessity rest in part upon conjecture. Still, we think the weight of evidence points to a dictyostelid not unlike *Dictyostelium* and *Polysphondylium* in much of its development, but quite unlike these genera in its later stages of morphogenesis. We do not for one moment question the reality of this slime mold or the accuracy of van Tieghem's report, as some individuals have suggested—we only regret that it has not been rediscovered.

Let us now consider some of the characteristics of *Coenonia* in the light of what has been learned about other cellular slime molds and about the dictyostelids in particular.

(1) The regeneration of entire sorocarps from fractional parts of the

pseudoplasmodium is precisely what one could expect in a species such as *Dictyostelium discoideum*, even from prespore cells while they remain amoeboid.

(2) The character of the stalk would conform with that of *D. discoideum* in many particulars, but not in all. A height of 2-3 mm seems quite reasonable, as does a stalk whose thickness varies in proportion to the cell mass; and, not least important, it is composed of swollen, polyhedral cells with walls (membranes) of secreted cellulose. Only in the terminal sorus-bearing cupule do we find a structure without counterpart in *Dictyostelium*.

(3) The crampon or holdfast, long regarded as unique, must surely resemble, if not essentially duplicate, that of *D. rhizopodium* and allied species, for it forms near the substrate surface and later serves as a base from which the stalk originates.

(4) Although rare among the dictyostelids, spherical spores with cellulose walls are found in *D. rosarium* and *D. lacteum*, although somewhat smaller than those of *C. denticulata*. What is unique about the latter are their yellow pigmented walls and the external gelatinous layer.

(5) We are provided no information concerning the myxamoebae, either in the trophic phase or during the fruiting process. We know not whether they had filose pseudopods (as in the dictyostelids) or lobose (as in the acrasids) when moving freely in liquid. Neither do we know the type of nucleus, whether it contained a lobed nucleolus (as in the dictyostelids) or a compact, centrally positioned nucleolus (as in the acrasids). Judging by the size of the globose spores, we would surmise that the myxamoebae were about 18-25 μm in major diameter.

(6) Information concerning cell aggregation is minimal. We are told that the myxamoebae flowed in continuously to enlarge the cell mass, but we do not know if streams were formed as in most dictyostelids, or if the cells converged individually as in some of the smaller species of *Dictyostelium* and in all of the acrasids. Once together, however, they formed in succession a crampon and an upright cone (sorogen) as in *D. rhizopodium*.

(7) The cupule or chalice which contained the spores is without parallel among cellular slime molds presently known, and it must have been constructed in a manner quite different from that with which we are familiar. Otherwise, one must assume that the prespore cells in some manner surmounted the cupule, representing an expanded terminus of the sorophore. This seems improbable—instead, we suspect that in some manner, still unknown, the prestalk and prespore cells differentiated synchronously to produce a completed whole. There is the faintest suggestion of a cupule in the mature sorus of *D. sphaerocephalum*, where debris (and sometimes a few vacuolated cells) form a persistent calyx-like structure at the base of the sorus.

(8) The presence of small, upturned teeth on the external cells of the stalk has no parallel in any dictyostelid known to us; even more bizarre

are the long, fine teeth that were said to extend between the spores from the uppermost cells of the cupule. Were it not for items 1-4 above we too might doubt the reality of *Coenonia denticulata*, but they cannot be dismissed.

(9) Disturbing also is the alkaline liquid surrounding the decomposing beans in which the myxamoebae had grown, as is the slightly ammoniacal urine used for cultivating the slime mold on microscope slides from germination to maturity. Whereas some isolates of *Dictyostelium* (e.g., *D. mucoroides*) can grow and fruit in cultures of pH 7.5-8.0, this is far from optimal and a range of 6.0-6.5 is generally preferred by the dictyostelids. In contrast some acrasids (e.g., *Guttulinopsis vulgaris*) thrive at the higher pH.

(10) Whether dictyostelid or acrasid or neither, *Coenonia denticulata* as isolated, observed, and reported by van Tieghem (1884) was, and we believe remains, a very singular slime mold. By any standard it is a genus and species worth seeking.

How should *Coenonia denticulata* be classified? Should it be included in the Dictyosteliaceae as was done by E. W. Olive some eighty years ago—and he was the last researcher to record an opinion—or should it be assigned to a new and different family, the *Coenoniaceae*, as we would be inclined to do? However, since we have no culture, and no illustrations of the original one exist, perhaps it is well to leave the decision to the fortunate biologist who at some future date rediscovers it. In any case it should no longer be ignored.

In ending this monograph on the dictyostelids, it is perhaps well to look again at some of the striking adaptations encountered among these slime molds and to consider how they redound to the organism's advantage, and to that of an investigator as well. Cell aggregation coupled with ensuing morphogenesis certainly enables the myxamoebae to achieve ends that are impossible to attain as independent, unprotected cells; and the disposition of *Dictyostelium purpureum* and other long-stalked forms to build sorocarps toward a weak source of light, a minimal increase in temperature, or a more open area with slightly reduced humidity must be significant for the dispersal of spores, for in nature these slime molds grow not on agar surfaces but in thin watery films within the recesses of soil and decaying vegetation. Any mechanism that transports the spore-forming cells away from the site of prior growth and into a situation more favorable to subsequent dispersal by water, insects, or other means must ensure a wider distribution for the species and a greater possibility of survival. *Dictyostelium discoideum* achieves the same end in a different and seemingly more efficient manner, for the whole cell mass migrates to a more favored site before culmination and sacrifices relatively fewer myxamoebae in constructing a supportive stalk. *Dictyostelium polycephalum* seems more haphazard in its approach, as the delicate migrating pseudoplasmodia are apparently insensitive to light. However, as if compensating for this deficiency, many of these structures arise from a single primary aggregate and some perforce find the proper balance of substrate, temperature, and humidity required for sorocarp formation. The additional characteristic of producing not one but several adherent sorocarps with divergent sori enhances spore dispersal. This tendency is even further developed in species of *Polysphondylium*, with their whorls of sorus-tipped branches that project outward from the main stems, which are themselves constructed toward the light and into the more open areas peripheral to their sites of origin. And finally, in *Acytostelium* a pattern has emerged wherein no cells are sacrificed in sterile supportive tissue but all myxamoebae are transformed into reproductive spores. Sensitive to their environment and capable of singular collaboration and differentiation at the cellular level, these slime molds seem to have mastered the problems of when to produce their spores and how to ensure maximal distribution.

There are in addition two other modes of survival, microcysts and macrocysts. The first represent unaggregated myxamoebae that encyst individually, hence are able to carry the slime molds through limited periods of adversity. The second are something special! They are formed under conditions often suboptimal for sorocarp formation and, insofar as we know, are without parallel in other organisms, since they arise from newly formed zygotes that first attract and then engulf neighboring myxamoebae and in the process lay down three protective walls, one of which is rich in cellulose. Unique in origin and structure, they are also the sites of sexual reproduction in the dictyostelids.

The cellular slime molds are simple organisms by any standard one may wish to apply, and the information to be gained from them may likewise have limited application in terms of more complex and highly organized creatures. Yet, their very simplicity and ease of cultivation commend them as models for the analysis of certain basic phenomena relating to (1) progresssive cellular interdependence, (2) intercellular communication and response, and (3) coordinated divergent cellular differentiation. We know that the environment can determine whether a pseudoplasmodium of *D. discoideum* will continue to migrate or cease movement preparatory to fructification, and whether the sorophore of *D. purpureum* will become quite long or remain relatively short before beginning the formation of a sorus. Additionally, we can determine and decide when successive stages in morphogenesis will occur and, in many strains, shift further development from the asexual (sorocarp) to the sexual (macrocyst) mode.

The completed sorocarp in most species consists essentially of two cell types: encysted resting cells, or spores, and strongly vacuolate, nonviable, parenchyma-like stalk cells. In some species one can predict which block of cells within the pseudoplasmodium or sorogen will differentiate into one type or the other, but relatively little is known concerning the really basic question of what happens among the associated cells to effect the necessary changes in behavior, structure, and function. What induces the critical realignment of the myxamoebae when a pseudoplasmodium of *D. discoideum* ceases migration and rounds up preparatory to fruiting? Of this we still have only limited information. And what really causes a sorogen of *D. purpureum* to begin the construction of its sorophore in a vertical direction as opposed to continued extension in a horizontal plane? What alterations in metabolism, influenced if not governed by the environment, determine the specific course to be followed by particular cells? We have far too few answers. It would seem, however, that such may be found in a system where growth and morphogenesis are separated in time and space, where cells normally differentiate in only one of two ways, and where the processes leading to such contrasting differentiation are clearly within the control of the investigator. Any multicellular structure exceeds the sum of its cellular parts, and this is as true of the pseudoplasmodial or sorogenic stage of a cellular slime mold as of any higher plant or animal. But there is an obvious and significant difference in degree! This difference, as expressed in these primitive microorganisms, may offer simple tools for finding answers to important questions relating to progressive interdependence, functional specialization, and structural differentiation among once equipotential cells, not of the dictyostelids alone but of higher organisms as well.

Agnihothrudu, V. 1956. Occurrence of Dictyosteliaceae in the rhizosphere of plants in southern India. Experientia *12*: 149-151.

Ainsworth, G. C. 1973. Introduction and keys to higher taxa, pp. 1-7. *In* G. C. Ainsworth, F. K. Sparrow, and A. S. Sussman, eds., The Fungi, Vol. IVA. Academic Press, New York.

Alcantara, F., and M. Monk. 1974. Signal propagation during aggregation in the slime mould *Dictyostelium discoideum*. J. Gen. Microbiol. *85*: 321-334.

Aldrich, H. C. 1967. The ultrastructure of meiosis in three species of *Physarum*. Mycologia *59*: 127-148.

Allen, J. R., S. H. Hutner, E. Goldstone, J. J. Lee, and M. Sussman. 1963. Culture of the acrasian *Polysphondylium pallidum* WS-320 in defined media. J. Protozool. *10* (Suppl.): 13.

Anderson, J. S., D. I. Fennell, and K. B. Raper. 1968. *Dictyostelium deminutivum*, a new cellular slime mold. Mycologia *60*: 49-64.

Arndt, A. 1937. Rhizopodenstudien III. Untersuchungen über *Dictyostelium mucoroides* Brefeld. Wilhelm Roux Archiv. f. Entwicklungsmechanik *136*: 681-747.

Bacon, C. W., and A. S. Sussman. 1973. Effects of the self-inhibitor of *Dictyostelium discoideum* on spore metabolism. J. Gen. Microbiol. *76*: 331-344.

Bacon, C. W., A. S. Sussman, and A. G. Paul. 1973. Identification of a self-inhibitor from spores of *Dictyostelium discoideum*. J. Bacteriol. *113*: 1061-1063.

Balamuth, W. 1964. Nutritional studies on axenic cultures of *Naegleria gruberi*. J. Protozool. *11*(Suppl.): 19-20.

Banerjee, S. D., J. R. Allen, E. M. Goldstone, S. H. Hutner, J. J. Lee, and J. H. Diamond. 1964. Subculture of the acrasian *Polysphondylium pallidum* WS-320 on defined media. J. Protozool. *11*(Suppl.): 20.

Barker, J. 1868. Proceedings of the Dublin Microscopal Club, Dec. 19, 1867. p. 123. *In* Quart. J. Microscop. Sci. Vol. 7.

Becker, J. E. 1959. A taxonomic and cultural study of the white-spored polysphondylia. M.S. thesis. Univ. of Wisconsin, Madison. 58 p.

Benson, M. R., and D. P. Mahoney. 1977. The distribution of dictyostelid cellular slime molds in southern California with taxonomic notes on selected species. Amer. J. Bot. *64*: 496-503.

Beug, H., F. E. Katz, and G. Gerisch. 1973. Dynamics of antigenic membrane sites relating to cell aggregation in *Dictyostelium discoideum*. J. Cell Biol. *56*: 647-658.

Blaskovics, J. C., and K. B. Raper. 1957. Encystment stages of *Dictyostelium*. Biol. Bull. *113*: 58-88.

Bonner, J. T. 1944. A descriptive study of the development of the slime mold *Dictyostelium discoideum*. Amer. J. Bot. *31*: 175-182.

———. 1947. Evidence for the formation of cell aggregates by chemotaxis in the development of the slime mold *Dictyostelium discoideum*. J. Exptl. Zool. *106*: 1-26.

―――. 1949. The demonstration of acrasin in the later stages of the development of the slime mold *Dictyostelium discoideum*. J. Exptl. Zool. *110*: 259-272.

―――. 1950. Observations on polarity in the slime mold *Dictyostelium discoideum*. Biol. Bull. *99*: 143-151.

―――. 1952. The pattern of differentiation in amoeboid slime molds. Amer. Naturalist *86*: 79-89.

―――. 1957. A theory of the control of differentiation in the cellular slime molds. Quart. Rev. Biol. *32*: 232-246.

―――. 1959a. The Cellular Slime Molds. Princeton Univ. Press, Princeton, N.J. 150 p.

―――. 1959b. Differentiation in social amoebae. Sci. Amer. *201*: 152-162.

―――. 1959c. Evidence for the sorting out of cells in the development of the cellular slime molds. Proc. Natl. Acad. Sci. USA *45*: 379-384.

―――. 1963. Epigenetic development in the cellular slime moulds. Symposium, Soc. Exptl. Biol. *17*: 342-358.

―――. 1967. The Cellular Slime Molds. Princeton Univ. Press, Princeton, N.J. 2nd ed. 205 p.

―――. 1970. Induction of stalk cell differentiation by cyclic AMP in the cellular slime mold *Dictyostelium discoideum*. Proc. Natl. Acad. Sci. USA *65*: 110-113.

―――. 1977. Some aspects of chemotaxis using the cellular slime molds as an example. Mycologia *69*: 443-459.

Bonner, J. T., and M. S. Adams. 1958. Cell mixtures of different species and strains of cellular slime moulds. J. Embryol. exp. Morph. *6*: 346-356.

Bonner, J. T., D. S. Barkley, E. M. Hall, T. M. Konijn, J. W. Mason, G. O'Keefe III, and P. B. Wolfe, 1969. Acrasin, acrasinase, and the sensitivity to acrasin in *Dictyostelium discoideum*. Develop. Biol. *20*: 72-87.

Bonner, J. T., A. D. Chiquoine, and M. Q. Kolderie. 1955. A histochemical study of differentiation in the cellular slime molds. J. Exptl. Zool. *130*: 133-157.

Bonner, J. T., W. W. Clarke Jr., C. L. Neely Jr., and M. K. Slifkin. 1950. The orientation to light and the extremely sensitive orientation to temperature gradients in the slime mold *Dictyostelium discoideum*. J. Cell and Comp. Physiol. *36*: 149-158.

Bonner, J. T., T. A. Davidowski, W.-L. Hsu, D. A. Lapeyrolerie, and H.L.B. Suthers. 1982. The role of surface water and light on differentiation in the cellular slime molds. Differentiation *21*: 123-126.

Bonner, J. T., and M. R. Dodd. 1962a. Aggregation territories in the cellular slime molds. Biol. Bull. *122*: 13-24.

―――. 1962b. Evidence for gas-induced orientation in the cellular slime molds. Develop. Biol. *5*: 344-361.

Bonner, J. T., and E. B. Frascella. 1952. Mitotic activity in relation to differentiation in the slime mold *Dictyostelium discoideum*. J. Exptl. Zool. *121*: 561-572.

Bonner, J. T., E. M. Hall, S. Noller, F. B. Oleson Jr., and A. B. Roberts. 1972. Synthesis of cyclic AMP and phosphodiesterase in various species of cellular slime molds and its bearing on chemotaxis and differentiation. Develop. Biol. *29*: 402-409.

Bonner, J. T., and M. E. Hoffman. 1963. Evidence for a substance responsible for the spacing pattern of aggregation and fruiting in the cellular slime molds. J. Embryol. exp. Morph. *11*: 571-589.

Bonner, J. T., A. P. Kelso, and R. G. Gillmor. 1966. A new approach to the problem of aggregation in the cellular slime molds. Biol. Bull. *130*: 28-42.

Bonner, J. T., P. G. Koontz Jr., and D. Paton. 1953. Size in relation to the rate of migration in the slime mold *Dictyostelium discoideum*. Mycologia *45*: 235-240.

Bonner, J. T., and M. J. Shaw. 1957. The role of humidity in the differentiation of the cellular slime molds. J. Cell. and Comp. Physiol. *50*: 145-154.

Bonner, J. T., T. W. Sieja, and E. M. Hall. 1971. Further evidence for the sorting out of cells in the differentiation of the cellular slime mold *Dictyostelium discoideum*. J. Embryol. exp. Morph. *25*: 457-465.

Bonner, J. T., and M. K. Slifkin. 1949. A study of the control of differentiation: the proportions of stalk and spore cells in the slime mold *Dictyostelium discoideum*. Amer. J. Bot. *36*: 727-734.

Bozzone, D. M., and J. T. Bonner. 1982. Macrocyst formation in *Dictyostelium discoideum*: mating or selfing? J. Exptl. Zool. *220*: 391-394.

Bradley, S. G., and M. Sussman. 1952. Growth of amoeboid slime molds in one-membered cultures. Arch. Biochem. Biophys. *39*: 462-463.

Brefeld, O. 1869. *Dictyostelium mucoroides*. Ein neuer Organismus und der Verwandschaft der Myxomyceten. Abh. Seckenberg. Naturforsch. Ges. *7*: 85-107.

————. 1884. *Polysphondylium violaceum* und *Dictyostelium mucoroides* nebst Bemerküngen zur Systematik der Schleimpilze. Unters. Gesammtgeb. Mykol. *6*: 1-34.

Brody, T., and K. L. Williams. 1974. Cytological analysis of the parasexual cycle in *Dictyostelium discoideum*. J. Gen. Microbiol. *82*: 371-383.

Buchanan, R. E., and N. E. Gibbons, eds. 1974. Bergey's Manual of Determinative Bacteriology. Williams and Wilkins Co., Baltimore. 8th ed. 1268 p.

Buell, C. B., and W. H. Weston. 1947. Application of the mineral oil conservation method to maintaining collections of fungous cultures. Amer. J. Bot. *34*: 555-561.

Butcher, R. W., and E. W. Sutherland. 1962. Adenosine-3',5'-phosphate in biological materials: I. Purification and properties of cyclic 3',5'-nucleotide phosphodiesterase and use of this enzyme to characterize -adenosine 3',5'-phosphate in human urine. J. Biol. Chem. *237*: 1244-1250.

Cappuccinelli, P., and J. M. Ashworth, eds. 1977. Development and Differentiation in the Cellular Slime Moulds. Elsevier/North-Holland Biomedical Press, Amsterdam, 317 p.

Cavender, J. C. 1969a. The occurrence and distribution of Acrasieae in forest soils. I. Europe. Amer. J. Bot. *56*: 989-992.

————. 1969b. The occurrence and distribution of Acrasiae in forest soils. II. East Africa. Amer. J. Bot. *56*: 993-998.

————. 1970. *Dictyostelium dimigraformum, Dictyostelium laterosorum* and *Acy-*

tostelium ellipticum: new Acrasiae from the American tropics, J. Gen. Microbiol. *62*: 113-123.

―――. 1972. Cellular slime molds in forest soils of eastern Canada. Can J. Bot. *50*: 1497-1501.

―――. 1973. Geographical distribution of Acrasiae. Mycologia *65*: 1044-1054.

―――. 1976a. Cellular slime molds of Southeast Asia. I. Description of new species. Amer. J. Bot. *63*: 60-70.

―――. 1976b. Cellular slime molds of Southeast Asia. II. Occurrence and distribution. Amer. J. Bot. *63*: 71-73.

―――. 1978. Cellular slime molds in tundra and forest soils of Alaska including a new species, *Dictyostelium septentrionalis*. Can. J. Bot. *56*: 1326-1332.

―――. 1980. Cellular slime molds of the southern Appalachians. Mycologia *72*: 55-63.

Cavender, J. C., and K. B. Raper. 1965a. The Acrasieae in nature. I. Isolation. Amer. J. Bot. *52*: 294-296.

―――. 1965b. The Acrasieae in nature. II. Forest soil as a primary habitat. Amer. J. Bot. *52*: 297-302.

―――. 1965c. The Acrasieae in nature. III. Occurrence and distribution in forests of eastern North America. Amer. J. Bot. *52*: 302-308.

―――. 1968. The occurrence and distribution of Acrasiae in forests of subtropical and tropical America. Amer. J. Bot. *55*: 504-513.

Cavender, J. C., K. B. Raper, and A. M. Norberg. 1979. *Dictyostelium aureostipes* and *Dictyostelium tenue*: new species of the Dictyosteliaceae. Amer. J. Bot. *66*: 207-217.

Cavender, J. C., A. C. Worley, and K. B. Raper. 1981. The yellow-pigmented Dictyostelia. Amer. J. Bot. *68*: 373-382.

Ceccarini, C., and A. Cohen. 1967. Germination inhibitor from the cellular slime mould *Dictyostelium discoideum*. Nature *214*: 1345-1346.

Chagla, A. H., K. E. Lewis, and D. H. O'Day. 1980. Ca^{2+} and cell fusion during sexual development in liquid cultures of *Dictyostelium discoideum*. Exptl. Cell Res. *126*: 501-505.

Chang, M. T. 1976. Macrocysts in *Dictyostelium rosarium*. M.S. thesis. Univ. of Wisconsin, Madison, 49 p.

Chang, M. T., and K. B. Raper. 1977. Macrocyst formation in *Dictyostelium rosarium*. Abstr. Ann. Mtg. Am. Soc. Microbiol. I-82, p. 168.

―――. 1981. Mating types and macrocyst formation in *Dictyostelium rosarium*. J. Bacteriol. *147*: 1049-1053.

Chang, M. T., K. B. Raper, and K. L. Poff. 1978. The environment and morphogenesis in *Dictyostelium*. Abstr. Ann. Mtg. Am. Soc. Microbiol. I-61, p. 91.

―――. 1983. The effect of light on morphogenesis in *Dictyostelium mucoroides*. Exptl. Cell. Res. *143*: 335-342.

Chang, Y. Y. 1968. Cyclic 3',5'-adenosine monophosphate phosphodiesterase produced by the slime mold *Dictyostelium discoideum*. Science *160*: 57-59.

Cienkowski, L. 1867. Über den Bau und die Entwicklung der Labyrinthuleen. Arch. mikr. Anat. *3*: 274-310.

―――. 1873. *Guttulina rosea*. Trans. Bot. Sect. 4th Mtg. Russian Naturalists at Kazan. [In Russian.]

Clark, M. A. 1974. Syngenic divisions of the cellular slime mold *Polysphondylium violaceum*. J. Protozool. *21*: 755-757.

Clark, M. A., D. Francis, and R. Eisenberg. 1973. Mating types in cellular slime molds. Biochem. Biophys. Res. Commun. *52*: 672-678.

Clark, M. A., and S. E. Speight. 1973. Macrocyst-forming ability of morphogenetic mutants of *Polysphondylium violaceum*. (Abstr. #44.) J. Protozool. *20*: 507.

Clark, R. L., and T. L. Steck. 1979. Morphogenesis in *Dictyostelium*: an orbital hypothesis. Science *204*: 1163-1168.

Clegg, J. S., and M. F. Filosa. 1961. Trelalose in the cellular slime mould *Dictyostelium discoideum*. Nature *192*: 1077-1078.

Coccuci, S. M., and M. Sussman. 1970. RNA in cytoplasmic and nuclear fractions of cellular slime mold amoebae. J. Cell Biol. *45*: 399-407.

Coemans, E. 1863. Recherches sur le polymorphisme et les différents appareils de reproduction chez les mucorinées—Deuxième partie. Bull. de l'Acad. Royale de Belgique *16*: 177-199.

Cohen, A. L. 1953. The isolation and culture of opsimorphic organisms. I. Occurrence and isolation of opsimorphic organisms from soil and culture of Acrasieae on a standard medium. Ann. N.Y. Acad. Sci. *56*: 938-943.

Cotter, D. A. 1975. Spores of the cellular slime mold *Dictyostelium discoideum*, pp. 61-72. *In* P. Gerhardt, R. N. Costilow, and H. L. Sadoff, eds., Spores VI. Amer. Soc. Microbiol., Washington, D.C.

———. 1977. The effects of osmotic pressure changes on the germination of *Dictyostelium discoideum* spores. Can J. Microbiol. *23*: 1170-1177.

———. 1979. Activation of *Dictyostelium discoideum* spores with guanidine and methyl derivatives of urea. Current Microbiol. *2*: 27-30.

———. 1981. Spore activation, pp. 385-411. *In* A. Turian and H. R. Hohl eds., The Fungal Spore. Academic Press, New York.

Cotter, D. A., and K. R. Dahlberg. 1977. Isolation and characterization of *Dictyostelium discoideum* spore mutants with altered activation requirements. Exptl. Mycol. *1*: 107-115.

Cotter, D. A., L. Y. Miura-Santo, and H. R. Hohl. 1969. Ultrastructural changes during germination of *Dictyostelium discoideum* spores. J. Bacteriol. *100*: 1020-1026.

Cotter, D. A., J. W. Morin, and R. W. O'Connell. 1976. Spore germination in *Dictyostelium discoideum*. II. Effects of dimethyl sulfoxide on postactivation lag as evidence for the multistate model of activation. Arch. Microbiol. *108*: 93-98.

Cotter, D. A., and R. W. O'Connell. 1976. Activation and killing of *Dictyostelium discoideum* spores with urea. Can J. Microbiol. *22*: 1751-1755.

Cotter, D. A., and K. R. Raper. 1966. Spore germination in *Dictyostelium discoideum*. Proc. Natl. Acad. Sci. USA *56*: 880-887.

———. 1968a. Factors affecting the rate of heat-induced spore germination in *Dictyostelium discoideum*. J. Bacteriol. *96*: 86-92.

———. 1968b. Properties of germinating spores of *Dictyostelium discoideum*. J. Bacteriol. *96*: 1680-1689.

———. 1968c. Spore germination in strains of *Dictyostelium discoideum* and other members of the Dictyosteliaceae. J. Bacteriol. *96*: 1690-1695.

Cotter, D. A. and K. R. Raper. 1970. Spore germination in *Dictyostelium dis-coideum*: trehalase and the requirement for protein synthesis. Develop. Biol. 22: 112-128.

Dahlberg, K. R., and D. A. Cotter. 1978. Autoactivation of spore germination in mutant and wild type strains of *Dictyostelium discoideum*. Microbios 23: 153-166.

Deasey, M. C. 1982. Spore formation by the cellular slime mold *Fonticula alba*. Mycologia 74: 607-613.

Deasey, M. C., and L. S. Olive. 1981. Role of Golgi apparatus in sorogenesis by the cellular slime mold *Fonticula alba*. Science 213: 561-563.

Dengler, R. E., M. F. Filosa, and Y. Y. Shao. 1970. Ultrastructural aspects of macrocyst production in *Dictyostelium mucoroides* (Abstr.) Amer. J. Bot. 57: 737.

Depraitère, C., and M. Darmon. 1978. Croissance de l'amibe sociale *Dictyostelium discoideum* sur différentes espèces bactériennes. Ann. Microbiol. (Inst. Pasteur) 129B: 451-461.

Dimond, R. L., M. Brenner, and W. F. Loomis. 1973. Mutations affecting n-acetylglucosaminidase in *Dictyostelium discoideum*. Proc. Natl. Acad. Sci. USA 70: 3356-3360.

Dowbenko, D. J., and H. L. Ennis. 1980. Regulation of protein synthesis during spore germination in *Dictyostelium discoideum*. Proc. Natl. Acad. Sci. USA 77: 1791-1795.

Durston, A. J. 1976. Tip formation is regulated by an inhibitory gradient in the *Dictyostelium discoideum* slug. Nature 263: 126-129.

Durston, A. J. and F. Vork. 1977. The control of morphogenesis and pattern in the *Dictyostelium discoideum* slug, pp. 17-26. *In* P. Cappuccinelli and J. M. Ashworth, eds., Development and Differentiation in the Cellular Slime Moulds. Elsevier/North-Holland Biomedical Press, Amsterdam.

———. 1979. A cinematographical study of the development of vitally stained *Dictyostelium discoideum*. J. Cell Sci. 36: 261-279.

Eagle, H. 1955. Nutrition needs of mammalian cells in tissue culture. Science 122: 501-504.

Eisenberg, R. M. 1976. Two dimensional microdistribution of cellular slime molds in forest soil. Ecology 57: 380-384.

Eisenberg, R. M., and D. Francis. 1977. The breeding system of *Polysphondylium pallidum*, a cellular slime mold. J. Protozool. 24: 182-183.

Ennis, H. L. an M. Sussman. 1958. The initiator cell for slime mold aggregation. Proc. Natl. Acad. Sci. USA 44: 401-411.

Erdos, G. W., A. W. Nickerson, and K. B. Raper. 1972. Fine structure of mac-rocysts in *Polysphondylium violaceum*. Cytobiol. 6: 351-366.

———. 1973. The fine structure of macrocyst germination in *Dictyostelium mu-coroides*. Develop. Biol. 32: 321-330.

Erdos, G. W., K. B. Raper, and L. K. Vogen. 1973. Mating types and macrocyst formation in *Dictyostelium discoideum*. Proc. Natl. Acad. Sci. USA 70: 1828-1830.

———. 1975. Sexuality in the cellular slime mold *Dictyostelium giganteum*. Proc. Natl. Acad. Sci. USA 72: 970-973.

———. 1976. Effects of light and temperature on macrocyst formation in paired mating types of *Dictyostelium discoideum*. J. Bacteriol. *128*: 495-497.

Eslava, A. P., M. I. Alvarez, P. V. Burke, and M. Delbrück. 1975. Genetic recombination in sexual crosses of *Phycomyces*. Genetics *80*: 445-462.

Farnsworth, P. 1973. Morphogenesis in the cellular slime mould *Dictyostelium discoideum*: the formation and regulation of aggregate tips and the specification of developmental axes. J. Embryol. exp. Morph. *29*: 253-266.

———. 1974. Experimentally induced aberrations in the pattern of differentiation in the cellular slime mould *Dictyostelium discoideum*. J. Embryol. exp. Morph. *31*: 435-451.

Farnsworth, P. A. 1975. Proportionality in the pattern of differentiation of the cellular slime mould *Dictyostelium discoideum* and the time of its determination. J. Embryol. exp. Morph. *33*: 869-877.

Farnsworth, P. A., and W. F. Loomis. 1975. A gradient in the thickness of the surface sheath in pseudoplasmodia of *Dictyostelium discoideum*. Develop. Biol. *46*: 349-357.

Farnsworth, P. A., and L. Wolpert. 1971. Absence of cell sorting out in the grex of the slime mould *Dictyostelium discoideum*. Nature *231*: 329-330.

Fayod, V. 1883. Beitrag zur Kenntniss niederer Myxomyceten. Botanische Zeitung *41*: 170-178.

Feinberg, A. P., W. R. Springer, and S. H. Barondes. 1979. Segregation of pre-stalk and pre-spore cells of *Dictyostelium discoideum*: observations consistent with selective cell cohesion. Proc. Natl. Acad. Sci. USA *76*: 3977-3981.

Fennell, D. I., K. B. Raper, and M. H. Flickinger. 1950. Further investigations on the preservation of mold cultures. Mycologia *42*: 135-147.

Filosa, M. F. 1962. Heterocystosis in cellular slime molds. Amer. Naturalist *96*: 79-92.

———. 1979. Macrocyst formation in the cellular slime mold *Dictyostelium mucoroides*: involvement of light and volatile morphogenetic substance(s). J. Expt. Zool. *207*: 491-495.

Filosa, M. F., and M. Chan. 1972. The isolation from soil of macrocyst-forming strains of the cellular slime mould *Dictyostelium mucoroides*. J. Gen. Microbiol. *71*: 413-414.

Filosa, M. F., and R. E. Dengler. 1972. Ultrastructure of macrocyst formation in the cellular slime mold, *Dictyostelium mucoroides*: extensive phagocytosis of amoebae by a specialized cell. Develop. Biol. *29*: 1-16.

Filosa, M. F., S. G. Kent, and M. U. Gillette. 1975. The developmental capacity of various stages of a macrocyst-forming strain of the cellular slime mold, *Dictyostelium mucoroides*. Develop. Biol. *46*: 49-55.

Firtel, R. A., and J. Bonner. 1972. Characterization of the genome of the cellular slime mold *Dictyostelium discoideum*. J. Mol. Biol. *66*: 339-361.

Forman, D., and D. R. Garrod. 1977a. Pattern formation in *Dictyostelium discoideum*. I. Development of prespore cells and its relationship to the pattern of the fruiting body. J. Embryol. exp. Morph. *40*: 215-228.

———. 1977b. Pattern formation in *Dictyostelium discoideum*. II. Differentiation and pattern formation in non-polar aggregates. J. Embryol. exp. Morph. *40*: 229-243.

Francis, D. W. 1962. Movement of pseudoplasmodia of *Dictyostelium discoideum*. Ph.D. diss. Univ. of Wisconsin, Madison. 90 p.

Francis, D. W. 1964. Some studies on phototaxis of *Dictyostelium*. J. Cell. Comp. Physiol. *64*: 131-138.

———. 1965. Acrasin and the development of *Polysphondylium pallidum*. Develop. Biol. *12*: 329-346.

———. 1975a. Cyclic AMP-induced changes in protein synthesis in a cellular slime mould, *Polysphondylium pallidum*. Nature *258*: 763-765.

———. 1975b. Macrocyst genetics in *Polysphondylium pallidum*, a cellular slime mould. J. Gen. Microbiol. *89*: 310-318.

———. 1977. Synthesis of developmental proteins in morphogenetic mutants of *Polysphondylium pallidum*. Develop. Biol. *55*: 339-346.

———. 1979. True divergent differentiation in a cellular slime mold, *Polysphondylium pallidum*. Differentiation *15*: 187-192.

———. 1980. Techniques and marker genes for use in macrocyst genetics with *Polysphondylium pallidum*. Genetics *96*: 125-136.

Francis, D. W., and D. H. O'Day. 1971. Sorting out in pseudoplasmodia of *Dictyostelium discoideum*. J. Exptl. Zool. *176*: 265-272.

Franke, J., and R. Kessin. 1977. A defined minimal medium for axenic strains of *Dictyostelium discoideum*. Proc. Natl. Acad. Sci. USA *74*: 2157-2161.

———. 1978. Auxotrophic mutants of *Dictyostelium discoideum*. Nature *272*: 537-538.

Frantz, C. E. 1980. P2: a behavioral mutant of *Dictyostelium discoideum*. J. Cell Sci. *43*: 341-366.

Free, S. J., and W. F. Loomis. 1974. Isolation of mutations in *Dictyostelium discoideum* affecting α-mannosidase. Biochimie *56*: 1525-1528.

Frischknecht-Tobler, v. U., F. Traub, and H. R. Hohl. 1979. Oekologische Beziehungen zwischen Zellulären Schleimpilzen und mikrobieller Aktivität eines Waldbodens in Jahresverlauf. Vierteljahresschrift der Naturforschenden Gesellschaft in Zürich *124*: 77-108.

Fukui, Y. 1976. Enzymatic dissociation of nascent macrocysts and partition of the liberated cytophagic giant cells in *Dictyostelium mucoroides*. Develop., Growth and Differ. *18*: 145-155.

Fuller, M. S., and R. M. Rakatansky. 1965. A preliminary study of the carotenoids in *Acrasis rosea*. Can. J. Bot. *44*: 269-274.

Fulton, C. 1974. Axenic cultivation of *Naegleria gruberi*. Exptl. Cell Res. *88*: 365-370.

Garrod, D. R. 1969. The cellular basis of movement of the migrating grex of the slime mould *Dictyostelium discoideum*. J. Cell Sci. *4*: 781-798.

Gauger, W. 1961. The germination of zygospores in *Rhizopus stolonifer*. Amer. J. Bot. *48*: 427-429.

George, R. P., R. M. Albrecht, K. B. Raper, I. B. Sachs, and A. P. MacKenzie. 1972. Scanning electron microscopy of spore germination in *Dictyostelium discoideum*. J. Bacteriol. *112*: 1383-1386.

George, R. P., H. R. Hohl, and K. B. Raper, 1972. Ultrastructural development of stalk-producing cells in *Dictyostelium discoideum*, a cellular slime mould. J. Gen. Microbiol. *70*: 477-489.

Gerisch, G. 1959. Ein Submerskulturverfahren für entwicklungsphysiologische Untersuchungen an *Dictyostelium discoideum*. Naturwissenschaften *46*: 654-656.

———. 1960. Zellfunktionen und Zellfunktionswechsel in der Entwicklung von *Dictyostelium discoideum* I. Zellagglutination und induktion der Fruchtkörperpolarität. Roux' Arch. f. Entwicklungsmechanik *152*: 632-654.

———. 1961a. Zellfunktionen und Zellfunktionswechsel in der Entwicklung von *Dictyostelium discoideum*. II. Aggregation homogener Zellpopulationen und Zentrenbildung. Develop. Biol. *3*: 685-724.

———. 1961b. Zellfunktionen und Zellfunktionswechsel in der Entwicklung von *Dictyostelium discoideum*. III. Getrennte beeinflussung von Zelldifferenzierung und morphogenese. Roux' Arch. f. Entwicklungsmechanik *153*: 158-167.

———. 1961c. Zellfunktionen und Zellfunktionswechsel in der Entwicklung von *Dictyostelium discoideum*. V. Stadienspezifische Zellkontaktbildung und ihre quantitative Erfassung. Exptl. Cell Res. *25*: 535-554.

———. 1961d. Zellkontakbildung vegetativer und aggregationsreifer Zellen von *Dictyostelium discoideum*. Naturwissenschaften *11*: 436-437.

———. 1962a. Zellfunktionen und Zellfunktionswechsel in der Entwicklung von *Dictyostelium discoideum*. IV. Der. Zeitplan der Entwicklung. Roux' Arch. f. Entwicklungsmechanik *153*: 603-620.

———. 1962b. Zellfunktionen und Zellfunktionswechsel in der Entwicklung von *Dictyostelium discoideum*. VI. Inhibitoren der Aggregation, ihr Einfluss auf Zellkontaktbildung und morphogenetische Bewegung. Exptl. Cell Res. *26*: 462-484.

———. 1962c. Die zellulären Schleimpilze als Objekte der Entwicklungsphysiologie. Bericht. der Deutsch. Bot. Gesellschaft. *75*: 82-89.

———. 1963a. Life history of *Dictyostelium*. Inst. Wiss. Film, Göttingen. C 87b, b & w, sound, 159 m, 14½ min.

———. 1963b. *Dictyostelium purpureum* (Acrasina)—propagation phase. Inst. Wiss. Film. Göttingen. E 629, b & w, silent, 134 m, 12½ min.

———. 1963c. *Dictyostelium purpureum* (Acrasina)—aggregation and sorophore formation. Inst. Wiss. Film, Göttingen. E 630, b & w, silent, 148 m, 13½ min.

———. 1963d. *Dictyostelium discoideum* (Acrasina)—aggregation and sorophore formation. Inst. Wiss. Film. Göttingen. E 631, b & w, silent, 141 m, 13 min.

———. 1963e. Eine für *Dictyostelium* ungewöhnliche Aggregationsweise. Naturwissenschaften *50*: 160.

———. 1964a. Die Bildung des Zellverbandes bei *Dictyostelium minutum*. I. Übersicht über die Aggregation und den Funktionswechsel der Zellen. Roux' Arch. f. Entwicklungsmechanik *155*: 342-357.

———. 1964b. *Dictyostelium minutum* (Acrasina)-aggregation. Inst. Wiss. Film, Göttingen. E673, b & w, silent, 160 m, 15 min.

———. 1965. Stadienspefische Aggregationsmuster bei *Dictyostelium discoideum*. Roux' Arch. f. Entwicklungsmechanik *156*: 127-144.

———. 1966. Die Bildung des zellverbandes bei *Dictyostelium minutum*. II. Analyse der Zentrengründung an Hand von Filmaufnahmen, Roux' Arch. f. Entwicklungsmechanik *157*: 174-189.

———. 1968. Cell aggregation and differentiation in *Dictyostelium*, pp. 157-197. *In* A. A. Moscona and A. Monroy, eds., Current Topics in Developmental Biology, Vol. 3. Academic Press, New York.

Gerisch, G. 1977. Membrane sites implicated in cell adhesion: their developmental control in *Dictyostelium discoideum*, pp. 36-42. *In* B. R. Brinkley and K. R. Porter, eds., International Cell Biology 1976-1977. Rockefeller University Press, New York.

————. 1979. Control circuits in cell aggregation and differentiation of *Dictyostelium discoideum*, pp. 225-239. *In* J. D. Ebert and T. Okada, eds., Mechanisms of Cell Change. John Wiley & Sons, New York.

————. 1980. Univalent antibody fragments as tools for the analysis of cell interactions in *Dictyostelium*, pp. 243-270. *In* M. Friedlander. A. Monroy, and A. A. Moscona, eds., Current Topics in Developmental Biology, Vol. 14. Academic Press, New York.

Gerisch, G., H. Beug, D. Malchow, H. Schwarz, and A. v. Stein. 1974. Receptors for intercellular signals in aggregating cells of the slime mold, *Dictyostelium discoideum*, pp. 49-66. *In* E.Y.C. Lee and E. E. Smith, eds., Biology and Chemistry of Eucaryotic Cell Surfaces, Miami Winter Symposia, Vol. 7. Academic Press, New York.

Gerisch, G., and B. Hess. 1974. Cyclic-AMP-controlled oscillations in suspended *Dictyostelium* cells: their relation to morphogenetic cell interactions. Proc. Natl. Acad. Sci. USA *71*: 2118-2122.

Gerisch, G., and A. Huesgen. 1976. Cell aggregation and sexual differentiation in pairs of aggregation-deficient mutants of *Dictyostelium discoideum*. J. Embryol. exp. Morph. *36*: 431-442.

Gerisch, G., D. Hülser, D. Malchow, and U. Wick. 1975. Cell communication by periodic cyclic-AMP pulses. Phil. Trans. R. Soc. Lond. B *272*: 181-192.

Gerisch, G., H. Krelle, S. Bozzaro, E. Eitle, and R. Guggenheim. 1980. Analysis of cell adhesion in *Dictyostelium* and *Polysphondylium* by the use of *Fab*, pp. 293-307. *In* A.S.G. Curtis and J. D. Pitts, eds., Cell Adhesion and Motility, Cambridge Univ. Press, Cambridge.

Gerisch, G., and U. Wick. 1975. Intracellular oscillations and release of cyclic AMP from *Dictyostelium* cells. Biochem. Biophys. Res. Commun. *65*: 364-370.

Gezelius, K. 1959. The ultrastructure of cells and cellulose membranes in Acrasiae. Exptl. Cell Res. *18*: 425-453.

————. 1962. Growth of the cellular slime mold *Dictyostelium discoideum* on dead bacteria in liquid media. Physiologia Plantarum *15*: 587-592.

Gezelius, K., and B. G. Ranby. 1957. Morphology and fine structure of the slime mold *Dictyostelium discoideum*. Exptl. Cell Res. *12*: 265-289.

Gillies, N. E., N. Hari-Ratnajothi, and C. N. Ong. 1976. Comparison of the sensitivity of spores and amoebae of *Dictyostelium discoideum* to γ-rays and ultraviolet light. J. Gen. Microbiol. *92*: 229-233.

Githens III, S., and M. L. Karnovsky. 1973. Phagocytosis by the cellular slime mold *Polysphondylium pallidum* during growth and development. J. Cell Biol. *58*: 536-548.

Glazer, P. M., and P. C. Newell. 1981. Initiation of aggregation by *Dictyostelium discoideum* in mutant populations lacking pulsatile signalling. J. Gen. Microbiol. *125*: 221-232.

Glynn, P. J. 1981. A quantitative study of the phagocytosis of *Escherichia coli* by the myxamoebae of the slime mould *Dictyostelium discoideum*. Cytobios *30*: 153-166.

Goldstone, E. M., S. D. Banerjee, J. R. Allen, J. J. Lee, S. H. Hutner, C. J. Bacchi, and J. F. Melville. 1966. Minimal defined media for vegetative growth of the acrasian *Polysphondylium pallidum* WS-320. J. Protozool. *13*: 171-174.

Gray, W. D., and C. J. Alexopoulos. 1968. Biology of the myxomycetes. Ronald Press Co., New York. 288 p.

Gregg, J. H. 1965. Regulation in the cellular slime molds. Develop. Biol. *12*: 377-393.

——. 1966. Organization and synthesis in the cellular slime molds, pp. 235-281. *In* G. C. Ainsworth and A. S. Sussman, eds. The Fungi, Vol. II. Academic Press, New York.

—— 1967. Cellular slime molds, pp. 359-376. *In* F. H. Wilt and N. K. Wessels, eds., Methods in Developmental Biology. T. Y. Crowell Co., New York.

Gregg, J. H., and W. S. Badman. 1970. Morphogenesis and ultrastructure in *Dictyostelium*. Develop. Biol. *22*: 96-111.

Gregg, J. H., and R. W. Davis. 1982. Dynamics of cell redifferentiation in *Dictyostelium mucoroides*. Differentiation *21*: 200-205.

Gregg, J. H., and G. C. Karp. 1978. Patterns of cell differentiation revealed by L-[³H]fucose incorporation in *Dictyostelium*. Exptl. Cell. Res. *112*: 31-46.

Grimm, von M. 1895. Ueber den Bau und die Entwickelungsgeschichte von *Dictyostelium mucoroides* Brefeld. Scripta Botanica Horti Universitalis Imperialis Petropolitanae *4*: 279-298 (renumbered 15-20).

Gross, J. D., C. D. Town, J. J. Brookman, K. A. Jermyn, M. J. Peacey, and R. R. Kay. 1981. Cell patterning in *Dictyostelium*. Phil. Trans. R. Soc. Lond. B *295*: 497-508.

Häder, D. P., and K. L. Poff. 1979. Inhibition of aggregation by light in the cellular slime mold *Dictyostelium discoideum*. Arch. Microbiol. *123*: 281-285.

Hagiwara, H. 1971. The Acrasiales in Japan. I. Bull. Natl. Sci. Museum *14*: 351-366.

——. 1972. Acrasiales. Mem. Natl. Sci. Museum *5*: 173-177.

——. 1973a. The Acrasiales in Japan. II. Rept. Tottori Mycol. Inst. *10*: 591-595.

——. 1973b. Enumeration of the Dictyosteliaceae. Mycological reports from New Guinea and the Solomon Islands. Bull. Natl. Sci. Museum *16*: 493-497.

——. 1974. The Acrasiales in Japan. III. On two species of the Dictyosteliaceae from the Yaeyama Islands, Okinawa. Mem. Natl. Sci. Museum *7*: 81-85.

——. 1976a. Cellular Slime Molds from Mount Margherita (Mts. Ruwenzori), East Africa. Bull. Natl. Sci. Museum Ser. B (Bot.) *2*: 53-62.

——. 1976b. Distribution of the Dictyosteliaceae (Cellular Slime Molds) in Mt. Ishizuchi, Shikoku. Trans. Mycol. Soc. Japan *17*: 226-237.

——. 1978. The Acrasiales in Japan. IV. Bull. Natl. Sci. Museum, Ser. B (Bot.) *4*: 27-32.

——. 1979. The Acrasiales in Japan. V. Bull. Natl. Sci. Museum Ser. B (Bot.) *5*: 67-72.

Hagiwara, H. 1982. Altitudinal distribution of Dictyostelid cellular slime molds in the Gosainkund region of Nepal. Reports on the Cryptogamic Study in Nepal, March 1982 (Miscellaneous Publication of the National Science Museum, Tokyo).

Harper, R. A. 1918. The evolution of cell types and contact and pressure responses in *Pediastrum*. Mem. Torrey Club *17*: 210-239.

————. 1926. Morphogenesis in *Dictyostelium*. Bull. Torrey Bot. Club *53*: 229-268.

————. 1929. Morphogenesis in *Polysphondylium*. Bull. Torrey Bot. Club *56*: 227-258.

————. 1932. Organization and light relations in *Polysphondylium*. Bull. Torrey Bot. Club *59*: 49-84.

Hashimoto. Y., Y. Tanaka, and T. Yamada. 1976. Spore germination promoter of *Dictyostelium discoideum* excreted by *Aerobacter aerogenes*. J. Cell Sci. *21*: 261-271.

Hashimoto, Y., and M. Wada. 1980. Comparative study of sensitivity of spores and amoebae of *Dictyostelium discoideum* to ultraviolet light. Radiation Res. *83*: 688-695.

Hashimoto, Y., and K. Yanagisawa. 1970. Effect of radiation on the spore germination of the cellular slime mold *Dictyostelium discoideum*. Radiation Res. *44*: 649-659.

Hayashi, M., and I. Takeuchi. 1981. Differentiation of various cell types during fruiting body formation of *Dictyostelium discoideum*. Develop., Growth and Differ. *23*: 533-542.

Heftmann, E., B. E. Wright, and G. U. Liddel. 1960. The isolation of Δ^{22}-Stigmasten-3β-ol from *Dictyostelium discoideum*. Arch. Biochem. Biophys. *91*: 266-270.

Hemmes, D. E., E. S. Kojima-Buddenhagen, and H. R. Hohl. 1972. Structural and enzymatic analysis of the spore wall layers in *Dictyostelium discoideum*. J. Ultrastructure Res. *41*: 406-417.

Hirschy, R. A. 1963. The macrocysts of *Dictyostelium purpureum*. M.S. thesis. Univ. of Wisconsin, Madison. 61 p.

Hirschy, R. A., and K. B. Raper. 1964. Light control of macrocyst formation in *Dictyostelium*. Bacteriol. Proc. G 76, p. 27.

Hohl, H. R. 1965. Nature and development of membrane systems in food vacuoles of cellular slime molds predatory upon bacteria. J. Bacteriol. *90*: 755-765.

————. 1976. Myxomycetes, pp. 463-500. *In* D. J. Weber and W. M. Hess, eds., The Fungal Spore: Form and Function. John Wiley & Sons, New York.

Hohl, H. R., and S. T. Hamamoto. 1969. Ultrastructure of spore differentiation in *Dictyostelium discoideum*: the prespore vacuole. J. Ultrastructure Res. *26*: 442-453.

Hohl, H. R., S. T. Hamamoto, and D. E. Hemmes. 1968. Ultrastructural aspects of cell elongation, cellulose synthesis, and spore differentiation in *Acytostelium leptosomum*, a cellular slime mold. Amer. J. Bot. *55*: 783-796.

Hohl, H. R., R. Honegger, F. Traub, M. Markwalder. 1977. Influence of cAMP on cell differentiation and morphogenesis in *Polysphondylium*, pp. 149-172. *In* P. Cappuccinelli and J. M. Ashworth, eds., Development and Differentiation in the Cellular Slime Moulds. Elsevier/North-Holland Biomedical Press, Amsterdam.

Hohl, H. R., and J. Jehli. 1973. The presence of cellulose microfibrils in the proteinaceous slime track of *Dictyostelium discoideum*. Arch. Mikrobiol. *92*: 179-187.

Hohl, H. R., L. Y. Miura-Santo, and D. A. Cotter. 1970. Ultrastructural changes during formation and germination of microcysts in *Polysphondylium pallidum*, a cellular slime mould. J. Cell Sci. *7*: 285-305.

Hohl, H. R., and K. B. Raper. 1962. The nutrition of cellular slime molds. Abstr. Am. Soc. Cell Biol. (2nd Ann. Mtg.), p. 72.

———. 1963a. Nutrition of cellular slime molds. I. Growth on living and dead bacteria. J. Bacteriol. *85*: 191-198.

———. 1963b. Nutrition of cellular slime molds. II. Growth of *Polysphondylium pallidum* in axenic culture. J. Bacteriol. *85*: 199-206.

———. 1963c. Nutrition of cellular slime molds. III. Specific growth requirements of *Polysphondylium pallidum*. J. Bacteriol. *86*: 1314-1320.

———. 1964. Control of sorocarp size in the cellular slime mold *Dictyostelium discoideum*. Develop. Biol. *9*: 137-153.

Horn, E. G. 1971. Food competition among the cellular slime molds. Ecology *52*: 475-484.

Hostak, M. 1960. Induced aggregation of myxamoebae in *Acytostelium*. M.S. thesis, Univ. of Wisconsin, Madison. 59 p.

Huffman, D. M., and L. S. Olive. 1963. A significant morphogenetic variant of *Dictyostelium mucoroides*. Mycologia *55*: 337-341.

———. 1964. Engulfment and anastomosis in the cellular slime molds (Acrasiales). Amer. J. Bot. *51*: 465-471.

Inouye, K., and I. Takeuchi. 1979. Analytical studies on migrating movement of the pseudoplasmodium of *Dictyostelium discoideum*. Protoplasma *99*: 289-304.

———. 1980. Motive force of the migrating pseudoplasmodium of the cellular slime mold *Dictyostelium discoideum*. J. Cell Sci. *41*: 53-64.

Jergenson, L. C., and W. G. Long. 1967. The Acrasiales of a South Dakota woodlot. Proc. S. D. Acad. Sci. *46*: 153-157.

Jones, W. E., and D. Francis. 1972. The action spectrum of light induced aggregation in *Polysphondylium pallidum* and a proposed general mechanism for light response in the cellular slime molds. Biol. Bull. *142*: 461-469.

Kahn, A. J. 1964. The influence of light on cell aggregation in *Polysphondylium pallidum*. Biol. Bull. *127*: 85-96.

———. 1968. An analysis of the spacing of aggregation centers in *Polysphondylium pallidum*. Develop. Biol. *18*: 149-162.

Kanda, F. 1981. Composition and density of dictyostelid cellular slime molds in Kushiro Moor, Hokkaido. Jap. J. Ecol. *31*: 329-333.

Kawabe, K. 1980. Occurrence and distribution of dictyostelid cellular slime molds in the southern Alps of Japan. Jap. J. Ecol. *30*: 183-188.

———1982. *Dictyostelium brunneum*: A new species of brown-pigmented cellular slime mold. Trans. Mycol. Soc. Japan *23*: 91-94.

Khoury, A. T., R. A. Deering, G. Levin, and G. Altman. 1970. Gamma-ray-induced spore germination of *Dictyostelium discoideum*. J. Bacteriol. *104*: 1022-1023.

Kitzke, E. D. 1950. Two members of the Acrasieae isolated in Milwaukee County, Wisconsin. Papers Mich. Acad. Sci., Arts, Letters *34*: 13-17.

―――. 1951. Some ecological aspects of the Acrasiales in and near Madison, Wisconsin. Papers Mich. Acad. Sci., Arts, Letters *35*: 25-32.

―――. 1952. A new method for isolating members of the Acrasieae from soil samples. Nature *170*: 284-285.

Konijn, T. M. 1961. Cell aggregation in *Dictyostelium discoideum*. Ph.D. diss. Univ. of Wisconsin, Madison. 186 p.

―――. 1965. Chemotaxis in the cellular slime molds. I. The effect of temperature. Develop. Biol. *12*: 487-497.

―――. 1968. Chemotaxis in the cellular slime molds. II. The effect of cell density. Biol. Bull. *134*: 298-304.

―――. 1969. Effect of bacteria on chemotaxis in the cellular sime molds. J. Bacteriol. *99*: 503-509.

―――. 1970. Microbiological assay of cyclic 3′, 5′-AMP. Experientia *26*: 367-369.

―――. 1972. Cyclic AMP as a first messenger, pp. 17-31. *In* P. Greengard, G. A. Robison, and R. Paoletti, eds., Advances in Cyclic Nucleotide Research, Vol. 1. Raven Press, New York.

Konijn, T. M., D. S. Barkley, Y. Y. Chang, and J. T. Bonner. 1968. Cyclic AMP: a naturally occurring acrasin in the cellular slime molds. Amer. Naturalist *102*: 225-233.

Konijn, T. M., Y. Y. Chang, and J. T. Bonner. 1969. Synthesis of cyclic AMP in *Dictyostelium discoideum* and *Polysphondylium pallidum*. Nature *224*: 1211-1212.

Konijn, T. M., and K. B. Raper. 1961. Cell aggregation in *Dictyostelium discoideum*. Develop. Biol. *3*: 725-756.

―――. 1965. The influence of light on the time of cell aggregation in the Dictyosteliaceae. Biol. Bull. *128*: 392-400.

―――. 1966. The influence of light on the size of aggregations in *Dictyostelium discoideum*. Biol. Bull. *131*: 446-456.

Konijn, T. M., J.G.C. van de Meene, J. T. Bonner, and D. S. Barkley. 1967. The acrasin activity of adenosine-3′,5′-cyclic phosphate. Proc. Natl. Acad. Sci. USA *58*: 1152-1154.

Konijn, T. M., J.G.C. van de Meene, Y. Y. Chang, D. S. Barkley, and J. T. Bonner. 1969. Identification of adenosine-3′,5′-monophosphate as the bacterial attractant for myxamoebae of *Dictyostelium discoideum*. J. Bacteriol. *99*: 510-512.

Kopachik, W. 1982a. Size regulation in *Dictyostelium*. J. Embryol. exp. Morph. *68*: 23-35.

―――. 1982b. Orientation of cells during slug formation. Roux's Archiv. Develop. Biol. *191*: 348-354.

Koshland, D. E. 1977. A response regulator model in a simple sensory system. Science *196*: 1055-1063.

Krzemieniewska, H. 1961. Acrasieae: Szkic monograficzny. Acta Microbiologica Polonica *10*: 3-79.

Krzemieniewska, H., and S. Krzemieniewski. 1927. Z mikroflory gleby w Polsce. (Contribution à la microflore du sol en Pologne.) Acta Soc. Bot. Poloniae *4*: 141-144.

Kubai, D. 1978. Mitosis and fungal phylogeny, p. 218. *In* I. B. Heath, ed., Nuclear Division in the Fungi. Academic Press, New York.

Kuserk, F. T. 1980. The relationship between cellular slime molds and bacteria in forest soil. Ecology *61*: 1474-1485.

Kuserk, F. T., R. M. Eisenberg, and A. M. Olsen. 1977. An examination of the methods for isolating cellular slime molds (Dictyostelida) from soil samples. J. Protozool. *24*: 297-299.

LaBudde, B. F. 1956. A cytological study of *Dictyostelium*. M.S. thesis. Univ. of Wisconsin, Madison. 50 p.

Laine, J., N. Roxby, and M. B. Coukell. 1975. A simple method for storing cellular slime mold amoebae. Can. J. Microbiol. *21*: 959-962.

Leach, C. K., J. M. Ashworth, and D. R. Garrod. 1973. Cell sorting out during the differentiation of mixtures of metabolically distinct populations of *Dictyostelium discoideum*. J. Embryol. exp. Morph. *29*: 647-661.

Lee, Y.-F. 1971. Notes on Japanese Acrasiales. I. Genus *Dictyostelium*. Trans. Mycol. Soc. Japan *12*: 142-150.

Loeblich, A. R., and H. Tappan. 1961. Suprageneric classification of the Rhizopodia. J. Paleontol. *35*: 245-329.

Lonski, J. 1976. The effect of ammonia on fruiting body size and microcyst formation in the cellular slime molds. Develop. Biol. *51*: 158-165.

Lonski, J., and N. Pesut. 1977. Induction of encystment of *Polysphondylium pallidum* ameba. Can. J. Microbiol. *23*: 518-522.

Loomis, W. F. 1971. Sensitivity of *Dictyostelium discoideum* to nucleic acid analogues. Exptl. Cell Res. *64*: 484-486.

––––. 1975. *Dictyostelium discoideum*: A Developmental System. Academic Press, New York, 214 p.

––––. 1979. Biochemistry of aggregation in *Dictyostelium*. Develop. Biol. *70*: 1-12.

––––, ed. 1982. The Development of *Dictyostelium discoideum*. Academic Press, New York, 551 p.

Machac, M. A., and J. T. Bonner. 1975. Evidence for a sex hormone in *Dictyostelium discoideum*. J. Bacteriol. *124*: 1624-1625.

Macinnes, M. A., and D. Francis. 1974. Meiosis in *Dictyostelium mucoroides*. Nature *251*: 321-324.

MacWilliams, H. K., and J. T. Bonner. 1979. The prestalk-prespore pattern in cellular slime molds. Differentiation *14*: 1-22.

Maeda, Y., and I. Takeuchi. 1969. Cell differentiation and fine structures in the development of the cellular slime molds. Develop., Growth and Differ. *11*: 232-245.

Marchal, E. 1885. Champignons Coprophiles de Belgique. Bull. Soc. Roy-Bot. Belg. *24:* 74, Pl. III.

Martin, G. W., and C. J. Alexopoulos. 1969. The Myxomycetes. Univ. Iowa Press, Iowa City, 560 p.

Mato, J. M., and T. M. Konijn. 1975. Chemotaxis and binding of cyclic AMP in cellular slime molds. Biochem. Biophys. Acta *385*: 173-179.

Mato. J. M., A. Losada, V. Nanjundiah, and T. M. Konijn. 1975. Signal input for a chemotactic response in the cellular slime mold *Dictyostelium discoideum*. Proc. Natl. Acad. Sci. USA *72*: 4991-4993.

Matsukuma, S., and A. J. Durston. 1979. Chemotactic cell sorting in *Dictyostelium discoideum*. J. Embryol. exp. Morph. *50*: 243-251.

Maxman, R. S., and E. R. Sutherland. 1965. Adenosine-3′,5′-phosphate in *Escherichia coli*. J. Biol. Chem. *240*: 1309-1314.

McQueen, D. J. 1971a. A components study of competition in two cellular slime mold species: *Dictyostelium discoideum* and *Polysphondylium pallidum*. Can. J. Zool. *49*: 1163-1177.

———. 1971b. Effects of continuous competition in two species of cellular slime mold: *Dictyostelium discoideum* and *Polysphondylium pallidum*. Can. J. Zool. *49*: 1305-1315.

Mishou, K. E., and E. F. Haskins. 1971. A survey of Acrasieae in the soils of Washington State. Syesis *4*: 179-184.

Moens, P. B. 1976. Spindle and kinetochore morphology of *Dictyostelium discoideum*. J. Cell Biol. *68*: 113-122.

Mouton, H. 1902. Recherches sur la digestion chez les amibes et leurs diastase intracellulaire. Ann. Inst. Pasteur *16*: 457-509.

Mühlethaler, K. 1956. Electron microscopic study of the slime mold *Dictyostelium discoideum*. Amer. J. Bot. *43*: 673-678.

Mullens, I. A., and P. C. Newell. 1978. cAMP binding to cell surface receptors of *Dictyostelium*. Differentiation *10*: 171-176.

Müller, K., and G. Gerisch. 1978. A specific glycoprotein as the target site of adhesion blocking Fab in aggregating *Dictyostelium* cells. Nature *274*: 445-449.

Müller, U., and H. R. Hohl. 1973. Pattern formation in *Dictyostelium discoideum*: temporal and spatial distribution of prespore vacuoles. Differentiation *1*: 267-276.

Nadson, G. A. 1899. Des cultures du *Dictyostelium mucoroides* Bref. et des cultures pures des amibes en général. Scripta Botanica Horti Universitalis Imperialis Petropolitanae *15*: 188-190.

Nanjundiah, V., K. Hara, and T. M. Konijn. 1976. Effect of temperature on morphogenetic oscillations in *Dictyostelium discoideum*. Nature *260*: 705.

Nelson, N., L. S. Olive, and C. Stoianovitch. 1967. A new species of *Dictyostelium* from Hawaii. Amer. J. Bot. *54*: 354-358.

Nesom, M., and L. S. Olive. 1972. *Copromyxa arborescens*, a new cellular slime mold. Mycologia *64*: 1359-1362.

Newell, P. C. 1977a. Aggregation and cell surface receptors in cellular slime molds, pp. 1-57. *In* J. L. Reissig, ed., Microbial Interactions, Ser. B, Vol. 3. Chapman and Hall, London.

———. 1977b. How cells communicate: the system used by slime moulds. Endeavour, New Ser. *1*: 63-68.

———. 1978. Cellular communication during aggregation of *Dictyostelium*. J. Gen. Microbiol. *104*: 1-13.

Newell, P. C., A. Telser, and M. Sussman. 1969. Alternative developmental pathways determined by environmental conditions in the cellular slime mold *Dictyostelium discoideum*. J. Bacteriol. *100*: 763-768.

Nickerson, A. W., and K. B. Raper. 1973a. Macrocysts in the life cycle of the Dictyosteliaceae. I. Formation of the macrocysts. Amer. J. Bot. *60*: 190-197.

———. 1973b. Macrocysts in the life cycle of the Dictyosteliaceae. II. Germination of the macrocysts. Amer. J. Bot. *60*: 247-254.

Norberg, A. M. 1971. The *Dictyostelium mucoroides* complex. Ph.D. diss. Univ. of Wisconsin, Madison. 127 p.

O'Day, D. H. 1979. Aggregation during sexual development in *Dictyostelium discoideum*. Can. J. Microbiol. *25*: 1416-1426.

O'Day, D. H., and A. J. Durston. 1979. Evidence for chemotaxis during sexual development in *Dictyostelium discoideum*. Can. J. Microbiol. *25*: 542-544.

———. 1980. Sorogen elongation and side branching during fruiting body development in *Polysphondylium pallidum*. Can. J. Microbiol. *26*: 959-964.

O'Day, D. H., and K. E. Lewis. 1975. Diffusible mating-type factors induce macrocyst development in *Dictyostelium discoideum*. Nature *254*: 431-432.

———. 1977. Sex hormone of *Dictyostelium discoideum* is volatile. Nature *268*: 730-731.

———. 1981. Pheremonal interactions during mating in *Dictyostelium discoideum*, pp. 199-221. *In* D. H. O'Day and P. A. Horgen, eds., Sexual Interactions in Eukaryotic Microbes. Academic Press, New York.

Oehler, R. 1922. Demonstration: *Dictyostelium mucoroides* (Brefeld). Centralblatt f. Bakt. und Parasitol. *89*: 155-156.

Olive, E. W. 1901. A preliminary enumeration of the sorophoreae. Proc. Amer. Acad. Arts Sci. *37*: 333-344.

———. 1902. Monograph of the Acrasieae. Proc. Boston Soc. Natur. Hist. *30*: 451-513.

Olive, L. S. 1960. Acrasiales of the West Indes. Mycologia *52*: 819-822.

———. 1967. The *Protostelida*—a new order of the mycetozoa. Mycologia *59*: 1-29.

———. 1974. A cellular slime mold with flagellate cells. Mycologia *66*: 685-690.

———. 1975. The Mycetozoans. Academic Press, New York. 293 p.

Olive, L. S., and C. Stoianovitch. 1960. Two new members of the Acrasiales. Bull. Torrey Bot. Club *87*: 1-20.

Oudemans, C.A.J.A. 1885. Anwisten voor de Flora Mycologica van Nederlands *9-10*: 39-40, pl. IV.

Pan, P., E. M. Hall, and J. T. Bonner. 1972. Folic acid as second chemotactic substance in the cellular slime moulds. Nature New Biology *237*: 181-182.

———. 1975. Determination of the active portion of the folic acid molecule in cellular slime mold chemotaxis. J. Bacteriol. *122*: 185-191.

Paul, J. 1960. Cell and Tissue Culture. Williams and Wilkins Co., Baltimore. 312 p.

Perkins, D. G. 1962. Preservation of *Neurospora* stock cultures with anhydrous silica gel. Can. J. Bot. *8*: 591-594.

Persoon, C. H. 1794. Neuer Versuch einer systematischen Eintheilung der Schwämme. Neues. Mag. Bot. *1*: 88.

Pescatore, R. F. 1972. Competition among species of *Dictyostelium*. M.S. thesis. Univ. of Wisconsin, Madison. 84 p.

Pfeffer, W. 1900-1906. The physiology of plants, 3 volumes. English trans. by A. J. Ewart Vol. III, pp. 344-45. Clarendon Press, Oxford.

Pfützner-Eckert, R. 1950. Entwicklungsphysiologische Untersuchungen an *Dic-*

tyostelium mucoroides Brefeld. Roux' Archiv f. Entwicklungsmechanik *144*: 381-409.

Pinoy, E. 1903. Nécessité d'une symbiose microbienne pour obtenir la culture des Myxomycètes. Compt. Rend. Acad. Sci. Paris *137*: 580-581.

———. 1907. Role des bactéries dans le développement de certains Myxomycètes. Ann. Inst. Pasteur *21*: 623-656, 686-700.

Poff, K. L., and W. L. Butler. 1974. Spectral characteristics of the photoreceptor pigment of phototaxis in *Dictyostelium discoideum*. Photochem. and Photobiol. *20*: 241-244.

Poff, K. L., and W. F. Loomis. 1973. Control of phototactic migration in *Dictyostelium discoideum*. Exptl. Cell Res. *82*: 236-240.

Poff, K. L., W. F. Loomis, and W. L. Butler. 1974. Isolation and purification of the photoreceptor pigment associated with phototaxis in *Dictyostelium discoideum*. J. Biol. Chem. *249*: 2164-2167.

Poff, K. L., and M. Skokut. 1977. Thermotaxis by pseudoplasmodia of *Dictyostelium discoideum*. Proc. Natl. Acad. Sci. USA *74*: 2007-2010.

Poff, K. L., and B. D. Whitaker. 1979. Movement of slime molds, pp. 356-382. *In* W. Haupt and M. E. Feinleib, eds., Encyclopedia of Plant Physiology, New Ser. Vol. 7. Springer-Verlag, Berlin.

Potts, G. 1902. Zur Physiologie des *Dictyostelium mucoroides*. Flora (Jena) *91*: 281-347.

Rai, J.N., and J.P. Tewari. 1961. Studies in cellular slime moulds from Indian soils. I. On the occurrence of *Dictyostelium mucoroides* Bref. and *Polysphondylium violaceum* Bref. Proc. Indian Acad. Sci. *53* 1-9.

Raper, K. B. 1935. *Dictyostelium discoideum*, a new species of slime mold from decaying forest leaves. J. Agr. Res. *50*: 135-147.

———. 1937. Growth and development of *Dictyostelium discoideum* with different bacterial associates. J. Agr. Res. *55*: 289-316.

———. 1939. Influence of culture conditions upon the growth and development of *Dictyostelium discoideum*. J. Agr. Res. *58*: 157-198.

———. 1940a. Pseudoplasmodium formation and organization in *Dictyostelium discoideum*. J. Elisha Mitchell Sci. Soc. *56*: 241-282.

———. 1940b. The communal nature of the fruiting process in the Acrasiae. Amer. J. Bot. *27*: 436-448.

———. 1941a. Developmental patterns in simple slime molds. Third Growth Symposium. Growth *5*: 41-76.

———. 1941b. *Dictyostelium minutum*, a second new species of slime mold from decaying forest leaves. Mycologia *33*: 633-649.

———. 1951. Isolation, cultivation and conservation of simple slime molds. Quart. Rev. Biol. *26*: 169-190.

———. 1956a. *Dictyostelium polycephalum* n. sp.: a new cellular slime mould with coremiform fructifications. J. Gen. Microbiol. *14*: 716-732.

———. 1956b. Factors affecting growth and differentiation in simple slime molds. Mycologia *48*: 169-205.

———. 1960. Levels of cellular interaction in amoeboid populations. Proc. Am. Phil. Soc. *104*: 579-604.

———. 1962. The environment and morphogenesis in cellular slime molds, pp. 111-141. *In* The Harvey Lectures, Ser. 57. Academic Press, New York.

———. 1973. Acrasiomycetes, pp. 9-36. *In* G. C. Ainsworth, F. K. Sparrow, and A. S. Sussman, eds., The Fungi, vol. IV B. Academic Press, New York.

Raper, K. B., and D. F. Alexander. 1945. Preservation of molds by the lyophil process. Mycologia *4*: 499-525.

Raper, K. B., and J. C. Cavender. 1968. *Dictyostelium rosarium*: a new cellular slime mold with beaded sorocarps. J. Elisha Mitchell Sci. Soc. *84*: 31-47.

Raper, K. B., and D. I. Fennell. 1952. Stalk formation in *Dictyostelium*. Bull. Torrey Bot. Club *79*: 25-51.

———. 1967. The crampon-based Dictyostelia. Amer. J. Bot. *54*: 515-528.

Raper, K. B., and M. S. Quinlan. 1958. *Acytostelium leptosomum*: a unique cellular slime mould with an acellular stalk. J. Gen. Microbiol. *18*: 16-32.

Raper, K. B., and N. R. Smith. 1939. The Growth of *Dictyostelium discoideum* upon pathogenic bacteria. J. Bacteriol. *38*: 431-445.

Raper, K. B., and C. Thom. 1932. The distribution of *Dictyostelium* and other slime molds in soils. J. Wash. Acad. Sci. *22*: 93-96.

———. 1941. Interspecific mixtures in the Dictyosteliaceae. Amer. J. Bot. *28*: 69-78.

Raper, K. B., A. C. Worley, and D. Kessler. 1977. Observations on *Guttulinopsis vulgaris* and *Guttulinopsis nivea*. Mycologia *69*: 1016-1030.

Raper, K. B., A. C. Worley, and T. A. Kurzynski. 1978. *Copromyxella*: a new genus of Acrasidae. Amer. J. Bot. *65*: 1011-1026.

Reinhardt, D. J. 1966. Silica gel as a preserving agent for the cellular slime mold *Acrasis rosea*. J. Protozool. *13*: 225-226.

———. 1975. Natural variants of the cellular slime mold *Acrasis rosea*. J. Protozool. *22*: 309-317.

Reinhardt, D. J., and A. L. Mancinelli. 1968. Developmental responses of *Acrasis rosea* to the visible light spectrum. Develop. Biol. *18*: 30-41.

Robertson, A., and D. J. Drage. 1975. Stimulation of late interphase *Dictyostelium discoideum* amoebae with an external cyclic AMP signal. Biophys. J. *15*: 765-775.

Robertson, A., D. J. Drage, and M. H. Cohen. 1972. Control of aggregation in *Dictyostelium discoideum* by an external periodic pulse of cyclic adenosine monophosphate. Science *175*: 333-335.

Robson, G. E., and K. L. Williams. 1979. Vegetative incompatibility and the mating-type locus in the cellular slime mold *Dictyostelium discoideum*. Genetics *93*: 861-875.

———. 1980. The mating system of the cellular slime mould *Dictyostelium discoideum*. Current Genetics *1*: 229-232.

———. 1981. Quantitative analysis of macrocyst formation in *Dictyostelium discoideum*. J. Gen. Microbiol. *125*: 463-467.

Roos, U.-P. 1975a. Mitosis in the cellular slime mold *Polysphondylium violaceum*. J. Cell Biol. *64*: 480-491.

———. 1975b. Fine structure of an organelle associated with the nucleus and cytoplasmic microtubules in the cellular slime mould *Polysphondylium violaceum*. J. Cell Sci. *18*: 315-326.

———. 1980. The spindle of cellular slime molds. *In* M. De Brabander and J. De May, eds., Microtubule and Microtubule Inhibitors. Janssen Research Foundation. Elsevier/North-Holland Biomedical Press, Amsterdam.

Roos, U.-P., and R. Camenzind. 1981. Spindle dynamics during mitosis in *Dictyostelium discoideum*. Europ. J. Cell Biol. *25*: 248-257.

Roos, W., and G. Gerisch. 1976. Receptor-mediated adenylate cyclase activation in *Dictyostelium discoideum*. Fed. Eur. Biochem. Soc. Letters *68*: 170-172.

Roos, W., V. Nanjundiah, D. Malchow, and G. Gerisch. 1975. Amplification of cyclic-AMP signals in aggregating cells of *Dictyostelium discoideum*. Fed. Eur. Biochem. Soc. Letters *53*: 139-142.

Roos, W., C. Scheidegger, and G. Gerisch. 1977. Adenylate cyclase activity oscillations as signals for cell aggregation in *Dictyostelium discoideum*. Nature *266*: 259-261.

Rosen, S. D., J. A. Kafka, D. L. Simpson, and S. H. Barondes. 1973. Developmentally regulated, carbohydrate-binding protein in *Dictyostelium discoideum*. Proc. Natl. Acad. Sci. USA *70*: 2554-2557.

Rosen, S. D., D. L. Simpson, J. E. Rose, and S. H. Barondes. 1974. Carbohydrate binding protein from *Polysphondylium pallidum* implicated in intercellular adhesion. Nature *252*: 128 and 149-151.

Ross, I. K. 1960. Studies on diploid strains of *Dictyostelium discoideum*. Amer. J. Bot. *47*: 54-59.

Rossomando, E. F., and M. Sussman. 1973. A 5'-adenosine monophosphate-dependent adenylate cyclase and an adenosine 3':5'-cyclic monophosphate-dependent adenosine triphosphate pyrophosphohydrolase in *Dictyostelium discoideum*. Proc. Natl. Acad. Sci. USA *70*: 1254-1257.

Rostafinski, J. 1875. Śluzowce (Mycetozoa) Monografia. Paris (with supplement, 1876).

Rubin, J., and A. Robertson. 1975. The tip of *Dictyostelium discoideum* pseudoplasmodium as an organizer. J. Embryol. exp. Morph. *33*: 227-241.

Runyon, E. H. 1942. Aggregation of separate cells of *Dictyostelium* to form a multicellular body. Collecting Net. *17*: 88.

Russell, G. K., and J. T. Bonner. 1960. A note on spore germination in the cellular slime mold *Dictyostelium mucoroides*. Bull. Torrey Bot. Club *87*: 187-191.

Saga, Y., and K. Yanagisawa. 1982. Macrocyst development in *Dictyostelium discoideum*. I. Induction of synchronous development by giant cells and biochemical analysis. J. Cell Sci. *55*: 341-352.

Saga, Y. and K. Yanagisawa. 1983. Macrocyst development in *Dictyostelium discoideum*. III. Cell-fusion inducing factor secreted by giant cells. J. Cell Sci. *62*: 237-248.

Saga, Y., H. Okada and K. Yanagisawa. 1983. Macrocyst development in *Dictyostelium discoideum*. II. Mating-type-specific cell fusion and acquisition of fusion-competence. J. Cell Sci. *60*: 157-168.

Sakai, Y. 1973. Cell type conversion in isolated prestalk and prespore fragments of the cellular slime mold *Dictyostelium discoideum*. Develop., Growth and Differ. *15*: 11-19.

Sakai, Y., and I. Takeuchi. 1971. Changes of the prespore specific structure during dedifferentiation and cell type conversion of a slime mold cell. Develop., Growth and Differ. *13*: 231-240.

Sampson, J. 1976. Cell patterning in migrating slugs of *Dictyostelium discoideum*. J. Empryol. exp. Morph. *36*: 663-668.

Samuel, E. W. 1961. Orientation and rate of locomotion of individual amebas in the life cycle of the cellular slime mold *Dictyostelium mucoroides*. Develop. Biol. *3*: 317-335.

Schaap P., L. van der Molen, and T. M. Konijn. 1981a. Development of the simple cellular slime mold *Dictyostelium minutum*. Develop. Biol. *85*: 171-179.

———. 1981b. The vacuolar apparatus of the simple cellular slime mold *Dictyostelium minutum*. Biol. Cell *41*: 133-142.

Schuckmann, W. von. 1924. Zur biologie von *Dictyostelium mucoroides* Bref. Centralblatt f. Bakteriol. und Parasitol. *91*: 302-308.

———. 1925. Zur morphologie und biologie von *Dictyostelium mucoroides* Bref. Archiv. Protistenk. *51*: 495-529.

Schuster, F. 1964. Electron microscope observations on spore formation in the true slime mold *Didymium nigripes*. J. Protozool. *11*: 207-216.

Schwalb, M., and R. Roth. 1970. Axenic growth and development of the cellular slime mould *Dictyostelium discoideum*. J. Gen. Microbiol. *60*: 283-286.

Shaffer, B. M. 1953. Aggregation in cellular slime moulds: *In vitro* isolation of acrasin. Nature *171*: 975-976.

———. 1956a. Acrasin, the chemotactic agent in cellular slime moulds. J. Exptl. Biol. *33*: 645-657.

———. 1956b. Properties of acrasin. Science *123*: 1172-1173.

———. 1957a. Aspects of aggregation in cellular slime moulds. I. Orientation and chemotaxis. Am. Naturalist *91*: 19-35.

———. 1957b. Properties of slime-mould amoebae of significance for aggregation. Quart. J. Microscop. Sci. *98*: 377-392.

———. 1957c. Variability of behavior of aggregating cellular slime moulds. Quart. J. Microscop. Sci. *98*: 393-405.

———. 1958. Integration in aggregating cellular slime moulds. Quart. J. Microscop. Sci. *99*: 103-121.

———. 1961a. Species differences in the aggregation of the Acrasieae, pp. 294-298. *In* Recent Advances in Botany. Proc. 9th Int. Bot. Cong. Univ. of Toronto Press, Toronto.

———. 1961b. The cells founding aggregation centres in the slime mould *Polysphondylium violaceum*. J. Exptl. Biol. *38*: 833-849.

———. 1962. The Acrasina, pp. 109-182 and 301-322. *In* M. Abercrombie and J. Brachet, eds., Advances in Morphogenesis, Vols. 2 and 3. Academic Press, New York.

———. 1965. Cell movement within aggregates of the slime mould *Dictyostelium discoideum* revealed by surface markers. J. Embryol. exp. Morph. *13*: 97-117.

———. 1975. Secretion of cyclic AMP induced by cyclic AMP in the cellular slime mould, *Dictyostelium discoideum*. Nature *255*: 549-552.

Shimomura, O., H.L.B. Suthers, and J. T. Bonner. 1982. The chemical identity of the acrasin of the cellular slime mold, *Polysphondylium violaceum*. Proc. Natl. Acad. Sci. USA *79*: 7376-7379.

Singh, B. N. 1946. Soil Acrasieae and their bacterial food supply. Nature *157*: 133.

Singh, B. N. 1947a. Studies on soil Acrasieae. 1. Distribution of species of *Dictyostelium* in soils of Great Britain and the effect of bacteria on their development. J. Gen. Microbiol. *1*: 11-21.

———. 1947b. Studies on soil Acrasieae. 2. The active life of species of *Dictyostelium* in soil and the influence thereon of soil moisture and bacterial food. J. Gen. Microbiol. *1*: 361-367.

Singh, B. N., and S. R. Das. 1970. Studies on pathogenic and non-pathogenic small free-living amoebae and the bearing of nuclear division on the classification of the order Amoebida. Phil. Trans. Roy. Soc. Lond. B. *259*: 435-476.

Skupienski, F. X. 1917. Sur la sexualité chez les champignons Myxomycètes. Compt. Rend. Acad. Sci. Paris *165*: 118-121.

———. 1981. Sur la sexualité chez une espèce de Myxomycète Acrasiée. Comp. Rend. Acad. Sci. Paris *167*: 960-962.

———. 1920. Recherches sur le cycle évolutif de certain myxomycètes. Published privately, Paris. 81 p.

Slifkin, M. K., and J. T. Bonner. 1952. The effect of salts and organic solutes on the migration time of the slime mold *Dictyostelium discoideum*. Biol. Bull. *102*: 273-277.

Smith, E., and K. L. Williams. 1980. Evidence for tip control of the 'slug/fruit' switch in slugs of *Dictyostelium discoideum*. J. Embryol. exp. Morph. *57*: 233-240.

Smith, K. L., and R. P. Keeling. 1968. Distribution of the Acrasiae in Kansas grasslands. Mycologia *60*: 711-712.

Snyder, H. M., and C. Ceccarini. 1966. Interspecific spore inhibition in the cellular slime moulds. Nature *209*: 1152.

Sobels, J. C., and A. L. Cohen. 1953. The isolation and culture of opsimorphic organisms. II. Notes on isolation, purification, and maintenance of myxomycete plasmodia. Ann. N.Y. Acad. Sci. *56*: 944-948.

Spiegel, F. W., and E. C. Cox. 1980. A one-dimensional pattern in the cellular slime mould *Polysphondylium pallidum*. Nature *286*: 806-807.

Spiegel, F. W., and L. S. Olive. 1978. New evidence for the validity of *Copromyxa protea*. Mycologia *70*: 843-847.

Spudich, J. A. 1974. Biochemical and structural studies of actomyosin-like proteins from non-muscle cells. II. Purification, properties, and membrane association of actin from amoebae of *Dictyostelium discoideum*. J. Biol. Chem. *249*: 6013-6020.

Stearn, W. T. 1966. Botanical Latin. Hafner, Publishing Co., New York. 566 p.

Stenhouse, F. O. and K. L. Williams. 1981. Investigation of cell patterning in the asexual fruiting body of *Dictyostelium discoideum* using haploid and isogenic diploid strains. Differentiation *18*: 1-9.

Sternfeld, J., and J. T. Bonner. 1977. Cell differentiation in *Dictyostelium* under submerged conditions. Proc. Natl. Acad. Sci. USA *74*: 268-271.

Sternfeld, J., and C. N. David. 1981a. Oxygen gradients cause pattern orientation in *Dictyostelium* cell clumps. J. Cell Sci. *50*: 9-17.

———. 1981b. Cell sorting during pattern formation in *Dictyostelium*. Differentiation *20*: 10-21.

Sumino, T. 1981. A cell-marking technique for a cellular slime mold. *Experientia* *37*: 1075.

Sussman, A. S., and H. A. Douthit. 1973. Dormancy in microbial spores. Ann. Rev. Plant Physiol. *24*: 311-352.

Sussman, M. 1951. The origin of cellular heterogeneity in the slime molds, Dictyosteliaceae. J. Exptl. Zool. *118*: 407-417.

———. 1952. An analysis of the aggregation stage in the development of the slime molds, Dictyosteliaceae. II. Aggregative center formation by mixtures of *Dictyostelium discoideum* wild-type and aggregateless variants. Biol. Bull. *103*: 446-457.

———. 1954. Synergistic and antagonistic interactions between morphogenetically deficient variants of the slime mold *Dictyostelium discoideum*. J. Gen. Microbiol. *10*: 110-120.

———. 1956. The biology of the cellular slime molds. Ann. Rev. Microbiol. *10*: 21-50.

———. 1958. A developmental analysis of cellular slime mold aggregation, pp. 264-295. *In* W. D. McElroy and B. Glass, eds., A Symposium on the Chemical Basis of Development. Johns Hopkins Press, Baltimore.

———. 1961. Cultivation and serial transfer of the slime mould *Dictyostelium discoideum* in liquid nutrient medium. J. Gen. Microbiol. *25*: 375-378.

———. 1963. Growth of the cellular slime mold *Polysphondylium pallidum* in a simple nutrient medium. Science *139*: 338.

———. 1966. Biochemical and genetic methods in the study of cellular slime mold development, pp. 397-410. *In* D. M. Prescott, ed., Methods in Cell Physiology, Vol. II. Academic Press, New York.

Sussman, M., and S. G. Bradley. 1954. A protein growth factor of bacterial origin required by the cellular slime molds. Arch. Biochem. Biophys. *51*: 428-435.

Sussman, M., and H. L. Ennis. 1959. The role of the initiator cell in slime mold aggregation. Biol. Bull. *116*: 304-317.

Sussman, M., and F. Lee. 1955. Interactions among variant and wild-type strains of cellular slime molds across thin agar membranes. Proc. Natl. Acad. Sci. USA *41*: 70-78.

Sussman, M., F. Lee, and N. S. Kerr. 1956. Fractionation of acrasin, a specific chemotactic agent for slime mold aggregation. Science *123*: 1171-1172.

Sussman, M., and E. Noel. 1952. An analysis of the aggregation stage in development of the slime molds, Dictyosteliaceae. I. The populational distribution of the capacity to initiate aggregation. Biol. Bull. *103*: 259-268.

Sussman, M., and R. R. Sussman. 1961. Aggregative performance. Exptl. Cell Res. Suppl. 8, 91-106.

———. 1962. Ploidal inheritance in *Dictyostelium discoideum*: stable haploid, stable diploid and metastable states. J. Gen. Microbiol. *28*: 417-429.

Sussman, R. R. 1961. A method for staining the chromosomes of *Dictyostelium discoideum* myxamoebae in the vegetative stage. Exptl. Cell Res. *24*: 154-155.

Sussman, R. R., and M. Sussman. 1963. Ploidal inheritance in the slime mould *Dictyostelium discoideum*: haploidization and genetic segregation of diploid strains. J. Gen. Microbiol. *30* 349-355.

Sussman, R. R. and M. Sussman. 1967. Cultivation of *Dictyostelium discoideum* in axenic medium. Biochem. Biophys. Res. Commun. *29*: 53-55.

Sutherland, J. B., and K. B. Raper. 1978. Distribution of cellular slime molds in Wisconsin prairie soils. Mycologia *70*: 1173-1180.

Szabo, S. P., D. H. O'Day, and A. H. Chagla. 1982. Cell fusion, nuclear fusion, and zygote differentiation during sexual development of *Dictyostelium discoideum*. Develop. Biol. *90*: 375-382.

Takeuchi, I. 1963. Immunochemical and immunohistochemical studies on the development of the slime mold *Dictyostelium mucoroides*. Develop. Biol. *8*: 1-26.

―――. 1969. Establishment of polar organization during slime mold development, pp. 297-304. *In* E. V. Cowdry and S. Seno, eds., Nucleic Acid Metabolism, Cell Differentiation and Cancer Growth. Pergamon Press, Elmsford, New York.

―――. 1972. Differentiation and dedifferentiation in cellular slime molds. Aspects Cell. Mol. Physiol., pp. 217-236.

Takeuchi, I., M. Hayashi, and M. Tasaka. 1977. Cell differentiation and pattern formation in *Dictyostelium*, pp. 1-16. *In* P. Cappuccinelli and J. M. Ashworth, eds., Development and Differentiation in the Cellular Slime Moulds. Elsevier/North-Holland Biomedical Press, Amsterdam.

Takeuchi, I., K. Okamoto, M. Tasaka, and S. Takemoto. 1978. Regulation of cell differentiation in slime mold development. Bot. Mag. Tokyo Special Issue *1*: 47-60.

Tasaka, M., I. Takeuchi. 1981. Role of cell sorting in pattern formation in *Dictyostelium discoideum*. Differentiation *18*: 191-196.

Toama, M. A., and K. B. Raper. 1967a. Microcysts of the cellular slime mold *Polysphondylium pallidum*. I. Factors influencing microcyst formation. J. Bacteriol. *94*: 1143-1149.

―――. 1967b. Microcysts of the cellular slime mold *Polysphondylium pallidum*. II. Chemistry of the microcyst walls. J. Bacteriol. *94*: 1150-1153.

Town, C., and J. Gross. 1978. The role of cyclic nucleotides and cell agglomeration in postaggregative enzyme synthesis in *Dictyostelium discoideum*. Develop. Biol. *63*: 412-420.

Town, C. D., J. D. Gross, and R. R. Kay. 1976. Cell differentiation without morphogenesis in *Dictyostelium discoideum*. Nature *262*: 717-719.

Town, C. D., and E. Stanford. 1977. Stalk cell differentiation by cells from migrating slugs of *Dictyostelium discoideum*: special properties of tip cells. J. Embryol. exp. Morph. *42*: 105-113.

―――. 1979. An oligosaccharide-containing factor that induces cell differentiation in *Dictyostelium discoideum*. Proc. Natl. Acad. Sci.. USA *76*: 308-312.

Traub, F. 1972. Acrasiales in Schweizer Waeldern. M.S. thesis. Univ. of Zurich, Switzerland. 76 p.

―――. 1977. Ein neues Konzept zur Taxonomie der Dictyostelia. Ph.D. diss. Univ. of Zurich, Switzerland. 113 p.

Traub, F., and H. R. Hohl. 1976. A new concept for the taxonomy of the family Dictyosteliaceae (cellular slime molds). Amer. J. Bot. *63*: 664-672.

Traub, F., H. R. Hohl, and J. C. Cavender. 1981a. Cellular slime molds of Switzerland. I. Description of new species. Amer. J. Bot. *68*: 162-171.

――――. 1981b. Cellular slime molds of Switzerland. II. Distribution in forest soils. Amer. J. Bot. *68*: 172-182.

van den Ende, H. 1976. Sexual Interactions in Plants: the Role of Specific Substances in Sexual Reproduction. *In* J. F. Sutcliffe and P. Mahlberg, eds. Experimental Botany. (An International Series of Monographs.) Vol 9. Academic Press, N.Y. 186 p.

van Haastert, P.J.M., R.J.W. DeWit, Y. Grijpma, and T. M. Konijn. 1982. Identification of a pterin as the acrasin of the cellular slime mold *Dictyostelium lacteum*. Proc. Natl. Acad. Sci. USA *79*: 6270-6274.

van Tieghem, Ph. 1880. Sur quelques Myxomycètes à plasmode agrégé. Bull. Soc. Bot. Fr. *27*: 317-322.

――――. *Coenonia,* genre nouveau de Myxomycètes à plasmode agrégé. Bull. Soc. Bot. Fr. *31*: 303-306.

Vogel, G., L. Thilo, H. Schwarz, and R. Steinhart. 1980. Mechanism of phagocytosis in *Dictyostelium discoideum*: phagocytosis is mediated by different recognition sites as disclosed by mutants with altered phagocytotic properties. J. Cell Biol. *86*: 456-465.

Vuillemin, P. 1903. Une Acrasiée bacteriophage. Compt. Rend. Acad. Sci. Paris *137*: 387-389.

Waddell, D. 1982. A predatory slime mould. Nature *298*: 464-466.

Wallace, M. A. 1977. Cultural and genetic studies of the macrocysts of *Dictyostelium discoideum*. Ph.D. diss. Univ. of Wisconsin, Madison. 204 p.

Wallace, M. A., and K. B. Raper. 1979. Genetic exchanges in the macrocysts of *Dictyostelium discoideum*. J. Gen. Microbiol. *113*: 327-337.

Watts, D. J. 1977. Vitamin requirements for growth of myxamoebae of *Dictyostelium discoideum* in a defined medium. J. Gen. Microbiol. *98*: 355-361.

Watts, D. J., and J. M. Ashworth. 1970. Growth of myxamoebae of the cellular slime mould *Dictyostelium discoideum*. Biochem. J. *119*: 171-174.

Weber, A. T., and K. B. Raper. 1971. Induction of fruiting in two aggregateless mutants of *Dictyostelium discoideum*. Develop. Biol. *26*: 606-615.

Weinkauff, A. M., and M. F. Filosa. 1965. Factors involved in the formation of macrocysts by the cellular slime mold *Dictyostelium mucoroides*. Can. J. Microbiol. *11*: 385-387.

Weiss, F. A. 1957. Maintenance and preservation of cultures, pp. 99-119. *In* Amer. Soc. Bacteriologists Manual of Microbiological Methods. McGraw-Hill, New York.

Whittingham, W. F., and K. B. Raper. 1956. Inhibition of normal pigment synthesis in spores of *Dictyostelium purpureum*. Amer. J. Bot. *43*: 703-708.

――――. 1957. Environmental factors influencing the growth and fructification of *Dictyostelium polycephalum*. Amer. J. Bot. *44*: 619-627.

――――. 1960. Non-viability of stalk cells in *Dictyostelium*. Proc. Natl. Acad. Sci. USA *46*: 643-649.

Wickerham, L. J., and M. H. Flickinger. 1946. Viability of yeasts preserved two years by the lyophil process. Brewers Digest *21*: 55-59.

Williams, K. L. 1980. Examination of the chromosomes of *Polysphondylium pallidum* following metaphase arrest by benzimidazole derivatives and colchicine. J. Gen. Microbiol. *116*: 409-415.

Wilson, C. M. 1952. Sexuality in the Acrasiales. 1952. Proc. Natl. Acad. Sci. USA *38*: 659-662.

———. 1953. Cytological study of the life cycle of *Dictyostelium*. Amer. J. Bot. *40*: 714-718.

Wilson, C. M., and I. K. Ross. 1957. Further cytological studies in the Acrasiales. Amer. J. Bot. *44*: 345-350.

Worley, A. C., K. B. Raper, and M. Hohl. 1979. *Fonticula alba*: a new cellular slime mold (Acrasiomycetes). Mycologia *71*: 746-760.

Wright, B. E. 1958. Effect of steroids on aggregation in the slime mold *Dictyostelium discoideum*. (Abstr.) Bact. Proc., pp. 115-116.

———. 1973. Critical Variables in Differentiation. Prentice-Hall, Englewood Cliffs, N.J. 109 p.

Wurster, B., P. Pan, G. G. Tyan, and J. T. Bonner. 1976. Preliminary characterization of the acrasin of the cellular slime mold *Polysphondylium violaceum*. Proc. Natl. Acad. Sci. USA *73*: 795-799.

Yamamoto, M. 1977. Some aspects of behavior of the migrating slug of the cellular slime mold *Dictyostelium discoideum*. Develop., Growth and Differ. *19*: 93-102.

Yamamoto, A., Y. Maeda, and I. Takeuchi. 1981. Development of an autophagic system in differentiating cells of the cellular slime mold *Dictyostelium discoideum*. Protoplasma *108*: 55-69.

Zopf, W. 1885. Die Pilzthiere oder Schleimpilze. Encykl. Naturwiss. *3*: 1-174.

INDEX

absolute density, 33

acellular slime molds (Myxomycetes), 228

acellular sorophores, 393, 395

Acrasidae (acrasids), 225

"acrasidiastase" of Vuillemin, 10

"Acrasiée Bacteriophage" of Vuillemin, 10

Acrasiées of van Tieghem, 5, 104, 226

acrasin, 107-116; choice of name, 108; definition of (Bonner), 108; early characterization (Shaffer), 110; flowing water experiments (Bonner), 107-109; proof of role in chemotaxis (Shaffer), 111

acrasin destruction by phosphodiesterase, 113

acrasin emission: by aggregation centers, 127; by inflowing streams of cells, 127

acrasin levels, detected by cells, 126

acrasins presently known: cyclic AMP in D. discoideum and some other species, 112; a derivative of pterin in D. lacteum, 115; glorin in P. violaceum and P. pallidum, 114

Acrasiomycetes: characteristics of, 225; classification of, 225, 236-238; commonly called "cellular slime molds," 226

Acrasis, derivation of generic name, 104

Acrasis granulata van Tieghem, 5, 227

Acrasis rosea Olive and Stoianovitch, 229-231

Acrasis van Tieghem, 5, 17, 227; as redescribed by Olive and Stoianovitch, 227, 229

activated charcoal, 177, 238, 293, 394, 402, 404

actomyosin-like proteins, 126

Acytosteliaceae Raper and Quinlan, 228, 393

Acytostelium Raper, 228, 393; description and comments, 393; key to species, 395-396; species descriptions:
A. ellipticum Cavender (1970): AE-2, type, 405; characterization, 405; illustrations, 406; sources, 405; unique sorogens and spores, 406;
A. irregularosporum Hagiwara (1971): characterization, 402; extremely delicate, 404; Hagiwara #28, type, 403; illustration, 403; very low nutrient substrates, 403;

A. leptosomum Raper (1956): acellular sorophores and globose spores, 396-400; characterization, 396; FG-12a, type, and other isolates, 396; illustrations, 397-398; sources, 396;
A. subglobosum Cavender (1976): characterization, 400; illustrations, 401; LB-1, type, 400; sources, 400; very delicate species, 400-402

adenosine-3', 5'-monophosphate (cyclic AMP or cAMP), 112-113; acrasin of D. discoideum and other large Dictyostelia, 112; degraded by phosphodiesterase, 113

Aerobacter aerogenes (see Klebsiella pneumoniae), 49

aggregateless mutants: fruiting by artificially aggregated cells, 121; synergistic development with other mutants (Weber), 121; or wild type (Sussman), 121

aggregation in Acytostelium, induction by alkaloids, 115

aggregation patterns: occasionally "halo"-form, 128; sometimes spiral, 127; typically radial, 127

aggregative stimulants for Acytostelium, preformed centers, alkaloids: methyl reserpate, 3-epiyohimbine, yohimbine tartrate, 116

Agnihotrudu, V., 24, 378

Ainsworth, G. C., 225

Alcantara and Monk, 123

alkaloids and cell aggregation in Acytostelium, 115

alternative morphogenesis, sorocarps or macrocysts, 181

altitude, influence of, 42, 43, 47

"amibodiastase" of Mouton, 10

aneuploids, in D. discoideum crosses, 215

Appalachian mountains, 374

arboretum, University of Wisconsin, 264, 273, 288, 333

Arndt, A., 15; first cinematography, 15; inflowing waves in aggregating streams, 105

Arndt's developmental stages, 15

arrested sori, 287, 290

Ashworth, J. M., 135

autoinhibition of spore germination, 89

axenic culture, 68-76

axenic strains, development of, 73

Library of Congress Cataloging in Publication Data

Raper, Kenneth B. (Kenneth Bryan), 1908-
The Dictyostelids.

Bibliography: p.   Includes index.
1. Dictyosteliales. I. Rahn, Ann Worley, II. Title.
QK635.D53R37 1984   589.2'9   83-43089
ISBN 0-691-08345-2 (alk. paper)